Biology from a
CHRISTIAN WORLDVIEW

Devotional BIOLOGY

Learning to Worship the Creator of Organisms

Kurt P. Wise, PhD

Copyright © 2018 Kurt P. Wise. Published by Compass Classroom.

All rights reserved. This book or any portion thereof may not be reproduced or used in any manner whatsoever without the express written permission of the publisher except for the use of brief quotations in a book review.

Printed in the United States of America

Second Printing, 2021, v9

ISBN-13: 978-0-9990409-2-8

Compass Classroom
605 West Iris Drive
Nashville, Tennessee 37204

CompassClassroom.com

CONTENTS

A NOTE TO TEACHERS IX

ACKNOWLEDGMENTS XI

INTRODUCTION 2
WHY THIS BIOLOGY TEXT IS UNIQUE
    Christian Theism vs. Naturalism [▶ 0.1] 3
    Holism vs. Reductionism 4
    Young-Age Creation vs. Evolution 6
    God-Centered vs. Biology-Centered 8
    Christian Responsibility 9
    Summary of Introduction 10
    Advanced Discussion Topic 11
    Test & Essay Questions 11

CHAPTER 1: BIOLOGY FOR THE BELIEVER 12
WHAT IS BIOLOGY?
    Creation and Revelation [▶ 1.1] 13
    The Origin of Modern Science [▶ 1.2] 14
    What is Science? [▶ 1.3] 20
        A Challenge to Define 20
        The Nature of Science 22
    What is Biology? 28
    Why Study Biology? [▶ 1.4] 29
    Biology and Our Priesthood 29
    Biology and Our Kingship [▶ 1.5] 33
    Biology and Our Image [▶ 1.6] 40
    Biology and Our Service to Others 44
    Summary of Chapter 46
    Advanced Discussion Topics 47
    Test & Essay Questions 48

CHAPTER 2: THE LIVING GOD 50
DIFFERENT TYPES OF LIFE & THE ORIGIN OF LIFE
    The Living God [▶ 2.1] 51
    Created Life 51
        Different Types of Life 51

Biblical Life [▸ 2.2] 54
The Spectrum of Perfection of Life [▸ 2.3] 57
The Nature of Life [▸ 2.4] 60
The Origin Of Life [▸ 2.5] 63
    The Law of Biogenesis 63
    How did Life Come to Be? 67
Life: Our Responsibility [▸ 2.6] 69
    Our Responsibility to God 69
    Our Responsibility to the Creation 74
Summary of Chapter 80
Advanced Discussion Topics 82
Test & Essay Questions 82

CHAPTER 3: THE GLORY OF GOD 84
BIOLOGICAL BEAUTY

The Beauty Of God [▸ 3.1] 85
Biological Beauty [▸ 3.2] 88
    Deep Beauty 89
    Ubiquitous Beauty 91
    Profound Beauty 93
    Multifaceted Beauty 94
    Sparkling Beauty 95
The Origin Of Beauty 96
    Human Appreciation of Beauty 97
Beauty: Our Responsibility 98
    Our Responsibility to God 98
    Our Responsibility to the Creation 101
Summary of Chapter 103
Advanced Discussion Topic 104
Test & Essay Questions 105

CHAPTER 4: GOD IS DISTINCT 106
CREATED KINDS AND BIOLOGICAL DISCONTINUITY

The Uniqueness Of God [▸ 4.1] 107
The Biblical Kind 107
Baraminology [▸ 4.2] 115
Biological Discontinuity [▸ 4.3] 116
    Extremely Shallow Discontinuity Within Species 116
    Very Shallow Discontinuity Between Varieties, Cultivars, and Breeds 117
    Shallow Discontinuity Between Species 118
    Semi-Deep Discontinuity Between Genera 121

Contents

Deep Discontinuity 121
    Deeper Discontinuity in Higher Groups 122
    Extremely Deep Discontinuity Surrounding Life 123
The Origin Of Discontinuity [▶ 4.4] 124
Discontinuity: Our Responsibility 128
    Our Responsibility to God 128
    Our Responsibility to the Creation 130
Summary Of Chapter 131
Test & Essay Questions 133

CHAPTER 5: GOD IS GOOD 136
BIOLOGICAL COOPERATION AND BIOLOGICAL EVIL

God is Good [▶ 5.1] 137
Biological Cooperation 137
    Symbiosis and Mutualism 137
The Origin of Mutualism 142
Biological Evil [▶ 5.2] 143
    The Origin of Biological Evil 144
    Negative Effects of the Curse [▶ 5.3] 145
    Evil-Minimizing Effects of the Curse 148
    Goodness of the Creation [▶ 5.4] 154
Mutualistic Symbiosis: Our Responsibility 156
    Our Responsibility to God 156
    Our Responsibility to the Creation 158
Summary of Chapter 162
Advanced Discussion Topics 164
Test & Essay Questions 164

CHAPTER 6: GOD IS PERSON 166
BIOLOGICAL PERSONALITY

God is a Person [▶ 6.1] 167
Personality in the Biological World 170
    Biological Uniqueness 171
    Biological Activity 172
    Biological Activity Continued [▶ 6.2] 175
    Biological Intelligence [▶ 6.3] 181
    Biological Intelligence Continued [▶ 6.4] 183
    Organismal Will [▶ 6.5] 186
    Biological Emotions [▶ 6.6] 188
    Self-Awareness 190
    Biological Relationship 191
The Origin of Personality [▶ 6.7] 192

Animal Behavior: Our Responsibility 194
   Our Responsibility to God 194
   Our Responsibility to the Creation 198
Summary of Chapter 200
Test & Essay Questions 201
References 202

CHAPTER 7: PROVIDER GOD 206
BIOPROVISION AND THE ANTHROPIC PRINCIPLE

God is Love [▶ 7.1] 207
Provision and the Anthropic Principle 207
   Necessary Large-Scale AP Characteristics 209
   Solar System Structure [▶ 7.2] 211
   Necessary Small-Scale AP Characteristics [▶ 7.3] 217
   Compounds [▶ 7.4] 223
   Unnecessary AP Characteristics [▶ 7.5] 230
The Origin of Bioprovision 231
Bioprovision: Our Responsibility 233
   Our Responsibility to God 233
   Our Responsibility to the Creation 233
Summary of Chapter 238
Test & Essay Questions 240
References 241

CHAPTER 8: THE SUSTAINING GOD 244
UNDERSTANDING THE BIOMATRIX

God is Sustainer [▶ 8.1] 245
Provision and the Biomatrix 245
   Food Providers 246
   Break Down Molecules 249
   Convert Molecules 250
   Provide Nutrients 253
   Protection 255
   Biogeochemical Cycles [▶ 8.2] 255
The Origin of Biological Sustenance 259
Biological Sustenance: Our Responsibility 262
   Our Responsibility to God 262
   Our Responsibility to the Creation [▶ 8.3] 262
Summary of Chapter 268
Test & Essay Questions 270
Reference 271

Contents

CHAPTER 9: GOD IS ONE 272
BIOCHEMISTRY AND SYSTEMS BIOLOGY

One God [▶ 9.1] 273
The Unity of Life 273
    Common Monomers 273
    Similar Structures [▶ 9.2] 278
    Systems Biology 279
The Origin of Biological Unity 285
Biological Unity: Our Responsibility [▶ 9.3] 287
    Our Responsibility to God 287
    Our Responsibility to the Creation 289
Summary of Chapter 294
Test & Essay Questions 295

CHAPTER 10: GOD IS THREE 296
CLIMATES, BIOMES, AND BIODIVERSITY

Three Persons of the Godhead [▶ 10.1] 297
Climactic Variety 298
    Diversity of Biological Communities 298
    Land Biome Variety [▶ 10.2] 299
    Water Biome Variety [▶ 10.3] 302
Biodiversity [▶ 10.4] 304
    Biological Diversity in Communities 305
    Diversity Amplified 307
    Disparity 308
The Origin of Disparity 309
Diversity: Our Responsibility [▶ 10.5] 311
    Our Responsibility to God 311
    Our Responsibility to the Creation 312
Summary of Chapter 319
Test & Essay Questions 320

CHAPTER 11: GOD OF HIERARCHY 322
MOLECULAR BIOLOGY: THE CELL

Divine Hierarchy [▶ 11.1] 323
Hierarchy of Biological Organization [▶ 11.2] 325
    The Unit of Biological Life 325
    Traits of All Cells [▶ 11.3] 329
    Traits of All Eukaryotic Cells 331
    Specialized Cell Traits [▶ 11.4] 334
    Why the Hierarchy of Biological Organization Exists 335
Netted Hierarchy of Biological Similarity [▶ 11.5] 336

The Origin of Biological Hierarchy 342
Biological Hierarchy: Our Responsibility 344
    Our Responsibility to God 344
Summary of Chapter 346
Test & Essay Questions 348

CHAPTER 12: THE ALMIGHTY GOD 350
CELLULAR ENERGY METABOLISM

Almighty God [▸12.1] 351
Physical World Energy As Illustration 351
Cellular Energy Metabolism: Biology's Power Source ... 352
    Photosynthesis [▸12.2] 353
    Light-Independent Reactions [▸12.3] 355
    Aerobic Respiration 359
The Origin of Cellular Metabolism 361
Cellular Metabolism: Our Responsibility 362
    Our Responsibility to God 362
Summary of Chapter 363
Test & Essay Questions 365

CHAPTER 13: GOD THE WORD 366
BIOLOGICAL COMMUNICATION AND DNA

The Communicating God [▸13.1] 367
Biological Communication 368
    Animal Communication 369
    The Language of Life [▸13.2] 371
    DNA as the Language of Life [▸13.3] 385
The Origin of Biological Communication 389
Biological Communication: Our Responsibility 391
    Our Responsibility to God 391
    Our Responsibility To The Creation 392
Summary of Chapter 394
Test & Essay Questions 396
References 396

CHAPTER 14: FULLNESS OF GOD 398
BIOLOGICAL REPRODUCTION

Fullness of God [▸14.1] 399
"Be Fruitful" 399
    Cellular Reproduction 401
    Meiosis [▸14.2] 406
    Non-Physical Reproduction 410
"Multiply" and "Breed" [▸14.3] 412

Contents

"Fill" [▶14.4] 417
The Origin Of Biological Reproduction 419
   The Origin of Modern Diversity 420
Biological Reproduction: Our Responsibility 422
   Our Responsibility to God 422
   Obeying God's Command to Multiply and Fill 424
   Our Responsibility to the Creation 427
Summary of Chapter 428
Test & Essay Questions 430

CHAPTER 15: HISTORY OF LIFE 432
THE NATURAL HISTORY OF LIVING ORGANISMS

The Communicating God [▶15.1] 433
The History of Organisms 434
   Biblical Time 434
   Creation Week [▶15.2] 439
   Edenian Epoch [▶15.3] 444
   Antediluvian Epoch 446
   Arphaxadian Epoch [▶15.4] 452
Summary of Chapter 463
Test & Essay Questions 464

APPENDIX: EVOLUTION 466
PERSPECTIVES ON EVOLUTION

Introduction 467
Abiogenesis 468
   Evidence for Abiogenesis 469
   Perspectives on Abiogenesis 470
Biological Evolution 472
   Microevolution Evidence 473
   Speciation Evidences 475
   Macroevolution Evidences 477
Perspectives on Biological Evolution 486
   Perspectives on Mutation 486
   Perspectives on Natural Selection 488
   Perspectives on Speciation 491
   Perspectives on Macroevolution 496
Summary of Appendix 503
Test & Essay Questions 503

GLOSSARY **506**

INDEX **518**

A NOTE TO TEACHERS

This text can be used in two ways. It can be read as a stand-alone book, or it can be used as a textbook alongside the Devotional Biology video curriculum taught by Dr. Kurt Wise (available for purchase at CompassClassroom.com).

A Unique Course

It is important for teachers familiar with traditional biology textbooks to realize this book is different.

It was written to look first at Christian theology as found in the Bible. It then applies that theology to what we see in living organisms. Finally, it examines what that means to how we live our lives. Each chapter will start with a section on God and His attributes, move into the scientific material, then end with a discussion about our responsibility to the creation.

The book also follows a different structure in terms of *when* certain topics are covered as well as *which* major topics are covered.

Finally, this book introduces many concepts that are simply not found in traditional biology textbooks. Many of these concepts are enlightening, but some can be a bit complex to understand on a first reading.

It is okay if a student does not comprehend everything: the goal of this class is to expand the way students think about God and His creation. Both are exceedingly complex. It is not a bad thing to struggle to understand; this is often where learning occurs.

Using Devotional Biology Video Lectures

If you are going to use this book with the video series, there are two ways to approach the material. You are welcome to try both to see which works better with your particular student:

1. Some students do better by watching the video lesson first, then reading the associated text.
2. Others may want to read the text first, then watch the associated video.

Either way, it could be useful to repeat the video or the text if a concept is difficult to grasp. This will help long-term retention, as well.

As you look over the Table of Contents, you will notice a small "play" symbol with a number after different sections, such as: "Creation and Revelation [▶ 1.1]."

A Note to Teachers

This means Devotional Biology Video 1.1 "God Desires to be Known" is associated with the textbook section starting at "Creation and Revelation" and continuing until the [▶1.2] appears. You would thus stop reading when you reach the next symbol and number in the text, such as "Origin of Modern Science [▶1.2]." This section is associated with Video 1.2 "Christian Foundation of Science and Biology."

Please note that the video titles are not always the same as the chapter sections because some video segments cover multiple chapter sections. As you are taking the class, simply ensure you are matching the video numbers such as 1.1, 5.3, or 14.2 to the markers [▶1.1], [▶5.3], or [▶14.2].

Advanced Discussion Topics And Test Questions

If you purchased the Devotional Biology video curriculum, look for a Teacher's Guide that provides a scope and sequence for scheduling the class as well as answers to Test Questions. (The Teacher's Guide is also available at CompassClassroom.com.)

You will find Advanced Discussion Topics at the end of most chapters, and Test & Essay Questions at the end of every chapter. Both of these sections were originally written for teachers using the textbook in a classroom environment. If, however, you would like to work from these lists, they could be incorporated in part or in whole as a discussion between parent and child, or as student projects.

Glossary and Index

Effort has been taken to enable the student and teacher to quickly locate key terms throughout the text. Definitions of terms appear in the margins as you read. Key terms have been collected in a Glossary at the end of the book.

Lab Manual and Lab Materials

Dr. Kurt Wise has written fourteen lab exercises to accompany Devotional Biology. These enhance the curriculum and provide one full credit for a high school lab science.

Purchase the available Devotional Biology Lab Materials and Lab Manual from CompassClassroom.com. Lab Materials include a microscope, DNA model kit, chromosome simulation kit, labware, microscopic slides, chemicals, and tools.

Questions?

Visit us at CompassClassroom.com for support by joining our Devotional Biology group or through our email help system.

ACKNOWLEDGMENTS

Innumerable people and sources have contributed to the concepts, ideas, and even words of this text. Teachers and mentors, student peers and students of mine, textbooks and reference works, popular works and primary literature—I have borrowed from them all.

Perhaps nothing has taught me more about biology than the challenge of teaching college biology to hundreds of students over two decades before attempting this work. Then, as the first draft of this text was composed and subsequent editions were drafted, numerous students have functioned as guinea pigs in the development of the text. Their comments, tears, test answers, and groans have all made their mark on these pages. Then there are the near-whole-text reviewers in both theology and biology who devoted such a large chunk of their time and attention pouring over early versions of the text. They include (in the order of their review) Gregg Allison, Tom Hennigan, Brad Reynolds, Gordon Wilson, Joe Francis, Stephanie Hartz, and Leonard Brand. Many others reviewed smaller sections of the text.

As valuable as these many contributions have been, they should only be blamed for whatever might be good about the text. The shortcomings of the volume are, of course, entirely my fault.

INTRODUCTION

Why This Biology Text is Unique

"For in six days the Lord made heaven and earth, the sea, and all that is in them, and rested on the seventh day."
Exodus 20:11, ESV

If you are a Christian and a student, this text has been designed for you. If you are typical of such students, you probably do not even like science, but must take a science class to fulfill the requirements of your academic program.

In fact, it is likely that you do not really know why a course like this is required. It may also be true that you chose Biology because it was the least undesirable of the science courses available. I pray that this text will convince you of the value of such a course. Even more, I hope that by the end of this course you even come to like science...at least a little!

0.1 | Christian Theism vs. Naturalism

Most college biology texts are written from the perspective of **naturalism**—the belief that only physical things[a] exist[b]. Since such a perspective or belief affects a person's view or understanding of all things—including the world—it is a **worldview**.

Naturalism is more specifically the naturalistic worldview, or the worldview of naturalism[c]. According to naturalism there is no God, there are no angels, and nothing non-physical exists, such as soul, spirit, good, evil, or even purpose. Naturalism accepts the existence of only that which can be detected directly by human senses[d].

naturalism—a worldview that accepts the existence of only physical things (vs. Christian theism)

worldview—a person's perspective (set of beliefs) that colors or influences the way that person interprets everything that person perceives (e.g. naturalism; Christian theism)

---

a Physical things are temporary, destructible things that can be detected in some way by our senses (touch, sight, smell, taste, or hearing) or by an enhancement of our senses (through microscopes and telescopes, rockets and satellites, amplifiers, etc.). Physical things not only include things made of matter (that takes up space and weighs something in the presence of gravity), but also includes physical energy (such as light), space, and physical time.

b Technically, there are two different types of naturalism. The belief that the physical is all that exists is known as 'philosophical naturalism', whereas the practice of studying the physical world as if the physical was all that exists (even if one believes non-physical things are real) is known as 'methodological naturalism'. However, whether the physical world is studied from philosophical or methodological naturalism, the description and explanation that results (i.e., the science) is the same—a naturalistic description that excludes the non-physical. Consequently, this text will refer to both forms of naturalism by the simpler term 'naturalism' and apply to that term the definition of philosophical naturalism.

c Although a person who believes in naturalism is called a naturalist, the word 'naturalist' can also refer to a person who merely focuses his or her study on the physical things in the environment, without regard to a particular belief about whether non-physical things also exist. Because of this ambiguity in meaning, we will avoid the use of the term 'naturalist' and refer instead to 'naturalism' or 'naturalistic worldview'—both of which are unambiguously referring to the *worldview* of naturalism. Also, since philosophical naturalism accepts the existence of only physical or material things, it is sometimes referred to as physicalism or materialism. Unfortunately, 'materialism' more commonly refers to a desire to amass material things and 'materialist' to a person who hoards or strives to acquire material things. To avoid misunderstandings which might arise from the use of such terms this text will not refer to materialism, materialists, or the materialistic worldview.

d Included among things that can be detected directly by human senses would be those things that cannot be detected with unaided senses but must be detected by means of one or more of thousands of different physical

Christian theism—a worldview that begins with the existence of the Christian God, Creator of the physical world and its organisms as well as the non-physical world and its non-physical beings

reductionism—a perspective that seeks to understand something by looking at its component parts (vs. holism)

atom—an electrically neutral arrangement of proton(s) and an equal number of orbiting electrons(s)

molecule—atoms bound together by covalent bonds

cell—tiny, membrane-bound structures that make up all organisms; smallest unit of biological life

The author of this text accepts the existence of God, so this text adopts a theistic perspective. Then, among the various theistic worldviews that exist, the author not only accepts the existence of God, but more specifically the existence of the God described in the Bible. Consequently, the worldview of this text is **Christian theism** (or the worldview of biblical theism as opposed to the worldview of naturalism)[a].

Biblical theism believes in one triune God Who defines good and evil, provides purpose, and created both the spirit world (*e.g.* angels, souls, spirits) *and* the physical world (*e.g.* the universe, astronomical objects, the earth, organisms). I pray that this text will nurture your own biblical worldview and preserve you from any and all non-biblical worldviews you may encounter in the future.

Holism vs. Reductionism

Most college biology texts are also written from the perspective of reductionism—a logical consequence of naturalism. To understand something completely, a person who believes in naturalism does not believe he needs to consider purpose, or an unseen God, or in soul or spirit (because he does not believe any of these things exist).

Since naturalism believes nothing exists beyond the physical, it should be possible to understand something completely by taking it apart and understanding its physical parts and how they fit together (*i.e.* 'the whole is the sum of its parts'). This belief is called **reductionism** and the perspective is reductionistic.

---

devices we have designed to amplify or extend our senses (such as microscopes, telescopes, magnetometers, seismometers, Geiger counters, infra-red cameras, x-ray machines, *etc., etc.*).

[a] Note that although naturalism is the most common worldview found in biology textbooks, it is not the only worldview alternative to Christian or biblical theism. Many other alternative worldviews are advocated around the world, and each provides an alternative worldview of biology. As an example, many environmental activists almost deify organisms, seeming to believe that humans have done so much damage as to be considered an evil. Another category of increasingly popular worldviews in the West are the transcendental worldviews. These worldviews are closer to an exact opposite of the naturalistic worldview, for they focus on the non-physical, tending to deny entirely the existence of the physical world. With all their diversity, what all of these alternative worldviews have in common is that each was designed to challenge the worldview that God wishes us to adopt. Naturalism, for example, denies the biblical claims of the existence of God, the creation of organisms, and the special status of humans. Other worldviews deny these biblical truths or others—such as the personhood of God, the existence of the physical world, or the goodness of matter. Although this text contrasts biblical theism with naturalism, it does so only because naturalism is the most common alternative to biblical theism in biology, not because it is the only worldview out there. The student is encouraged to embrace a biblical worldview and beware of *all* false alternatives (not just naturalism).

Most biology texts are written from the perspective of naturalism, and thus tend to adopt a reductionistic perspective. Consequently, it is common for most biology texts to arrange their chapters from small things (the *micro*) to large things (the *macro*)—in a stepwise fashion dealing with **atoms** then **molecules**, **cells**, **organs**, **organisms**, and finally ecosystems.

The author of this text not only explicitly rejects naturalism, he also explicitly rejects reductionism, believing instead that the creation contains **emergent properties** (those that cannot be explained by the parts that make it up). He also believes that **life** cannot be understood without considering divine intent, or the 'big picture'. He believes that the biblical perspective is **holistic**, not reductionistic.

The Bible begins with God ("In the beginning God…"), then relates the creation of all things (Genesis chapter one), then relates God's interaction with mankind in general (Genesis 2-11), then with a chosen people group, and finally to us individually. The author believes a holistic perspective of biology is the proper one. He also believes it is a better teaching strategy to begin the study of biology with things you as a student are already familiar with—like the critters and plants themselves.

Consequently, the chapters in this text begin with the macro and deal with the micro when it is most appropriate to do so—namely when it actually does help understand the whole. This text starts with organisms and deals with molecules along the way. As a side benefit, the

organ—biological system of tissues in multi-cellular organisms; make up organ systems in large organisms

organism—an animal, plant, or single-celled life form

emergent properties—properties of an entity unaccounted for by the entity's component parts

life—non-physical source of vitality. Different types of life: divine life is part of the natural essence of God and creature life is created by God in spirit beings and organisms; biblical life is possessed by God, spirit beings, and living humans and animals; nephesh life is possessed by living humans and animals; biological life is possessed by all organisms.

holism—a perspective that seeks to understand something by looking at its larger context, and discovers more to something than is accounted for by a sum of its parts (vs. reductionism)

large-to-small approach is opposite that of the reductionist approach, so it implicitly reinforces a (holistic) biblical worldview[a].

Young-Age Creation vs. Evolution

Since a person who believes in naturalism rejects a creator God, he or she believes everything came into being without help (*i.e.* spontaneously or 'naturally'), changing—'evolving'—from previously existing physical things[b]. With a naturalistic worldview perspective, a person has no choice but to believe that everything came to be by some sort of **naturalistic evolution**—the idea that all physical things originate by spontaneous or natural change from previously existing physical things[c].

According to naturalistic evolution, life has been developing over billions of years, it has always been subject to **natural evil** (degenerative aging, animal **death**, suffering), and the diversity of human languages has been developed over thousands of years, and there never was a global flood on this planet.

However, since the God of the Bible is the Creator of all things (Ex. 20:11; Col. 1:16), the author accepts creationism[d] rather than evolutionism (*i.e.* he is a creationist, rather than an evolutionist, and

---

evolution, naturalistic—the belief that all things originate by spontaneous or natural change from previously existing physical things (vs. young-age creationism)

natural evil—anything in the physical world that causes suffering of humans or animals

death, biblical—cessation of biological life in animals & humans as an evil-minimizing effect of the curse

---

a   Because there is no external organizer in naturalistic evolution, in many cases evolutionary theory also operates from the micro to the macro. Typically, subatomic particles are formed first, then atoms, and then molecules. Molecules are formed before cells, and single-celled organisms are formed before multi-celled organisms, organisms before species, and species before communities. Pursuing a macro-to-micro course also implicitly argues against naturalistic evolution.

b   In naturalism this was true up until recent times. Because of the naturalistic belief that everything arose from previously existing physical things, most naturalists from the time of Aristotle to the middle of the 20th Century believed that the physical universe has always existed—that it did not have a beginning. Naturalists consistently believed that everything physical came from something physical that preceded it. It is only with the extraordinary success of the big bang theory in the 1960's that naturalists reluctantly accepted that the universe itself had to have had a beginning. Not only is this original belief of biblical theism a rather recent belief of science, but it stands as a substantial challenge to the naturalistic worldview.

c   Many variations on evolution have been proposed, including forms of theistic evolution that involve God helping or accelerating the physical transformations. Since biology is dominated by naturalism, the dominant form of evolution in biology is naturalistic evolution, where physical changes are unaided, and are thus spontaneous or 'natural'. Consequently, the text will contrast creationism with *naturalistic* evolution. Unless otherwise indicated, when the text uses the word 'evolution' it is referring to naturalistic evolution, not any of the various forms of theistic evolution that have been proposed over the years.

d   For philosophical clarity, note that 'creation' and 'evolution', 'creationism' and 'evolutionism', and 'creationist' and 'evolutionist' refer to claims about the *origin* and modification of things, not the on-going existence of things. Although there is a now <u>un</u>popular theological theory that claimed that things were being continually re-created, most people today—whether creationists or evolutionists—accept that things continue to exist for reasons *other* than creation or evolution. They commonly believe either that physical things have an inherent property of continued existence or that they are being held together by a process unrelated to their origin or modification.

believes that ultimately, all physical things came to be *super*naturally). More particularly, the author believes that the Bible speaks with truth on all matters that it addresses, including its claims about the physical world[a].

As result, he affirms the creation of a complete, un-cursed universe in the course of six days (Gen. 1) only thousands of years ago[b], a curse in response to Adam's sin that introduced natural evil into the world (Gen. 3), a global **Flood** in the days of Noah a millennium and a half later that destroyed all living things on the land with the exception of those in the ark (Gen. 6-9), and a couple centuries later, a judgment on humans at Babel that was the source of the diversity of human language (Gen. 11:1-10)[c].

Thus, not only does this text present a creationist perspective of the world, it more specifically presents a **young-age creationist** perspective of the world[d]. Nearly every chapter relates the success of young-age creation and the failure of naturalistic evolution to explain

Flood—unique, global, year-long catastrophe that destroyed all land animals except those on Noah's ark

creation, young-age—the belief that God created the entire universe about 6000 years ago (vs. naturalistic evolution)

---

a An alternative perspective of Scripture that was first popularized by Galileo, suggests that the Bible speaks authoritatively on spiritual matters only. Various degrees of application of this concept to the Bible has generated a number of alternative perspectives of earth history, including a variety of old-age creationist views that accept the millions and billions of years of conventional science.

b As per the chronogenealogies of Genesis 5 and 11, the time from the creation to Abraham was approximately 2000 years. Since Abraham lived approximately 2000 B.C. (derived both from Biblical chronology and from conventional archaeology), the creation was roughly 6000 years ago. Alternative numbers (*e.g.* those of the Septuagint) and interpretations of Scripture can add as much as 1500 years to the pre-Abraham chronology, so young-age creationists accept an origin of the universe within the last 6000-8000 years.

c See Chapter 15 for more detail on the young-age creationist perspective of history.

d Note that the young-age creationist perspective of earth history is the position of a very small minority (<<1%) of scientists. It is even a small minority of college-educated evangelical Christians (perhaps <5%?). Of the hundreds of universities and colleges with accredited biology majors, less than ten accept a young-age creationist perspective of origins.

some important aspect of biology[a]. The last chapter presents a summary of earth history from a young-age creationist perspective. Since the purpose of the text is to present a creationist perspective of the world, no systematic presentation or critique of evolution is presented in the chapters of the book, so it is offered instead in an appendix.

God-Centered vs. Biology-Centered

Biology is such an enormous field that authors of biology textbooks must select a few topics and ignore most of the others. Since most biology textbooks are written from the perspective of naturalism, it seems only natural that most biology textbooks include the things naturalistic biologists believe they understand about the biological world and avoid those things that continue to mystify them.

Consequently, most textbooks focus on the accomplishments of biologists, the nature of the biological world, and unify the topics with naturalistic evolution. As brilliant as scientists may be, as impressive as their accomplishments are, and as awesome as the biological world is, the author believes the focus of a textbook of biology should be on neither humans nor organisms, but rather upon the One Who created them. The chief end of man is to bring glory to God, so we ought to continually glorify and worship Him.

Scripture also tells us that God created the physical world to show us the invisible God and His invisible qualities or attributes (Rom. 1:20). It stands to reason that in some sense

---

[a] Young-age creationism and naturalistic evolution are explicitly referenced in the text because the former is the perspective of the author and the latter is the dominant perspective of textbooks of biology. Other versions of creation and evolution are not mentioned in the text only because of lack of space (not, for example, because they are unimportant). Also, cases in the text where young-age creationism has been successful at explaining the biological world may also be cases where other variations on creation and/or evolution have been successful at explanation. And, conversely, cases in the text where naturalistic evolution has been *unsuccessful* at explanation may also be cases where other variations on creation and/or evolution have been unsuccessful at explanation.

the obverse of this statement must also be true—that the attributes of God can help us better understand the creation. Consequently, this text focuses on how the major attributes of God illustrate the nature of God. Thirteen of the fifteen chapters open with a description of a characteristic of God and introduce that particular aspect of the biological creation that God created in order to physically illustrate that characteristic .

Christian Responsibility

Another distinctive feature of this text is its emphasis on personal responsibility. Most biology texts engage in very little discussion of ethics and personal responsibility, probably because of the wide diversity of opinions that exist in our society, and the fact that in strict adherence to naturalism there is no such thing as right and wrong. Christians, however, cannot avoid personal responsibility. We have an obligation to learn about the biological creation so that we grow in our understanding of how to know God more intimately, how to share God more effectively, and how to obey God's commands more completely. Then, once we have acquired that understanding we have a responsibility to use that knowledge wisely.

Thus, we have a responsibility to worship God, to share God with others, to guard and keep the creation He gave us, and to enhance the divinity-illustrating characteristics of the creation so as to bring God more glory. With this thought in mind the first chapter of the text focuses on what kinds of responsibility a believer has in regards to the biological creation. Nearly every chapter thereafter concludes with comments about specific responsibilities that the believer has with respect to the biological creation.

In summary, this text opens with a chapter on Christian responsibility in the biological creation and ends with a chapter reviewing young-age creation earth history. Between those bookends are thirteen chapters devoted to the characteristics of God. Each chapter opens with a brief discussion of an attribute of God, then discusses that part of the biological world that God created to illustrate that attribute. This is followed in each case by a short discussion on how the origin of that aspect of the biological world is better explained by young-age creation than by naturalistic evolution. The last part of each chapter deals with our responsibility to that biological creation—first our responsibility to God (to worship Him), then our responsibility to others (to share Him), and finally our responsibility to the creation (to care for it and enhance it to His glory).

It is my hope that this text will change you. After reading and studying this book previous students have claimed it has helped them grow in their relationship with God. Biology has helped them better understand things they have long known about God but struggled to understand. They have come to see God in those things He made, to be awed by God in ways they have never been before, and to know God more intimately. Students have also commented that studying this material has helped them in their Christian walk. They learned new ways to worship, to share their faith with others, to stand for what they believe, and to glorify God.

Finally, students claim that issues in this text have changed their perspective on who they are. They have come to understand their purpose, recognized their roles as priests and rulers of the creation, and learned additional ways to do what is right. I pray all this for you. I pray that this course will initiate a life-long journey of worshipping and glorifying the One Who made all things and testifying that truth to others.

Kurt P. Wise, PhD
July 2015

Summary of Introduction

A. Whereas most biology texts are written from a naturalistic worldview perspective, *Devotional Biology* is written from a Christian theistic worldview perspective.

   a. Worldview is a belief or perspective that affects the way a person understands all things.

   b. Naturalism, or a naturalistic worldview, is the belief that physical things are the only things that exist (i.e. rejects non-physical things like God, souls, spirits, good, evil, purpose).

   c. The Christian theistic (or biblical) worldview believes in one triune God Who defines good and evil, provides purpose, and created both the spirit world (e.g. angels, souls, spirits) and the physical world (e.g. the universe, astronomical objects, the earth, organisms).

   d. Although there are many alternative worldviews, because biology is dominated by that of naturalism, *Devotional Biology* contrasts the Christian theistic worldview with naturalism.

B. Whereas most biology texts are written from a reductionistic perspective (reductionism), *Devotional Biology* is written from a holistic perspective (holism).

   a. Reductionism is a logical consequence of naturalism; holism is a logical consequence of Christian theism.

   b. Reductionism is the belief that the whole can be fully understood by understanding the component parts and how they fit together ("the whole is the sum of its parts"). Holism is the belief that there is more to the whole than can be understood from the component parts (i.e. the whole has emergent properties not found in the component parts).

   c. Reductionistic biology begins with molecules (chemistry) and moves from micro to macro (or from the components of organisms to the interactions among organisms). (Christian) holistic biology begins with God and moves from the macro to micro (or from the interactions among organisms to the parts of organisms).

C. Whereas most biology texts are written from a naturalistic evolutionary perspective, *Devotional Biology* is written from a young-age creation perspective.

   a. Naturalistic evolution (i.e. all physical things come to be by spontaneous or natural change from previously existing physical things) is a logical consequence of naturalism; creation (i.e. the physical world was created supernaturally by God) is a logical consequence of Christian theism.

   b. Young-age creation is a biblically-based claim that the universe was created in six days 6,000-8,000 years ago, in an un-cursed condition (no decay, death, suffering), the universe was cursed shortly thereafter in response to man's sin, life on earth was judged about a millennium and a half later with a global Flood, and most of the diversity of human languages was created at Babel a couple of centuries after that. In contrast, according to naturalistic evolution, life has been developing over billions of years, it has always been subject to decay, death, and suffering, there never was a global flood on this planet, and the diversity of human languages has been developed over thousands of years.

c. *Devotional Biology* argues that biology is better explained by young-age creationism than naturalistic evolution.

D. Whereas most biology texts focus on the achievements of biologists and the awesomeness of the biological world, *Devotional Biology* focuses on the attributes of God and how they are illustrated in the biological world.

E. Whereas most biology texts are weak in human responsibility and ethics (a logical consequence of naturalism), *Devotional Biology* stresses human responsibility and ethical behavior.

Advanced Discussion Topic

The author of this text met with considerable resistance in both the production of this textbook and in the implementation of the associated course from which a video curriculum was developed, and the opposition came from believing science professors at Christian colleges. Discuss why this might be. Things to ponder:

A. the impact of early education on philosophy of education

B. what must be taken out of a science course in order to "add" comments about God

C. the comfort of thinking about the world the way someone in another discipline thinks

D. the record of past success in the integration of science and theology

E. the commonness of this text's perspective in the church

Test & Essay Questions

1. Define worldview / naturalism / theism / Christian theism / reductionism / holism / naturalistic evolution / young-age creation / emergent properties.

2. Compare and contrast Christian theism and naturalism / holism and reductionism / young-age creationism and naturalistic evolution / 'micro to macro' and 'macro to micro' approaches to biology / ethics discussion in most textbooks and ethics discussion in this textbook.

3. Explain why, of all the worldviews that exist, the author chose the contrast the worldviews of naturalism and Christian theism.

4. Which of the following is logical consequence of naturalism / Christian theism, and explain why it is a logical consequence of that worldview: holism or reductionism / creation or evolution / ethics or lack of ethics?

5. Which of the following is true of most biology textbooks / this textbook: naturalism or Christian theism / holism or reductionism / young-age creation or naturalistic evolution / God-centered or scientific achievement-centered / lack of ethics or ethics?

6. Explain what 'the whole is the sum of its parts' / 'the whole is more than a sum of its parts' means.

CHAPTER 1

BIOLOGY FOR THE BELIEVER

What is Biology?

"Praise God, O heaven and earth, seas and all creatures in them."
Psa. 69:34, GNT

1.1 | Creation and Revelation

God identified Himself to Moses and Israel as 'I am' (Ex. 3:14) because it is part of His very nature to exist. It is impossible for God not to exist. Consequently, God is eternal ('the King eternal': I Tim. 1:17[a]). He had no beginning. He always was, is, and always will be. God and *only* God is eternal and uncreated. "…by Him were all things created…" (Col. 1:16). He created both the physical world (everything detectable, or potentially detectable, with our senses of sight, smell, hearing, touch, and taste) and the non-physical world (everything that is not detectable with our senses).

"God is spirit" (John 4:24). Therefore, unless He chooses to reveal Himself, human eyes cannot see Him, ears cannot hear Him, tongues cannot taste Him, noses cannot smell Him, and skin cannot feel Him. Unless He wanted us to perceive Him, God would be undetectable and unknowable. He would not have to 'hide' to be unknown to us. He would not have to do anything at all.

In fact, considering the awesomeness of God and how far we fall short of His glory (Rom. 3:23), we do not deserve to know Him. It seems only 'natural' that such a God *should* be unknowable to us[b]. However, astonishingly enough, this is *not* the God of the Bible. Instead, the God of Scripture desires to be known.

Before man's rebellion, God apparently made it a habit to walk and talk with Adam and Eve in the cool of the day (the implication in Gen. 3:8a). Even after the Fall of man, Enoch 'walked with God' until God took him directly into heaven (Gen. 5:24), Abraham 'was called the friend of God' (James 2:23), Moses spoke with God face to face 'as a man speaks to his friend' (Exo. 33:11), David was chosen by God as a man 'after His own heart' (I Sam. 13:14), Israel was cherished as the 'the apple of His eye' (Deu. 32:10), New Testament believers are adopted children who can call Him 'Abba' (Rom. 8:15), and the church is cherished by God in the way a bride is cherished by her bridegroom (*e.g.* Song of Solomon).

---

a   Other references that God is eternal: Deu. 33:27 & Heb. 9:14.

b   Thus, it is only reasonable that human reasoning would modify the Truth (as in modern Judaism), or create an alternative to the Truth (as in Islam), to envision God as distant and unapproachable. Such a god is too great to be known personally, and for a human to know such a god personally would be disrespectful, impious, or even sacrilegious. Although this might make sense to human reason, such is not the God of the Bible.

From the very beginning God has sought out man so that we could know Him. To do so, God has condescended to reveal Himself to man. Although He could create the entire universe and its components in an instant and still not need rest, He condescended to create over the course of six days and rested on the seventh day as an example to man (Mark 2:27; Exo. 20:8-11).

A couple thousand years ago "…the Word was made flesh, and dwelt among us…" (John 1:14). God went so far as to humble Himself and take on the form of a servant (Philippians 2:6-7)—all as an example to us. Jesus Christ even permitted a greater abasement than that. He allowed our sin to be placed upon Him, and He allowed Himself to take the full measure of punishment for our sin. He actually received on Himself His Father's anger towards our sin and paid for an eternity of suffering for the sins we committed. And He made Himself 'to be sin' for us, so that we might enjoy an eternal relationship with the Almighty Holy God (II Cor. 5:21).

He did so much for us that all we have to do in return is believe in what Christ has already done (Acts 16:31)—to trust that He has done all that is necessary for us to be acceptable in God's sight[a].

As part of the revelation of Himself to man, God created the physical world so that humans could see His invisible qualities and attributes (Rom. 1:20a). This was true of 'even His eternal power and Godhead' (Rom. 1:20b). According to the larger passage (Rom. 1:18ff), God has so convincingly used His creation to show His attributes that every person has actually already come to 'know God' (Gen. 1:21a). Every person did not just come to know *about* God; every person has come to *know* God. God's revelation through His creation is so effective that no person is left with an excuse. No one will be able to stand before God and say that he or she never knew God.

1.2 | The Origin of Modern Science

Because He intended the creation to illustrate His attributes, God designed the creation—and humans—in a very deliberate and special

---

a   If before this moment you have never trusted in what Christ did for you, would you like to? The Bible says that if we understand that we are sinners (Rom. 3:23), deserving judgment for our sins (Rom. 6:23), but that Jesus died on the cross to pay the penalty for our sins (I Cor. 15:1-4), then to receive eternal life we need only believe in what He has already done (John 3:16; Acts 16:31). If you have so trusted in what Jesus has done for you, then the Bible says you HAVE eternal life that no one can take from you (John 10:27-29)—otherwise, in fact, it would not be *eternal* life!

manner. A number of things had to be true about the universe and human beings for humans to recognize the illustrations and infer from them the nature of the invisible God. These might be called the 'knowability traits' of the creation, and they include the following :

The physical world actually exists. Although the existence of the physical world might seem quite 'obvious' to most of us in the Western world, many of the worldviews of the Eastern world believe the physical world to be an illusion[a]. Consequently, if people believe consistently with their declared worldview, a majority of the world's population rejects the claim that a physical world actually exists in some place other than the imagination of the human mind.

But, believers know the physical world exists because God created it to illustrate His attributes. An actual physical world not only allows inferences to be made from it, but also allows verification of those inferences.

---

a   Even in the western world, Plato (c426-c348 BC) pictured the physical world as shadows dancing on the wall of a cave—mere silhouettes of those things which were truly real—the invisible world of perfect ideas, concepts, and forms.

Human senses are reliable. Even though we know of instances where our senses can lead us astray (*e.g.* with mirages) and we know that we can use the senses to lead others astray (*e.g.* as illusionists), human senses must be generally reliable so that humans can correctly perceive revelation from His creation. From this we can then conclude that the physical world that we perceive is the physical world that actually exists.

The creation is ordered. For the creation to illustrate something there must be some sort of structure or order to carry that illustration.

The order of creation is simple enough to be understood by humans and the human brain is complicated enough to understand creation's order. Although a full understanding of God would certainly be outside the capability of any finite being, God still wants us to understand something about Him. This means He chose to illustrate understandable things, created the universe in such a way as to illustrate those concepts in understandable ways, and constructed our brains in such a way as to comprehend those concepts.

We can thus conclude that the order we perceive in the creation is the actual order that is there, and not un-naturally imposed upon the creation by our minds.

Regularities of the creation can be detected and understood in the course of individual human lifetimes. Humans have long recognized that there are consistent patterns in the creation. It is almost as if there were rules or laws that things in the universe must obey. This actually seems quite reasonable, for the God of Scripture is a personal God with reliable and consistent behavior, Who desires very consistent behavior of others.

If the creation mirrors God's nature, the creation might be expected to have regular behaviors of its own. And, since God desires for each individual human to know Him, it is reasonable to assume that He created at least some of that order and some of those regularities in such a way as to be detectable in the span of one individual's lifetime.

Regularities of the creation are consistent across the entire universe for all time. Since God is an unchanging God, creation's order and regularities should be consistent through time. And, since God desires to be known by all people, no matter where they are and no matter when they live, the law-like patterns of the creation ought to be true across all space for all of time.

The order of the creation, including all its regularities, is unified. Since God desires us to recognize the one true God in the creation, all its illustrations will point to the same God. Because there are many and various facets to God's character, there are probably many and various illustrations of His character in the creation. As our understanding of these illustrations increases, we would expect that they will weave together as threads in a tapestry into a coherent, interlocking picture of the God of creation.

We can gain truth and understanding by studying the creation. Since God has created the physical world to teach us something about Him, it is reasonable to assume that studying the creation will lead to truth and

understanding. At the very least we will gain truth and understanding about God.

There is intrinsic value in studying the physical world. There is perhaps no other endeavor as glorious and fulfilling as seeking God. The more we know of Him, the more awesome we realize Him to be, the more awed we are in knowing Him, and more benefited we are in becoming like Him.

Since He created the physical world to teach us about Himself, better knowing the creation leads to better knowledge of God. There is great value in studying the physical world to better know its Creator.

Truth is advanced by continual study of the creation. Since God is so much greater than we can understand or imagine, it is likely that He put enough truth in the creation to keep all humans challenged for all time. God has authored His word in such a way as to provide simple truths for children, deeper truths for those older in the faith, deeper truths still for those who diligently study, and even deeper truths to challenge those who spend lifetimes in study.

Likewise, God has designed His creation with simple truths for all, deeper truths for those who seek, deeper truths still for those who diligently study, and even deeper truths to challenge those who spend lifetimes in study of His creation.

Truth about the unseen can be inferred from the study of observable things. God created observable things so that we could understand those attributes of God that we cannot see. We infer from this that there is

probably much that exists that we cannot see, and we are justified in inferring the nature of those unseen things from the things we can see.

For every event there is a cause. God is a God of cause and effect. He wills and it occurs, He speaks and it is done, He promises and it is true, He redeems and we are His. The universe itself is an effect which resulted from His creation, and He has built the law of cause and effect into the creation as one of its regularities.

Just as we are supposed to infer the cause of the universe's regularities (as illustrations of His character), so also we are justified in seeking cause for effects that we observe in the creation.

Human language is capable of describing, understanding, and teaching truths about the creation. The same God Who created the creation so we could recognize His character in it, gave us language.

It is reasonable to assume that the language He gave us has been created in such a way as to make it possible to describe the creation and the attributes of God illustrated in that creation. It is reasonable to assume that that same language has been designed so that we can reliably pass on information about that creation to others.

Furthermore, the same God Who gave us language, spoke the universe into being by the word of His mouth. Not only should human language be capable of describing the universe, but human language itself should be similar to the structure of the creation. This would explain why mathematics—a language created by humans—has been so successful at approximating the very structure of the creation.

God's desire to illustrate His nature in those things He made yields quite an astonishing creation. Such a world is not the expectation of naturalism. Although our survival might suggest that our senses must be somewhat reliable, naturalism gives us no good reason to believe that the physical world's order just *happens* to be simple enough for us to understand[a] and our brains just *happen* to have evolved enough to be sufficiently complex to understand it.

Nor in naturalism is there reason to believe that the order of the universe should be comprehensible in the course of a single human lifetime, or that the order is unifiable. In naturalism there is no good reason to believe that the regularities of the physical world are unchanging—let alone consistent across all time and space. For naturalism it is even more incredible—or even presumptuous—to

---
a Einstein is said to have said "The most incomprehensible thing about the universe is its comprehensibility."

believe that language humans invented happens to correspond with the structure of the entire universe.

Yet, after centuries of studying the physical world, it does seem as if the universe is understandable. It does seem as if the regularities of the universe are consistent across time and space. And more and more of the regularities of the universe have been unified. And it does seem as if mathematics comes astonishingly close to mimicking the very structure of the universe. The universe seems to be designed just as the Bible intimates—in such a way as to illustrate the very nature of its Creator.

In fact, belief in such a universe led to the origin of science itself. When people acknowledged that the universe not only existed, but was understandable and worthy to be understood, and, additionally, that humans were capable of understanding it, humans began studying the universe in order to understand it. This was the birth of what is called 'modern science'.

It is no accident that this occurred in the wake of the Reformation in Western Europe, among people who were freshly re-acquainted with the truths of Scripture. No worldview aside from a Christian (or biblical) worldview generates such an understanding of the physical world as to spawn the birth of modern science. It is unlikely that modern science could have been birthed in any other worldview. In fact, the birth and subsequent success of modern science is implicit confirmation of the truth of the biblical worldview[a].

1.3 | What is Science?

A Challenge to Define

Most people are under the impression that science is easy to define. After all, the word has some sort of a definition in the dictionary and even grade school textbooks offer definitions. Furthermore, since the word is used somewhat commonly, one might infer that it should be easy to determine whether something is science or not a science. Finally, since most of the definitions floating around in our society—

---

a   The characteristics of the universe that permit it to illustrate the nature of God turn out to be what philosophers of science call the 'presuppositions' of science. Presuppositions of science are those things that are assumed so that science can be done and those things that must be true for the pursuit of science to be reasonable. The biblical worldview is the only worldview known to provide reason to believe the presuppositions of science.

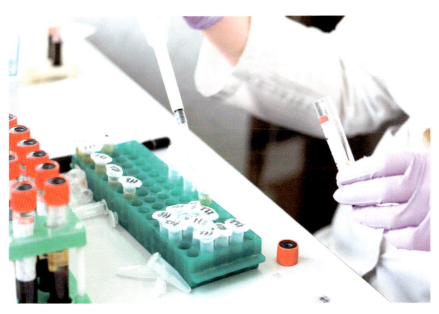

especially those which students are called upon to memorize—link science to 'the scientific method', there must be a procedure that all scientists use and no one outside of science uses.

In fact, none of these things are true. First of all, science is *not* easy to define. Even experts[a] struggle to define it, and thus far no single definition has been agreed upon. Secondly, it is not always clear whether or not something is a science. There are certain things that everyone is satisfied labeling as science. There are many other things that everyone agrees should not be labeled as science. But there are still other things that some people label as science and other people label as non-science. Finally, there is, in fact, no single 'scientific method' that all scientists use, and most of the methods that are used by scientists are also commonly used by people outside of science.

The whole story on how our society came to confuse the definition of science is long and complicated. Some of the confusion comes from the rather typical changes that occur in language, where words change meanings and words assume additional—often figurative—meanings. Some of the confusion also comes from the nature of science itself. After all, such a large variety of things are studied by science, and such a huge variety of people are scientists, that a simple definition might not be possible.

I suspect that the single most important cause for the confusion, however, has to do with the fallen nature of man, rather than the

---

a   Philosophers of science are those whose job would include creating a definition of science.

nature of science itself. In our society, scientists are held in very high regard. Many people find the esteem that comes with science to be an irresistible temptation. Some outside of science want the esteem for themselves, so they stretch the definition of science so they can call themselves scientists. Some within science want to preserve or increase the esteem they already receive, so they modify the definition of science in such a way that science is even more respectable than it really is. Some even want to exclude others from the esteem of science, so they modify the definition of science so that those other people fall outside the definition and can then be called *non*-scientists.

All this has produced a variety of inaccurate definitions of science. Not only has this led to overall confusion about the nature of science, but our society has come to cultivate (and teach our young people) an inaccurate understanding of the true nature of science[a].

The Nature of Science

Simple definitions of science are probably not possible. After all, scientists come from a wide variety of cultural backgrounds and they study many different things (*e.g.* from the structure of the universe to the makeup of electrons, from minerals of rocks to the workings of the human brain, from the behavior of extinct dinosaurs to the cause of polio, from the cause of gravity to the process of evolution). Furthermore, there might actually be human activities that lie on the

---

[a] In most educational materials (at all grade levels) science is defined improperly. Since students are often required to memorize these inaccurate definitions, most people have adopted inaccurate definitions of science.

edge of science, being (validly) defined as science by some and excluded from science by others.

On the other hand, though the fringes of science might be difficult to define precisely, the vast percentage of science is universally accepted as science. And, though science is a broad discipline, there are characteristics that seem to be found across its entire breadth. Let us now consider what seems to be the most important of the characteristics found across all of science.

Science is <u>something humans do</u> to understand the physical world by proposing tentative truths as theories of explanation and valuing fit with the physical world.

Science is a human activity. One implication of this statement is that science is done by humans. Chimpanzees do not do science. But neither does God… nor angels. Chimpanzees do not do science because they are incapable of doing so. God does not do science because He does not have to. Angels do <u>not</u> do science because it is not what they are called to do. Humans invented science and humans do science[a].

A second implication of the statement is that science is an activity. Many students might think of science as a bunch of things a person has to memorize. In contrast, scientists themselves tend to understand science as something scientists *do*—almost as if 'science' was a verb. A third implication is that since science is performed by humans, human nature plays an important role in science. For example, in our society, scientists are commonly portrayed as emotionless, unbiased seekers of truth. In fact, science is done by humans, and emotions are an essential element of being human.

Furthermore, bias has also been a part of every human who has ever lived. A person's bias might be a correct one or an incorrect one, but there is no way that any human can have no bias at all. Scientists are not only emotional beings, they are also fallible and fallen. Scientists do make mistakes. And, although it would be nice if every scientist was seeking the truth, human nature being what it is, a fair bit of science is done for less than the best motives.

Science is something humans do <u>to understand the physical world</u> by proposing tentative truths as theories of explanation and valuing fit with the physical world.

---

a As shall be discussed later in the text, humans invented and do science in order to fulfill the task that God assigned humans in the creation.

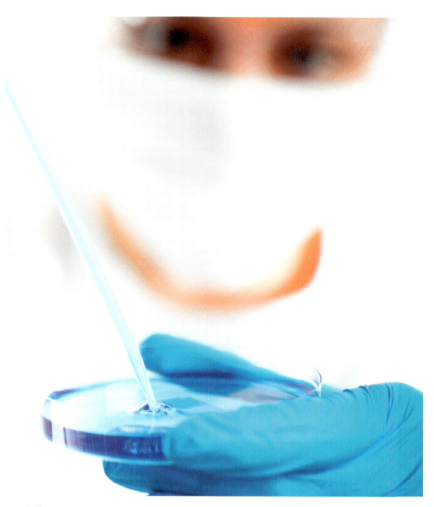

The purpose or goal of science is to understand the physical world. One implication of this is that pure science[a] does not generate anything useful. Unlike many in our society may understand, science did not provide us with light bulbs, air conditioners, or cars. It is *not* used to build bridges or computers or cure disease. Pure science only attempts to understand *how* the physical world works. Inventions, engineering feats, and medical cures are examples of applied science. The applied sciences seek to modify the physical world for the benefit of humanity. The applied scientist may or may not base his or her inventions on knowledge derived from the pure sciences.

A second implication is that science focuses its attention on the physical world. Whereas another discipline, theology, seeks to understand God; science seeks to understand the physical world. This does not mean that science rejects the existence of anything else—even

---
[a] Pure science is also known as 'natural science', or even 'modern science'.

though naturalists (and many scientists) do believe that the physical world is all there is. It merely means that science focuses its attention on the objects of the physical world, many times ignoring everything else. Thus, whereas understanding is the *purpose* of science; the physical world is the *object* of science—or said another way, the physical world is what science studies.

Science is something humans do to understand the physical world <u>by proposing tentative truths as theories of explanation</u> and valuing fit with the physical world.

A common *misconception* of science is that it has something to do with proof and certainty[a]. Rather, a better motto for science would be 'you never know for sure'. In an effort to understand the physical world, a scientist makes an educated guess called a scientific theory. Any attempt to *understand* the physical world cannot be known for sure. We'll never be able to test theories in every possible situation and at all possible times[b].

Furthermore, you can never know for sure that something might be discovered which shows the idea to be wrong. As the late paleontologist Stephen Jay Gould (1941-2002) used to say, honest scientists must always fear the 'mouse in Michigan'—that apocryphal mouse (that lived in the backyard of an elderly gentleman from Michigan) which falsified a highly celebrated theory about mouse behavior and humiliated the scientist who proposed it.

Reinforcing the idea of the tenuous nature of scientific theories is the turnover rate of scientific theories. New and better scientific theories are being suggested all the time, replacing older theories that are not as successful at explaining the world. Very few theories survive this process of modification for more than a few decades. None have survived for more than a few centuries. These short life spans for scientific theories suggest that every *current* theory of science may be wrong—in large or small part—and every scientist knows it. It may not be too far from the truth to say that every scientist prays that his or her theory is not shown wrong before he or she gets famous for proposing it... and if the theory *is* disproved that he or she will be the one to

---

a   The commonness of the phrase 'proven by science' illustrates the association of science and proof.

b   It is possible for a scientific theory to be shown to be false (if data of the physical world is contrary to the theory), but since a theory cannot be tested everywhere, at all times, under all circumstances, a scientific theory cannot be *proven* to be true.

show it wrong and become famous for its replacement theory! As a result of this, scientific theories of all types (*e.g.* hypotheses[a], historical scenarios, mechanisms, laws[b]) must be accepted as only *possible* truths.

Theories remain theories for their entire existence—science itself cannot finally declare any scientific theory to be certainly true. Although a given theory might actually be true, there is nothing in science that alerts the scientist that that particular theory is actually true[c].

Science is something humans do to understand the physical world by proposing tentative truths as theories of explanation and <u>valuing fit with the physical world.</u>

Any process seeking to discover truth requires some standard to determine which ideas should be held onto and which ideas should be rejected. In conservative Christianity, for example, the Bible is considered absolute truth because the God of Truth authored it.

Consequently, anything which does not compare favorably with the Bible is considered untrue. In science, the physical world is the standard of evaluating scientific theories. Scientific theories are created in an effort to explain the physical world. So, if a particular scientific theory is correct, or nearly correct, then the physical world should 'behave' in the particular manner expected by that theory.

Other ways this might be described is that the theory should 'fit' the physical world, or align with the physical world, or 'explain' the physical world. If the physical world is not the way the theory expects (*i.e.* the theory does not fit the physical world), then the theory

---

natural law—a regularity of the universe

---

a Whereas it is commonly taught that a 'scientific hypothesis' is an untested theory, the term 'scientific hypothesis' is rarely used by scientists, and when it is used it is not restricted entirely to untested theories. Actually, hypothesis and theory are used interchangeably by most practicing scientists.

b Whereas it is commonly taught that a scientific theory becomes a 'scientific law' with enough testing, scientific (or 'natural') laws are actually one very specific type of theory. A **natural law** is a regularity of the universe—something that is supposed to operate across all space and time. A theory that suggests no regularity for the entire universe can *never* become a law. Furthermore, a claim about a regularity of the universe is especially impossible to test over the entire universe for all of time. Natural laws cannot be proven and must always retain the status of tentative truths. Scientific or natural laws are thus nothing more than a particular type of scientific theory and must always remain a particular type of scientific theory.

c It is even questionable whether scientific claims can be legitimately arranged in order of increasing likelihood of being true. It is often believed that theories that have successfully survived many tests are somehow more likely true. For example, it is commonly taught that a scientist first makes an educated guess known as a scientific hypothesis, which, upon successfully testing, becomes a scientific theory, which in turn, upon more successfully testing, becomes a scientific law. There is an implicit suggestion that hypotheses, theories, and laws have increasing likelihoods of being true. However, if one 'mouse in Michigan' can prove a theory wrong, then just before the mouse was discovered was the theory more likely to be true just because the theory had successfully survived many tests before that? No, testing does not make a theory more or less true. It is either true or false and testing will not change that.

should be considered *un*true. Some scientists may consult additional standards of truth (Christians, for example, may consult Scripture), but the physical world is the standard that scientists accept across all the disciplines of science.

To restate our definition,

Science is something humans do to understand the physical world by proposing tentative truths as theories of explanation and valuing fit with the physical world.

When multiple theories are available to explain something in the physical world, theories are *preferred* that match the physical world best. Theories that explain more of the physical world are preferred over theories that explain less. Theories that can be tested (in other words, compared with the physical world) are preferred over theories that cannot be tested. Theories that have 'passed' more tests or more severe tests are preferred over theories that have only a few or easier tests.

Theories that fit better with other accepted theories are preferred over those that do not relate to anything else or conflict with other theories. Theories that have fewer internal problems (*e.g.* logical problems) are preferred over those with more. Even theories that lead to further research into the physical world are preferred over those that do not lead to other investigations.

What is Biology?

In modern science there are many different natural sciences, distinguished by studying different aspects of the physical world. Biology, for example, is one of the natural sciences. Derived from the Greek words *bios* ('life') and *logos* ('word' or 'discussion'), 'bio-logy' was originally understood to be the discussion or study of life and is often defined as 'the study of life'. And, if you believe that the physical world is all that exists (the worldview of naturalism), this would be an accurate definition of biology. If, however, the biblical worldview is true, then there is more to the world (and more to organisms) than the physical.

God is called the living God (*e.g.* Jer. 10:10; Matt. 16:16), but He is also spirit—not physical. Since biology is a natural science and can only study the physical, God is at least one living being Who cannot be studied by biology. Consequently, biology does not study *all* of life—but only living beings with physical bodies. The Bible also refers to life itself as something different from the physical—or at the very least something more than physical. For example, God first formed man from the dust of the earth (Gen. 2:7a). It would seem that man's complete physical being was formed at this time. But it was only after God breathed into man's nostrils the 'breath of life' that 'man became a living soul' (Gen. 2:7b).

Our physical body is an essential part of being human (after all, God creates physical bodies even for believers in heaven). But the body alone is not alive. Adam's body needed enlivening, so God gave it 'life', just as He did for all of us and even does for all animals (Psa. 104:29-30). Since it enlivens physical bodies (and is not itself the physical body), life involves something beyond the physical. And if life is non-physical, or something more than physical, then biology, as a natural science, may not be able to study life itself—at least not completely[a].

Thus, although the nature of the word itself would suggest 'biology' is the 'the study of life', the science of biology may not actually be able to study life! At best, science can study only organisms—the bodies of those physical things that possess life. So, rather than say that biology is the study of life (a common definition and an acceptable one in naturalism), it would be more accurate to say that **biology** is the study of organisms (physical beings having life).

biology—a science that studies organisms

---

a   More discussion on the non-physical nature of life is found in Chapter 2.

1.4 | Why Study Biology?

Whereas the first half of this chapter focused on what biology is, the second half of this chapter focuses on why we should study biology. Why should a student majoring in music, English, history, or a host of other non-science fields take a course in biology? Since a pure science seeks to understand the physical world rather than produce something of value to humanity, what use is there then in studying science at all? And is not science dominated by unbelievers who attack Christianity? Did not biologists come up with evolution, and does not evolution contradict the Bible? Why should a Christian young person have to suffer through memorizing a bunch of useless facts, especially when he or she dislikes it, and it may even be an enemy to his or her faith?

I would suggest that there are a host of reasons why a Christian should study biology. A dozen of them are introduced below. It should be noted that because of the holistic nature of God and His truth, most of these reasons are strongly interrelated and often difficult to clearly distinguish.

Biology and Our Priesthood

Theologians have pointed out that humans have an important **priestly** function. There are similarities between the Garden of Eden as described in Genesis 2 and Ezekiel 28:13 and the future abode of believers described in Revelation 21 and 22 (*e.g.* gems, gold, tree of

priest—divinely-assigned role of humans to use the creation to better know God, worship God, and bring others into the worship of God

life, God living with man, no curse, and no death). Both heaven and Eden are specially designed as places where man and God were to live together.

There are also similarities between the Garden of Eden and the tabernacle built by Moses and the temple built by Solomon (*e.g.* cherubim, a single entrance, God interacting directly with man). Between the dismissal of humans from Eden and admittance of humans to heaven, the tabernacle and temple were to provide a picture of both Eden and heaven.

In both the tabernacle and the temple, God descended to earth to sit between the cherubim of the ark, and there to dwell among priests. These priests in Numbers 3:7-8 were called upon to 'keep' and 'serve' (Hebrew words *shâmar* and *'âbad* respectively). First, of course, they were to 'keep' or look out for the physical care of the temple. At the same time, however, these priests were called upon to maintain a pure relationship with God which would compel them to worship God. In this manner they were also to 'keep' the temple a place of continual worship. Furthermore, this relationship with God was to compel them to share that relationship with others, thereby cultivating a similar love of God in others. In this way they were to 'serve', or meet the spiritual needs, of the remainder of the people by bringing them into temple worship. And all this was to bring glory to God.

When God charged Adam (and thus all humanity) in Gen. 2:15, He used the same Hebrew words, to 'dress' (*'âbad*) and 'keep' (*shâmar*) that He would use later for the temple priests. Man has been called upon from the very beginning to be a creation priest, or a priest of God in the creation. He is to preserve or care for the creation as the abode of God (as the temple priest was to care for the temple). He is to make it a place of continual worship by deepening his own relationship with God and using the creation as a stimulus to worship (as the temple priest was to do for the temple).

Finally, he is to serve others by bringing them into that worship (as the temple priest was to do for the people of Israel). In fact, man should bring the entire creation into the worship of its Creator. Towards that end, the study of biology can help a Christian be a better creation priest in at least three different ways.

Through Biology We Can Better Understand God. The more familiar a person is with an artist's work, the more intimate is his understanding of the artist. A careful study of God's handiwork (such as a careful study of organisms) has the potential of providing an intimate understanding of God Himself[a]. This would be the case even if God was not *trying* to teach us about Himself. However, Romans 1:18-20 indicates that God does intend to teach us about Himself. He carefully designed the physical world to provide physical illustrations of His invisible qualities and He carefully designed humans to recognize those qualities. The passage indicates that God reveals these truths to all people, even to unbelievers.

The Bible, especially as the Holy Spirit reveals its truths to believers, provides much more specific information about God than is revealed in His Creation. However, physical illustrations from the creation can help us understand abstract truths. Witnessing a pattern throughout the entire creation can also confirm and deepen our understanding of biblical claims. The deeper our understanding of God the more effectively we can fulfill our responsibilities as priests of the creation. A study of biology can thus deepen our understanding of God and help us be better creation priests.

Through Biology We Can Better Worship God. Once you recognize how a biological feature teaches something about God, I believe you

---

a   Noting here that, tragically, because of man's rebellion, some can come to an intimate knowledge of the creation, but reject its Creator ("...the invisible things of Him from the creation of the world are clearly seen... so that they are without excuse... [But]...when they *knew* God, they glorified Him not as God...": Rom. 1:20-21) will be forever changed. I pray that for the rest of your life, every time you see that biological feature you will be reminded of God.

Sometimes seeing such things can even cause us to spontaneously erupt into worship. Psalm 19:1-6, which begins with "The heavens declare the glory of God...," is an example of the worship that David experienced in response to viewing God's creation. In fact, following David's example, there is a long tradition among the Hebrews and in the church to produce psalms, and hymns and praises to God based upon the creation. A few of many examples would include *All Creatures of Our God and King* (1225 A.D.), "Fairest Lord Jesus, Ruler of all nature... Fair are the meadows, Fairer still the woodlands..." (*Fairest Lord Jesus*, 1677), "...There's not a plant or flower below But makes Thy glories known..." (*We Sing the Greatness of Our God*, 1715), "...Birds in song His glories show..." (*I Am His and He is Mine*, 1876), "...When through the woods and forest glades I wander... Then sings my soul..." (*How Great Thou Art*, 1885), "All nature sings... This is my Father's world..." (*This is My Father's World*, 1901), "...Join with all nature in manifold witness..." (*Great is Thy Faithfulness*, 1923).

Once we learn to recognize how God shows Himself through His creation, the biological creation can be used to stimulate us to worship. Sharing these experiences with others, such as David did in Psalm 19, can then stimulate others to worship. As this text introduces biological illustrations of God's nature, the author will be reminding

you of the responsibility we have to respond in worship[a]. I hope that you are compelled to worship in response to first learning these things. Even more, though, I pray this begins a life-long experience of priestly activity—using His creation to worship our God and bringing others into that same worship.

Through Biology We Can Better Glorify God. God created the physical world for his own pleasure (Rev. 4:11). The chief end of man is to glorify Him ("…whatsoever you do, do all to the glory of God": I Cor. 10:31). One way a person can glorify God is by recognizing that a specific aspect of the creation was placed there as an illustration of His nature. God is further glorified when we share those truths with others. In short, as we fulfill our priestly function, we bring glory to God. Every time this text reveals how the biological world illustrates a characteristic of God, you will learn a tool for glorifying God that you can use for the rest of your life.

1.5 | Biology and Our Kingship

We are not only priests, we are kings. Only after God created the sky, land, and sea (creation days 1-3a), then filled them with plants, lights, and animals (creation days 3b-6), did God create man. God then gave man '**dominion**' over all the things that He had created (Gen. 1:26-28)—over the 'birds of the air, the fish of the sea, the animals of the land, and over the earth itself'. The Hebrew words used in this passage translated 'dominion' and 'subdue' are words commonly used for kings. Psalm 8:4-8, understood by theologians to be an inspired commentary or elaboration on this passage, uses the word 'crown' (Psa. 8:5) and clarifies that that rule is over 'all things'.

The New Testament commentary on this same passage, I Cor. 15:24-28, uses the words 'kingdom', 'rule', 'authority', and 'reign'[b]. God is certainly the One Who deserves to rule over the creation. After all, He is the Creator of all those things and the only One Who has

dominion—divinely-assigned role of humans to rule over the creation, enhancing its glorification of God

---

a This reminder to worship is found in multiple chapters under the 'Our Responsibility' section of each chapter.

b Some people feel that the Psalm 8 and I Corinthians 15 passages are *only* making reference to Jesus Christ. In Psalm 8:4, the 'son of man' probably has reference to Jesus Christ, but the passage revolves about the question 'What is man that Thou art mindful of him?'—referring to humanity. Because humans failed in many of their tasks that God took on the form of man to fulfill those purposes. I Corinthians 15 is Jesus Christ's fulfillment of the role God gave to man.

absolute power over those things. Nonetheless, God chose man to be a ruler over the creation in His place.

God did not look at everything He had made and then determine which being would be most suited to be ruler. The "Let us make man…" statement in Genesis 1:26 indicates instead that He planned for humans to be rulers *before* He created them. This distinction is important because we are rulers according to divine appointment, not because of our ability. He designated us to be rulers *then* gave us the tools we would need to accomplish this task. We have no reason to boast of our role as rulers, for both the role and the abilities were given to us by God so that we can obey and glorify Him.

This also means that we fill the role of ruler even if we lack the ability (analogous to the case of a very young boy who becomes king over a country). Unfortunately, this also means that we *must* rule (we were made rulers—Gen.1:26—and we were commanded to rule—Gen. 1:28) and we *must* bear the responsibility of reigning, even if we turn out to be very poor rulers. The study of biology can help a Christian be a better ruler over the creation in at least three different ways.

Through Biology We Can be Better Shepherd Kings. So many rulers have so abused the authority given to them by God that many people cannot see **kingship** as something good. However, God's standard of the ideal ruler is the shepherd king. God is described as the shepherd of Israel (*e.g.* Psa. 23:1; 80:1; Isa. 40:11), and Jesus calls Himself 'the good shepherd' (John 10:11-16). God commanded the rulers of Israel

kingship—divinely-assigned role of humans to rule over the creation, serving both God and creation

to rule as good shepherds over the flock of God (Ezekiel 34). Just as the shepherd is willing to lay down his life for his sheep, so the shepherd king is to promote his subjects over his own pleasure and desires. He is sensitive to, and meets the needs of, his subjects and at the same time is subject to God. The shepherd king serves both God and subjects. The shepherd king is one example of the servant leadership advocated throughout Scripture.

Consider that when people choose rulers, they tend to choose men who have somehow proved their power. The first king mentioned in the Bible, for example, was Nimrod (Gen. 10:10). The Bible describes him as 'a mighty one on the earth' (Gen. 10:8) and 'a great and mighty hunter before the Lord' (Gen. 10:9). Although people choose such 'mighty men' so that such a person will protect them, many of the same mighty men abuse their power for their own gain. Likewise, when Samuel was seeking to anoint a king over Israel, he thought he had found such a person in Eliab, the strong, first-born son of Jesse (I Sam. 16:6). However, God did not choose Him. God passed over the older, stronger-looking sons of Jesse and chose Jesse's youngest son, the shepherd David. David had proven himself faithful to the care of his sheep and he had proven himself to have a heart for God (I Sam. 16:7).

We should follow the example of David as we rule over the creation of God. Rather than exploit, plunder, and abuse the creation for our own end, we should seek God with all our heart and serve the creation

He created. We should rule over the creation *as God would rule*[a]. One evidence of God's concern for the creation is the Sabbath rest He required the Israelites to provide to their *land* ("For six years you shall sow your land and gather in its fruits, but the seventh year you shall let it rest and lie fallow, that the poor of the people may eat; and what they leave the beasts of the field may eat. You shall do likewise with your vineyard, and with your olive orchard.": Exo. 23:10-11, and similarly in Lev. 25:2-7).

It is instructive to note that God did not assign us to be rulers without giving us the ability to rule. As God created humans, He gave humans the authority and power to rule. That we have the power can be seen in the incredible ways in which humans have directly altered the creation. That we have the authority can be seen in the massive impact our sin has had on the creation. When Satan fell, there was apparently no change in the physical creation. The fall of man was another matter entirely. According to Romans 8:18-20, the pain, suffering, bondage, and corruption to which the *entire* universe is subject is a consequence of man's sin. God cursed His own creation because of man's fall. Then the sin of man in the days of Noah brought about a global flood that destroyed all living things found on the land. Even when we think

---

a   It is common to say that we should not 'play God', and we should certainly not take a role God reserves for Himself or attempt to take such a role. However, God has called us—He has commanded us—to rule in His place, to act in His interest, and to rule as He would rule. If we do not do as God would have us to, then we are sinning against the Almighty God. There is a sense in which we *must* play God (*i.e.* rule as He has called us to, to act as He would act).

we are abdicating our responsibility to rule, we are still rulers. Our decisions—good or bad—substantially impact the creation.

In order to serve the creation most effectively we need to better understand it. If we do not know what plants need to thrive, or what critters need to stay healthy, or what the earth requires to best support the life on it, then we will do a poor job of caring for the creation. We will rule poorly even if we desperately wish to rule well. The more we understand the creation, the better we can serve it. The study of biology allows us to better understand the biological creation, and thus better rule over it. I pray that his text will encourage you to whole-heartedly serve God and rule over the biological creation more effectively.

Through Biology We Can be Better Stewards. As rulers over the creation, we are also called to be stewards. A steward in Scripture is placed in charge of someone else's belongings (*e.g.* Gen. 44:2; I Chr. 28:1). Though he is not the owner, the steward is made 'ruler' over those things (Luke 12:42, 44). Since we do not own the creation that we rule over, we are stewards of that creation.

The steward over a household serves that household, paying employees (*e.g.* Mat. 20:8), providing for the needs of the people in the house (*e.g.* Luke 12:42), and taking care of visitors (*e.g.* Gen. 43:19, 24). The steward is given his responsibility for the purpose of serving others (I Pet. 4:10). A steward who abuses others or serves himself rather than others is specifically condemned (Luke 12:45). In both our

roles—as shepherd kings and as stewards—we are expected to provide for the needs of the creation, serving the creation and not abusing it.

At the choice of the actual owner of the goods, stewards even enjoy the same kind of authority that the owner possesses. Biblical examples of stewards had the authority to buy and sell needed provisions (*e.g.* Gen. 44:1), the authority to capture thieves (Gen. 44:4-5) or release prisoners (*e.g.* Gen. 43:23), and the authority to renegotiate the debts of the owners (Luke 16:5-7). Likewise, God has granted us the authority over the creation to do with it as we like. At the same time, the steward was not to perform his own will, but the will of his master (Luke 12:45, 47). So likewise, we are to perform the will of God in the creation and not our own will. As in the case of our role as ruler, as stewards we are to oversee the creation *as God would*.

Finally, because the owner might ask for an accounting at any time, the biblical steward is to be ready at all times (Luke 12:43) to give an account of how well he has taken care of the owner's possessions (I Cor. 4:1; *e.g.* Luke 16:1-2). Since Christ could return at any time, as stewards of His creation, we must always be ready to give an account of how well we have taken care of those things He made. Applying the **stewardship** principle to our role as rulers of the creation, we must realize that the authority we have over the creation is not something earned or even deserved but granted by God. So, it is something to be thankful for. We should care for the creation because it is not ours, but God's. It was made for His pleasure and is His handiwork and we must treat it as important to God[a].

We should realize that someday we will have to give an account of everything that we did, including our role as rulers over His creation. Did we rule over it as we were supposed to? Did we care for it as if it were important to God? We are rulers whether we like it or not, and if we do not rule well, or if we refuse to rule at all, we are subject to judgment. The more we learn about God and the biological creation the more we learn about how to be a good steward of the biological creation.

I pray this course will better prepare you for the Lord's return by helping you care for His creation more effectively. As this text introduces biological illustrations of God's nature, the author will be

stewardship—divinely-assigned role of humans to be stewards of God's creation, preserving and enhancing its glorification of God, being ready always to give an account to God

---

[a] God is not said to 'love' His creation (His love seems to be reserved for fellow members of the Godhead and His children, *i.e.* humans), but as His creation and possession, He certainly values it.

reminding you of the stewardship responsibility we have towards the biological creation[a]. I hope that through this, you will unite with others in the church in creating a more constructive perspective of creation stewardship than has been common in the past.

Through Biology We Can Enhance Creation's Glorification of God. The parable of the talents in Matthew 25 introduces a concept too infrequently heeded by the church. In that parable (Mat. 25:14-30) the lord gave talents to three servants—or stewards. When the lord returned after a time he rewarded those who invested *and* increased their talents but condemned the servant who did not take a risk and merely preserved the talent he was given. It seems that the lord expected each servant not just to keep what he had, but also to take some risk and work hard to use the talent and multiply it. The lesson we get from this is that God desires for us to take those things which we have been given and *increase* or *enhance* them.

As rulers and stewards, we have been given charge of the creation. Applying the principle of increase or enhancement to our role as stewards and rulers, we are not just to preserve the creation, we are to take some risk and work hard to increase or enhance the creation. How do we do that? God created to bring glory to Himself, and it is the purpose of man to glorify God (I Cor. 10:31). Consequently, the creation is increased or enhanced when it brings God more glory. We glorify God as we fulfill our priestly responsibilities (*i.e.* as we

---

[a] This reminder of creation stewardship is found in the 'Our Responsibility' section of multiple chapters.

recognize the illustrations of God's nature in His creation, worship God as a result, and bring others into that worship). But we can *increase* the creation's glorification of God if we can increase how much glory the creation actually brings to God.

The good ruler *enhances* the creation. He brings out more evidence of the nature of God so as to reveal more and more of the glory of God. At first thought this would seem impossible. God created the world to bring glory to Himself. Can man even think to make the world better than God created it? Actually, we can, *if* God created it in such a way that we could. And He did. God revealed many illustrations of His nature in the creation, but it seems that he hid even more. God glorified Himself in the creation but included in that same creation the potential for even more glory. And He gave humans the ability to reveal that hidden potential and thus increase or enhance the glory that creation brings to God.

This may be analogous to parents who wrap gifts for their children, enjoy their children's excitement as they unwrap the presents, and revel in the thanks their children give them with the receipt of the gifts. It is as if God wrapped gifts of His glory for us to reveal, and God enjoys our excitement as we unwrap the gifts and He revels in the thanks we give Him for those gifts. Through biology we can discover the potential—the gifts of glory—God hid in the biological creation and find ways to reveal it. In our role as rulers, we can enhance the biological creation to the increased glory of God. As the text introduces more ways to enhance the attributes of God in the biological creation, I pray that you will learn how to glorify God more effectively[a].

1.6 | Biology and Our Image

Humans were created as the **image of God** (Gen. 1:26). We are not to make images of God (Ex. 20:4) because God has already made images of Himself, and they are us! From before the creation of the world, God planned a physical representation of Himself in the world. This representation was not to look like God, for God is spirit and not physical. In a certain sense this representation was not even to have God's attributes, because God is infinite and any created being could at most only have a tiny, limited measure of any of His attributes.

image of God—the divinely-declared, special status of each and every human—from conception to death—as a representative of God

---

a Specific suggestions on how to enhance the creation for the glory of God are provided under the 'Our Responsibility' section of multiple chapters.

Rather, humans were to be His representation because God *declared* them His representation. As in the case of our rulership status, our image status is not because we deserve it, but because He decided to make us His image. He did in fact grant us some tiny measure of His attributes, but only so that we could better image Him—so that we could bring more glory to Him. Just as a king of Egypt might design a ring to represent his power or have a statue or a building constructed to represent his person or kingdom, God designed and fashioned humans to be a representation of His person, kingship, and kingdom. We have the status of image whether we're a good image or a bad one, and whether we want to be an image or not. Nonetheless, it would be better if we imaged Him well. The study of biology can help a Christian be a better image of God in at least three different ways.

Through Biology We Can Better Care for Our Bodies. Because God is spirit we cannot 'look' like God, but our bodies are visual representations of God, nonetheless. The way in which we appear to others is the representation of God we offer to them. For as long as we are alive and in whatever we do, we are a visible representation of God. This is reason enough for humans to care for their bodies, so as to illustrate God well. For believers, however, there is an *additional* reason. The body of a believer is also home to the Holy Spirit—thus the very temple of God (I Cor. 3:16-17; 6:19-20; II Cor. 6:16).

If God lived in a particular building we would take great care to see that that building was well cared for. In fact, God does live in a

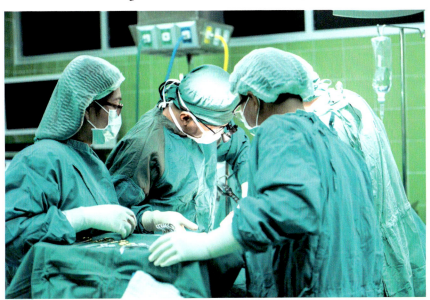

building—not one built by human hands—but a building God created: the body of the believer. Christians should take great care to see that their bodies are well cared for. Biology gives us insight into the workings of our bodies. The more we know about our bodies, the better we can maintain our bodies and care for the temple of God. Although the structure of the human body is not the focus of this text, I pray that what *is* said about the human body will help you better maintain the temple He has given you. In this way you can better image God.

Through Biology We Can Make Better Ethical Decisions. As the image of God, Our bodies are a physical representation of God. At the same time our decisions are a moral representation of God. We are to do right—to make the decisions that God would make. To be the best possible image of God, we should be perfect as He is perfect. So we are commanded (*e.g.* "Be holy, for I am holy": Lev. 11:45 & I Pet. 1:15-16; "…let us …perfect holiness…": II Cor. 7:1).

Part of that command to be perfect is fulfilled as we stop wrongdoing and do the right things instead. Yet, to do right in a given situation we need to understand enough about the situation to understand what is right and wrong. For example, if a person does not know that most IVF (in-vitro fertilization) procedures involve abortion, he or she may think that all IVF procedures serve only to promote life. Do the different birth control methods differ in ethics, or risk of abortion, or danger to the body? What about genetic engineering, or cloning? Are some methods displeasing to God? Are some honoring to God? Human technology permits us to do things we never dreamed of before, and human technology of the future will allow us to do even more. Which of these things are right and which are wrong? What about the changes in global climate, pollution, extinction, and human population? Which should be stopped and which should be allowed to continue? And what is the best way of doing what is right?

Learning biology and the language of biology can help us to know what is right and wrong in a host of bioethical situations. Biology can help us obey God's command to do what is right, and thus be better images of God. I pray that the ethics comments scattered through the text will help you to be holy as He is holy.

Through Biology We Can Learn to Respect God's Word Over Man's. We have been commanded to worship God and only God (*e.g.* Exo. 20:3-6), and we are to fear God and only God (Deu. 13:4). We are not to worship or fear the creation, we are not to worship or fear the idols that we make, and we are not to worship or fear other humans. In part this is because we are the image of God. It would not be appropriate for God to worship or fear anything He had created, so it is not appropriate that His image should be seen worshiping or fearing anything God created. Our awe and worship should be directed only towards God.

Down through the centuries, a number of challenges to Christianity and the Bible have come from the field of biology. Even many Christians have listened to the critics rather than God. As a result, they have questioned Christianity, they have doubted God,

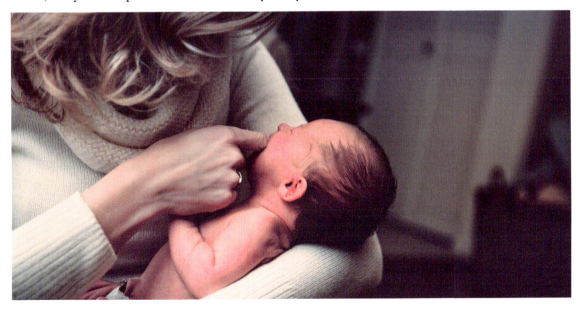

and they have questioned God's word. And this is *within* the ranks of Christianity! Such things should never happen. The conclusions of science are tentative truths, and this can be demonstrated by a study of science. God's Word carries more weight than biological challenges to it, and this can be demonstrated by a study of biology. For example, because biological evolution is a substantial challenge to Scripture in our day, critiques of evolutionary theory are found throughout the text. I pray these critiques will encourage you to embrace the claims of God over the claims of man, and thus better image God. I also pray that the overall effect of this course would lead to the strengthening of your faith, and that you would wholeheartedly embrace the claims of Scripture, becoming thereby immune to any efforts to undermine the authority of the Bible.

Biology and Our Service to Others

To a large extent, the reasons for studying biology that we have considered so far are a fulfillment of the first great commandment (Mat. 22:37-8). As we function as priests and kings of the creation and images of God, we love the Lord our God with all our heart, soul, and mind. The second commandment is to love others (Mat. 22:39). The study of biology can help a Christian better minister to the needs of others in at least three different ways.

Through Biology We Can Better Meet the Physical Needs of Others. One of the ways we love others is by serving them (Gal. 5:13-14). Studying the biological world has the potential of providing knowledge which we can use to meet the needs of others (*e.g.* to treat disease, provide food crops, and advise in bioethical dilemmas). An accurate understanding of the human body can allow a better understanding of the physical responses and needs of others. Even an understanding of the theories of science which oppose Scripture can permit us to more effectively minister to those who accept those theories. I pray this course will improve your overall ministry to others, and thus allow you to better obey the second great commandment.

Through Biology We Can Better Meet the Spiritual Needs of Others. Rom. 1:19-20 tells us that God uses His creation to introduce Himself to humans. If God can use His creation to illustrate Himself to humans, we should be able to use His creation to lead unbelievers to

Him and believers closer to Him. As we learn more about the creation, we should be able to discover more ways to share God with others. At the very least, an understanding of science will broaden and round a person's general knowledge, which in turn will allow discussion of a wider range of topics with others.

Learning biology should provide more opportunities for common ground in discussions and should make the Christian a more interesting person to others. And, as we become familiar with more biological illustrations of God's attributes we also learn more ways to introduce God to others. In this fashion, a study of biology can help us better love others by ministering more effectively to their spiritual needs. I pray that this text increases the effectiveness of your own evangelistic efforts.

Through Biology Christian Biologists Can Be Salt and Light to Other Professionals. Unbelievers everywhere need a Christian example. Christians are called upon to be salt and light in our world (Mat. 5:13-16) and to evangelize the world (Mat. 28:18-20). In whatever profession, in whatever position (*e.g.* friendship), with whatever authority (*e.g.* advising or voting) that believers possess, believers should introduce the world to the Gospel, a Christ-like lifestyle, and to Biblical morality and ethics. Unbelieving scientists, for example, need the testimony of believers. God has called some of his children to pursue the science of biology and be the salt and light that that discipline needs. I would be thankful indeed if this text were to stimulate someone to follow God's calling into professional biology.

Summary of Chapter

A. By His very nature, God is undetectable and unknowable. However, God wishes for us to know Him, so He condescended and created the physical world full of physical illustrations of His nature.

B. For the creation to illustrate the invisible attributes of God, the universe and man had to be created with certain 'knowability traits'.

C. The 'knowability traits' of the creation include:

   a. The physical world
      i. exists;
      ii. is ordered;
      iii. has an order simple enough for us to understand;
      iv. For every physical world event there is a cause;
   b. The regularities of the creation
      i. can be detected and understood in a single human lifetime;
      ii. are consistent across the entire universe for all time;
      iii. are ultimately unifiable;
   c. Human
      i. senses are reliable;
      ii. brains are complex enough to understand the order of the creation;
      iii. language (including mathematics) is capable of describing, understanding and teaching about the creation;
   d. Truth and understanding can be gained by studying the creation;
   e. There is intrinsic value in studying the physical world; and
   f. Truth can be advanced by a continual study of the creation.

D. These 'knowability traits' of the universe not only seem to be true, but they are readily explained by the Christian worldview (and no other worldview).

E. Since modern science assumes all the 'knowability traits' of the creation to be true, modern science was birthed in—and probably could only have been birthed in—a Christian worldview of the creation. The success of science is implicit confirmation of the truth of the biblical worldview.

F. Modern/Natural/Pure science

   a. is difficult to define and impossible to define precisely because 1) the word 'science' refers to several very different things; 2) so many different kinds of people do science and so many different things are studied by science, and 3) scientists are held in high regard, so people have distorted the meaning of science to make themselves look better or to make others look worse.
   b. is not defined by a 'scientific method' (because there is no single method used throughout all of science) and is not associated with proof (because theories of science cannot be known with certainty).
   c. is something humans do (a 'verb'; a distinctively human activity, done by fallen, biased humans) to understand the physical world (science's purpose is non-pragmatic; science's object is the physical world) by proposing tentative truths as theories of explanation (scientific theories are suggestions of understanding that can never be proven) and valuing fit with the physical world (science's standard of truth is the physical world).
   d. prefers theories that 1) explain more of the physical world, and/or 2) have more often been successfully compared with the physical world, and/or 3) fit better with other accepted theories, and/or 4) have fewer mismatches with the physical world, and/or 5) stimulate more science.

G. Biology is a science that studies organisms.

H. Humans are to serve as priests of the creation by (1) caring for the creation and using the creation to (2) better know God, (3) worship God and (4) bring others into the worship of God. It is good that all Christians study biology so that the believer can be a better priest of the creation by (1) learning enough about the creation to take better care of it, and (2) becoming familiar with more illustrations of God's nature so to deepen the believer's understanding of God's nature, (3) stimulate the believer into worship of God, and (4) give us opportunity to glorify God before others and include them in the believer's worship of God.

I. Humans are to rule as kings over the creation, (1) looking out for the best for the creation, all the while (2) ruling as God would rule, accountable (as stewards) to God for our actions, and (3) searching for ways (as stewards) to further enhance the glory of God in the creation. It is good that Christians study biology so that the believer can be a better ruler over the creation by (1) better understanding the creation so that the believer can better care for it and protect it (as a shepherd king), (2) learning more about God through the creation so that the believer can better rule over it as God would (as a steward), and (3) discover and reveal illustrations of God's nature that God hid within it, and thereby enhance the glory of God in the creation (as a good steward).

J. Humans are the image of God because God declared us His image. It is good that Christians study biology so that the believer can image God better by learning (1) how to take better care of the believer's body (as the temple of the Holy Spirit), (2) how to make better ethical decisions, and (3) how to respect God's word over man's.

K. Christians are to serve others. It is good that Christians study biology so that the believer can better serve others (1) by learning things about the human body and organisms that interact with it to allow the believer to better meet the physical needs of others, (2) by learning illustrations of the nature of God that allow the believer to better meet the spiritual needs of others. The believer can even (3) pursue biology professionally and thereby introduce other professional biologists to God.

Advanced Discussion Topics

A. There was a period of time when it was popular to derive spiritual truths from the creation (described by some as 'spiritualizing'). Most historians of both science and the church look uncharitably on this period, considering the entire endeavor not only a failure but also wrong-headed. What was wrong with this approach? How is spiritualizing different from the approach of this text (in finding illustrations of divine attributes in the creation)? What safeguards should be followed to avoid the same mistakes? Things to consider: compare and contrast the direction of inference (theology to creation vs. creation to theology); compare with proper and improper methods in the analogous project of biblical typology; and observe where Scripture is explicit vs. non-explicit.

B. Modern science was apparently not invented for the first 6000 years of human history. Why was modern science not invented until then? Things to consider: the socio-economic conditions that would support full-time jobs for people who do nothing useful, but merely study the creation; the importance of widespread literacy/higher education; and the relationship between the presuppositions of science and cultural/religious worldviews.

Test & Essay Questions

1. Define the physical world / non-physical world / scientific theory / science / knowability traits / biology.

2. Compare and contrast physical and non-physical / natural science and applied science / popular and biblical understandings of kingship / stewardship and dominion.

3. According to Romans 1:20, what is one of the ways that God reveals Himself?

4. Short Essay: Explain how Romans 1:20 / God's creation of illustrations logically leads to [any one of the 'knowability traits' of the creation]

5. Short Essay: Explain how [the understandability of the universe / the knowability of universe regularities in one human lifetime / the consistency of universe regularities in space and time / the unification of universe regularities / the adequacy of human language to describe, understand, and teach about the universe / the close approximation of the universe by mathematics] is *not* expected in naturalism.

6. Essay: Explain how a Christian worldview [leads to the 'knowability traits' of the creation / provides philosophical foundation for the presuppositions of science].

7. Short Essay: Explain why science is so hard to define.

8. Who does science? / What is the purpose of science? / What does science study? / What does science use as a standard of truth?

9. Short Essay: What does it mean to say that scientists often understand science to be a verb?

10. Short Essay: Can science be used to study God / angels / the soul / the human spirit? Why or why not?

11. Short Essay: Comment on the statement 'Science gave us the light bulb and computers and automobiles and planes'.

12. Short Essay: What does it mean to say that a scientific theory is tentative truth / possible truth? / What does the statement 'In science you can never know for sure" mean?

13. Short Essay: Comment on the statement 'Science has proven [insert any claim here]'.

14. List two different ways in which one scientific theory may be preferred over another.

15. What is the popular definition of biology and what is wrong with that definition?

16. The popular definition of biology is acceptable in what worldview? / Short Essay: In what worldview is the popular definition of biology unacceptable, and why is it unacceptable in that worldview?

17. List two evidences that suggest that life might not be physical.

18. Short Essay: Why is it that biology may not be able to study life? / Why is it ironic that biology may not be able to study life?

19. Essay: How would you explain to a Christian who is not majoring in science and dislikes science why he or she should take a course in biology?

20. Essay: If science merely seeks to understand the physical world, rather than produce something of value to humanity, what use is there in studying the physical world?

21. Essay: What does it mean to say that we are priests / kings, and what does that have to do with biology?

22. Short Essay: How can we use biology to know God / to glorify God / to worship God?

23. Short Essay: How can we use biology to be better kings / to be better stewards / to better execute dominion?

24. List two different ways that we can glorify God.

25. Short Essay: If God created the creation to glorify Himself, how do we increase God's glory in the creation?

26. Short Essay: How can we use biology to better care for the temple / to do right / to not fear man?

27. Short Essay: Why should we take care of our bodies?

28. Short Essay: What does it mean to say that many Christians fear science, and why should we not fear science?

29. Essay: How can we use biology to obey the second greatest commandment?

30. Short Essay: How can we use biology to be salt and light / to witness to others / to minister to others?

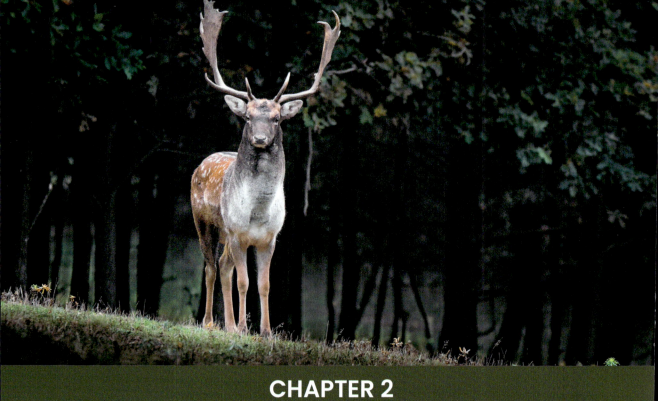

CHAPTER 2

THE LIVING GOD

Different Types of Life & The Origin of Life

"...worship Him Who lives forever and ever..."
Rev. 4:10, NKJV

2.1 | The Living God

In the Bible, God is repeatedly described as the living God, the God who lives, He Who lives forever, and He Who lives forever and ever[a]. God Himself claims He lives and that He lives forever[b]. He is described as the living redeemer and the living Father[c]. Jesus is called the bread of life, the light of life, the Prince of life, the Word of life, the living stone, the ever-living Word, life, eternal life, the One Who was dead but now lives, and the One Who was dead but now lives and lives forever[d]. Life is clearly one of the attributes of God Himself.

In creating the physical world to illustrate His invisible attributes (Rom. 1:20), God created organisms to illustrate His attribute of life. God created plants (Gen. 1:11-12) and living 'creatures' ('created' beings) in the sea (Gen. 1:20-21), the air (Gen. 1:20-21), and the land (Gen. 1:24). Even now, all living things are in His hands (Job 12:9-10) and are held together by 'the word of His power' (Heb. 1:3). God created and holds together everything that biologists study (Col. 1:16-17).

> VIDEO 2.1

Created Life

Different Types of Life

Intuitively, most people seem to have a sense that there are fundamental differences between **plants** and **animals** and further distinctions between the life of animals and the life of humans. Animal life seems to deserve more respect than plant life, and human life seems to deserve even more respect than animal life. Even aside from biblically-defined morality, there are also legal distinctions between killing a

plant—(multi-cellular) eukaryotic organism whose cells have cell walls of cellulose

animal—(multi-cellular) eukaryotic organism whose cells lack cell walls (also possesses biblical life)

---

a   Biblical references for the living God: Deut. 5:26; Josh. 3:10; I Sam. 17:26, 36; II Kings 19:4, 16; Isa. 37:4, 17; Psa. 42:2; 84:2; Jer. 10:10; 23:36; Dan. 6:20, 26; Hos. 1:10; John 6:69; Acts 14:15; Rom. 9:26; II Cor. 3:3; 6:16; I Thess. 1:9; I Tim. 3:15; 4:10; 6:17; Heb. 3:12; 9:14; 10:31; 12:22; Rev. 7:2. Biblical references for the God who lives: Jud. 8:19; Ruth 3:13; I Sam. 14:39, 45; 19:6; 20:3, 21; 25:26, 34; 26:10, 16; 28:10; 29:6; II Sam. 2:27; 4:9; 12:5; 14:11; 15:21; 22:47; I Kings 1:29; 2:24; 17:1, 12; 18:10, 15; 22:14; II Kings 2:2, 4, 6; 3:14; 4:30; 5:16, 20; II Chr. 18:13; Job 27:2; Jer. 4:2; 12:16; 16:14-15; 23:7-8; 38:16; 44:26; Psa. 18:46; Amos 8:14; Hos. 4:15. Biblical references for He Who lives forever: Dan. 4:34; Dan.; 12:7. Biblical references for He Who lives forever and ever: Rev. 4:9-10; 5:14; 10:6; 15:7.

b   Biblical references to God claiming that lives: Num. 14:28; Isa. 49:18; Jer. 22:24; 46:18; Eze. 5:11; 14:16, 18, 20; 16:48; 17:16, 19; 18:3; 20:3, 31, 33; 33:11, 27; 34:8; 35:6, 11; Zeph. 2:9; Rom. 14:11. Biblical reference where God claims that He lives forever: Deu. 32:40.

c   God is the living redeemer (Job 19:25) and the living Father (John 6:57).

d   References for the bread of life (John 6:33-35, 47-48, 51), the light of life (John 8:12), the Prince of life (Acts 3:15), the Word of life (I John 1:1-2), the living stone (I Pet. 2:4), the ever-living Word (I Pet. 1:23), life (John 11:25; 14:6), eternal life (I John 5:20), and the One Who was dead but now lives (Rev. 2:8) and lives forever (Rev. 1:18).

plant, killing a cheetah and killing a human. Could this be because plants and animals and humans differ in some fundamental manner?

Divine Life

Because life is part of the very nature of God, He has always lived and will always live. Life is an attribute of God. It is also important to remember that God is completely distinct from His creation and not to be confused or considered equal with it. Likewise, His attributes are distinct from similar qualities in His creation. The life of God is uncaused. It had no beginning, no cause, and no creator. The life of God is also autonomous. It is dependent upon nothing else for its continued existence. And, because God Himself is infinite, the life of God is probably infinite and unbounded in quality.

Consequently, the first great division of life is divine life on the one hand, possessed only by God, and creature life on the other hand, which is all other life. Whereas divine life is uncaused, autonomous, and infinite, creature life was created by God, is completely dependent upon God for its continued existence, and is limited in quality. This distinction is our first indication that not all life is the same, and a hint that even among creature life there might be different types of life.

The Life of Spirit Creatures

God created both physical beings and non-physical (or spiritual) beings ("…by Him were all things created… visible and invisible…": Col. 1:16). Among the invisible things, the Bible describes spirit creatures that God created (*e.g.* angels, cherubim, seraphim). Among them, only the cherubim are explicitly described as living (Ezekiel 1:5-25; 3:13; 10:15-22). The other spirit creatures are nowhere explicitly described as 'alive' or 'dead' or 'dying'.

However, the activity of these spirit creatures, as well as other similarities to what we know to be living beings, would suggest that these spirit creatures are living beings as well. Since no reference is made to the death of any of these spirit creatures (even in the lake of fire), the life of spirit creatures is probably unending. And, since these are spirit creatures—not having (or having need of) physical bodies—it is likely that they possess a different kind of life from living creatures with bodies, so it is listed here as a distinct type of creature life.

Nephesh Life

Creatures that do have physical bodies were brought into being in four different creation events in Genesis 1: plants on Day Three; sea creatures and birds on Day Five; land animals on Day Six; and humans later on Day Six. Of these, the Bible explicitly describes several of them as living—and dying[a]. Animals of the sea are described as living and dying[b]. Animals of the air, such as birds and bats are described as living and dying[c]. Animals of the land are described as living and dying[d]. Among the animals of the land, even creeping things, like insects, amphibians, and rodents are described as living and dying[e]. Humans are described in various places as alive, dead, or dying, even those newly born (II Sam. 12:18-23; I Kings 3:18-27) and those in the womb (Ex. 21:22-25).

More specifically each category of beings is described in the Hebrew as *nephesh hayim* (translated as 'living creature' for water creatures and birds in Gen. 1:20-21, 'living creature' for land animals in Gen. 1:24-25,

---

a   Presumably only something that was once alive can die, so references to something dead or dying suggest that that something was once alive.

b   Biblical references for animals of the sea being alive are Gen. 1:21; 9:2-4; Lev. 11:10, 46; Eze. 47:9; Rev. 8:8-9; and 16:3, and biblical references for animals of the sea dying are Exo. 7:18 and 21.

c   Biblical references for animals of the air being alive are Gen. 1:30; 2:19; 6:17-20; 7:2-3, 14-15, 21-23; 8:17; 9:2-4, 10; Lev. 14:6; 16:10, and 20-21, and a biblical reference for animals of the air dying is Lev. 14:6.

d   Biblical references for animals of the land being alive are Gen. 1:24, 30; 2:19; 6:17-20; 7:2-3, 21-23; 8:17; 9:2-4, 10; Ex. 19:13; 21:34-36; 22:4-7, 51-53; Lev. 11:30-32; Num. 22:33; I Kings 18:5; Prov. 12:10; and Eccl. 9:4, and biblical references for animals of the land dying are Gen. 33:13; Ex. 9:4, 6, 19; 21:33-36; 22:10-14; Lev. 7:24; 11:39; and Jer. 21:6.

e   Biblical references for creeping things, like insects, amphibians, and rodents being alive are Gen. 1:24, 30; 6:19-20; 7:14-15, 21-23; and Lev. 20:25, and biblical references for creeping things, like insects, amphibians, and rodents dying are Ex. 8:13; and Eccl. 10:1.

and 'living soul' for humans in Gen. 2:7). Thus, according to Scripture, humans and animals possess what might be described as nephesh life (or soul life). Nephesh life would then be a type of creature life found in creatures that by nature have both bodies and souls.

It is also likely that there is a different life possessed by humans and animals, for the Bible treats them as *very* different types of organisms. Some of the differences include: animals and humans were created in separate creation events; humans were formed directly by God, rather than by divine command (compare Gen. 2:7 with Gen. 1:20 and 24); God breathed life into humans directly (Gen. 2:7b); humans, and only humans, are the image of God; humans were made rulers over the animals; and the human soul continues forever, whereas the soul of animals apparently dies at the death of the animal's body (Eccl. 3:21). So, it is likely that there are at least two very different types of nephesh life: human life and animal life.

2.2 | Biblical Life

In the case of plants, Scripture does not seem to consider them alive—or at the very least they are not described as having nephesh or soul life. When God creates animals, He refers to each category of animal as living (Gen. 1:21 & 24), but when He creates plants (Gen. 1:11-12), He does not describe them as living. In fact, at the end of the **Creation Week**, plants were given as food to everything 'wherein there is life' (Gen. 1:29-30), strongly suggesting that plants themselves are not living (at least in the biblical sense of life).

A millennium and a half later, *after* requiring that 'all living things' of the land be included on the ark of Noah, God refers to plants only inferentially when He required that the food of the animals be included as well (Gen. 6:19-21). Again, the strong suggestion is that plants are not considered living beings in Scripture. In fact, nowhere in Scripture are plants referred to as living. Most commonly, the Bible refers to plants as 'green things' (*e.g.* Ex. 10:15; Rev. 9:4[a]), as if the Bible refers to thriving plants as photosynthesizing, rather than 'living'. As for whether plants can die, plants in the Bible can be 'cut down', 'cut off',

Creation Week—first six days (~6000 years ago) of time, during which God created the physical world

---

[a] Further biblical references where plants are referred to as green things include Gen. 1:30; 9:3; Deu. 12:2; I Ki. 14:23; II Ki. 16:4; 17:10; 19:26; II Chr. 28:4; Job 15:32; Job 39:8; Psa. 37:2; 37:35; 52:8; Isa. 15:6; 37:27; 57:5; Jer. 2:20; 3:6; 3:13; 11:16; 17:2; 17:8; Eze. 6:13; Hos. 14:8; Mark 6:39; and Rev. 8:7. In each case, Hebrew or Greek words are used that mean (the color) 'green' or are derived from words that mean (the color) 'green'.

'choked out', 'felled', 'hewed', 'smitten', 'broken', 'burned', 'plucked', and 'pulled up', but there is no biblical example of a plant being 'killed'[a].

Most of these phrases when applied to animals or humans would involve death (*e.g.* for a human to be 'cut down'). However, when these words are applied to plants, they do not necessarily refer to the complete end of a plant, since "there is hope in a tree when it is cut down, that it will sprout again" (Job 14:7) and "…if a seed dies, it brings forth much fruit." (John 12:24)[b]. The Bible seems to ignore whatever 'life' a plant might have and refer instead to what the plant produces (*e.g.* seeds, fruit, leaves, wood). This probably reflects the *purpose* of plants—which is to provide food for animals. The Bible also indicates that a plant can 'fade', 'languish', 'wither', 'dry up', 'fail', 'fall', and 'wax old'[c]. Of the scores of references to the termination of the vitality of

---

a Biblical references that indicate plants can be: (1) 'cut down' are Ex. 34:13; Deu. 7:5; 19:5; Jud. 6:25-30; II Ki. 6:4; 18:4; 19:23; 23:14; II Chr. 14:3; 31:1; 34:4; Job 8:12; 14:2, 7; Psa. 37:2; 80:16; 90:6; Isa. 9:10; 10:34; 37:24; Jer. 22:7; 46:23; and Luke 13:7-9; (2) 'cut off' are Job 18:16; 24:24; Eze. 17:9; and Dan. 4:14; (3) 'choked out' are Mat. 13:7; Mark 4:7; and Luke 8:7; (4) 'felled' are II Ki. 3:19, 25; 6:5; and Isa. 14:8; (5) 'hewn' down are Deu. 19:5; II Chr. 2:10; I Ki. 5:6; Isa. 33:9; 44:14; Jer. 6:6; 46:22; Dan. 4:14, 23; Mat. 3:10; 7:19; and Luke 3:9; (6) 'smitten' are Ex. 9:25, 31; Psa. 105:33; and Jonah 4:7; (7) 'broken' are Ex. 9:25; Job 24:20; Psa. 89:40; 105:33; and Isa. 27:11; (8) 'burned' are Deu. 12:3; II Ki. 23:15; Psa. 80:16; 83:14; Jer. 51:32; Joel 1:19; and Rev. 8:7; (9) 'plucked' are II Chr. 7:20; Eccl. 3:2; Jer. 24:6; 42:10; 45:4; Eze. 17:9; 19:12; Micah 5:14; Luke 17:6; and Jude 12; and (10) 'pulled up' are Eze. 17:9 and Amos 9:15.

b An example of a plant sprouting after being 'cut down' is given in Isa. 10:34-11:1, and an example of a plant sprouting after being 'hewn down' is given in Dan. 4:23f.

c Biblical reference that indicate plants can: (1) 'fade' are Isa. 1:30; 28:1, 4; 40:7-8; 64:6; Jer. 8:13; and Eze. 47:12; (2) 'languish' include Joel 1:12; (3) 'wither' are Gen. 41:23; Job 8:12; Psa. 1:3; 37:2; 90:6; 102:4, 11; 129:6; Isa. 15:6; 19:6, 7; 40:7, 24; Jer. 12:4; Eze. 17:9-10; Joel 1:12, 17; Amos 4:7; Jonah 4:7; Mat. 13:6; 21:19-20; Mark 4:6; 11:21; Luke 8:6; John 15:6; James 1:11; and I Pe. 1:24; (4) 'dry up' are Isa. 42:15; Job 18:16; Eze.

plants, only four passages might suggest that plants can perish or die, but each passage can also be interpreted to refer to the cessation of a plant's productivity, not its actual ceasing to exist[a].

Plants, then, not only seem to lack nephesh or soul life, but they seem to lack anything at all that the Bible refers to as life. Consequently, what might be called biblical life—or that which is considered 'life' in the Bible—is possessed by God, spirit creatures, humans, and animals, but it is not possessed by plants. Nephesh life, since it is possessed by animals and humans, but not God and spirit creatures, would then be a type or subset of biblical life. Other things, like **fungi**, protozoa, **algae**, and bacteria are not specifically mentioned in Scripture. However, their similarity with plants would suggest they, like plants, do not possess biblical life.

Biological Life

God and spirit creatures are not studied by biologists because they are not physical beings. This suggests a limitation of biology. Biology cannot study all beings that possess life. On the other hand, biologists study and consider living a number of things that may not be living as the Bible uses the term (*e.g.* bacteria, protozoa, algae, fungi, and plants). This text will refer to all being studied by biologists (humans, animals, plants, fungi, algae, protozoa, bacteria) as organisms, and the type of life possessed by organisms will be referred to as biological life.

Even More Types of Life

A summary of the different types of life we have considered is illustrated in Figure 2.1. Since biblical life, creature life, nephesh life, and biological life are not the same, but do overlap, it is likely that there are a number of different types of life and that biblical, creature, nephesh, and biological life are different categories of life—each including multiple, distinct types of life. Different types of spirit

---

fungi—(generally multi-cellular) eukaryotic organism whose cells have cell walls of chitin

alga (pl. *algae*)—eukaryotic organism that is generally a one-celled, aquatic producer

---

17:24; Hos. 9:16; Joel 1:12; and Mark 11:20; (5) 'fail' include Isa. 15:6; (6) 'fall' are Eccl. 11:3; James 1:11; and I Pe. 1:24; and (7) 'wax old' include Job 14:8).

a   The first passage refers to the death of a plant (Job 14:8), but this poetic passage continues by indicating that it will bud if it is watered (Job 14:9), meaning that the plant was not actually dead, but cut down and no longer bearing. The second passage indicates that plants can 'perish' (Psa. 80:16), but another poetic passage in the Psalms also indicates that the heavens will 'perish' (Psa. 102:25-26), so 'perish' in the Psalms is not restricted to once living things. The third passage (as part of one of Christ's parables) refers to the death of a seed (John 12:24) but then indicates that the seed bears fruit only if it dies, so death here seems to refer simply to burial (which if done to a human would lead to death!), not to the actual death of the seed. The fourth passage is part of an analogy and refers to plants 'twice dead' (Jude 12)–*i.e.* those 'with withered fruit' and 'without fruit', suggesting one 'death' is losing fruit and the second 'death' may refer to being unable to bear fruit at all.

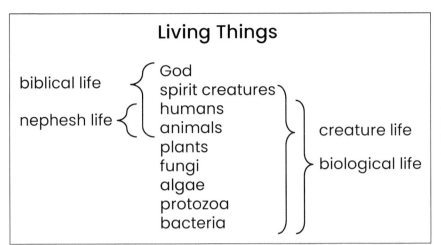

Figure 2.1
Different Types of Life

creatures, different types of animals (I Cor. 15:39), different types of microorganisms, and different types of plants may each contain distinct types of life.

2.3 | The Spectrum of Perfection of Life

Why are there so many different kinds of life? If there is only one God, and He wanted us to understand the invisible attribute of life, why did He provide more than one physical illustration of life? Part of the answer probably has something to do with the fact that there is more than one person of the Godhead (see Chapter 8). But another reason is that God wanted us to know more than the simple fact that God has the attribute of life.

God also wanted us to understand that He has *infinite* life. The finite brains of humans cannot fully comprehend anything that is infinite. So how does God teach a finite being that He possesses an infinite measure of a particular trait? Besides making the claim in His word, I suggest that God has created both the creation and the human mind in such a way that humans automatically conclude that God possesses an infinite or unlimited measure of a particular trait.

I suggest that God endowed different parts of His creation with a suite or spectrum of different degrees or perfections of the trait in question. Some things in the creation lack the trait entirely. Other things have only a miniscule measure of the trait. Other things have the trait in small measure, others still possess intermediate amounts of the trait, and still others possess the trait in large measure.

spectrum of perfection—a range of degrees of perfection of a God-illustrating trait, created in a suite of different physical things so that humans extrapolate to God's infinite manifestation of that trait

When the different measures of the trait found in different objects are arranged in order (from objects lacking the trait to objects having the trait in large measure), what results is a **spectrum of perfection** of the trait. At the same time, I suggest that God endowed humans with three automatic responses to this spectrum of perfection: 1) a tendency to recognize the different perfections in the creation; 2) a compulsion to order these different perfections into a spectrum; and 3) a tendency to extrapolate the spectrum beyond what is observed. Each occurrence of the trait in the creation provides an illustration to help us understand the trait. The spectrum of perfection of that trait causes us to extrapolate beyond what is actually observed towards an unlimited or infinite manifestation of that trait.

In Figure 2.2, the four circles represent four different examples of a given trait in the creation. The smallest circle represents something with a miniscule measure of that trait, the next circle something with a small measure, the next circle something with an intermediate measure, and the largest circle something with a large measure of that trait. It is a human tendency to arrange the circles in order of increasing size and then (with the dashed lines) extrapolate from what is seen towards a circle of unlimited or infinite diameter.

If humans possess some measure of the trait along the spectrum (as represented by the stick figure in the third circle), then humans also conclude that whatever possesses the trait is very much greater than humans. This spectrum of perfection aids our minds to envision an

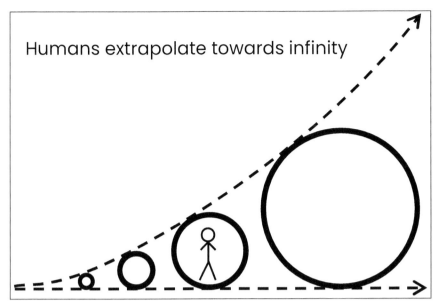

Figure 2.2
Spectrum of Perfection Graphical representation of human design in a spectrum of perfection—namely that human tendency of arranging items—from small to large—and then extrapolating from that towards the infinite.

infinite form of that trait. Though we will never *fully* comprehend such an infinity, our minds are lifted in that direction.

In the case of life, there are objects in the universe that lack life altogether. Such things as stars, planets, moons, and the substances of which they are made do not seem to possess life at all (represented by the intersection of the dashed lines on the left side of Figure 2.2). A whole suite of other things (bacteria, protozoa, algae, fungi, and plants) have so many physical attributes of life that they are considered alive even when they are not considered living in Scripture (represented by the smallest circle in Figure 2.2).

God even created some things between non-living things and biological life. Viruses have DNA (or the very similar RNA) but they are not made of cells and they can only do the other things characteristic of life by taking over the cells of organisms. So, there is a spectrum of perfection of life from non-living things, to viruses, to biologically living things. There is also a sense in which this spectrum can be continued among the biologically living things. An animal seems to be able to experience a fuller or more perfect life than a plant (represented by the second circle from the left in Figure 2.2). And, though the cells that make up the bodies of plants or animals or humans are living, such cells seem to possess a less perfect form or manifestation of life than the organisms they make up.

Even the same organism in different stages of development or different states of health seems to possess life in more or less perfect forms. A newly fertilized human egg appears to us to be less vibrant and active ('alive') than a swimmer during an Olympic event. A person in a coma shows fewer evidences of 'life' than a child laughing in the midst of play. A human, who has the *capacity* to know God has a more perfect form or manifestation of life than an animal (represented by the third circle from the left in Figure 2.2). Jesus even claimed that He had come not just to give life, but to give *abundant* life (John 10:10). This refers to a fullness of life that can be experienced by humans who know God and know Him intimately. A person who is in intimate relationship with God is in a sense more alive than a person who knows God more casually. And one who knows God is more alive than a human who does not (represented by the circle farthest to the right in Figure 2.2).

This creates a spectrum of perfection of life. From the spectrum of perfection of life, God's design of humans causes them to automatically extrapolate (the dashed lines in Figure 2.2) towards something more alive, dynamic, and exciting than the fullest life we have either experienced or imagined. This gives us insight and elevates our understanding towards the infinite measure of life possessed by God.

2.4 | The Nature of Life

Because life is part of the very nature of God, He has always lived and always will live. All other life was created by God, whether it be spiritual or physical ("...by Him were all things created... visible and invisible...": Col. 1:16). Since God created living organisms to illustrate His attribute of life, the more we learn about the nature of that life, the more we should be able to learn about God. Beyond that, a number of important issues in our society and in our lives are dependent upon a proper understanding of life. What life is and when life begins and ends, impacts our understanding of such things as birth control, abortion, in-vitro fertilization, fetal tissue research, cloning, and euthanasia.

There are several Biblical reasons to believe that life is far more than physical. First, since God is alive, but is also spirit (John 4:24), God's life—the source of all life—is not physical. Second, cherubim are alive, but they are spirit beings (having no body), so the life of spirit beings is not physical either. Third, after creating Adam, it was not until after God breathed the 'breath of life' into him that 'man' became a living soul (Gen. 2:7).

Elsewhere in Scripture human death comes with the departure of the ghost or spirit or soul (*e.g.* Job 14:10, Acts 12:23[a]) and "...the body without the spirit is dead..." (James 2:26). Ghost, spirit, and soul are all contrasted with the body (*e.g.* "...a spirit does not have flesh and bones...": Luke 24:39[b]), suggesting ghosts, spirits, and souls are not physical. Life, which also leaves the body at death would seem to be non-physical also. Fourth, each time God created animals, He called *nephesh hayim* into being (translated as 'living creatures' in Gen. 1:20-21, 24-25). *Nephesh hayim* is the same Hebrew phrase that is translated 'living soul' in reference to Adam in Genesis 2:7[c]. If human souls are something more than physical, then there is an implication that animal life is also something more than physical.

There are also *non*-biblical reasons to believe that life is more than just physical. First, even when studied carefully in a scientific laboratory with many sophisticated instruments, the death of an organism involves no change in mass or volume. When life departs, nothing is detected

---

a Further biblical references to giving up the ghost or spirit or soul: Gen. 25:8, 17; 35:18, 29; 49:33; Job 3:11; 10:18; 11:20; 13:19; Jer. 15:9; Eccl. 3:21; Matt. 27:50; Mark 15:37, 39; Luke 23:46; John 19:30; Acts 5:5, 10.

b Other biblical references referring separately to the physical and ghost or soul or spirit or Deu. 6:5; Mark 12:30; Luke 10:27; I Thess. 5:23; Heb. 4:12.

c The point here is not to equate the human soul with the animal soul, for it is clear that there is something quite distinct about the spirit of man (*e.g.* Eccl. 3:21), but only to say that the Bible indicates there is more to an animal than a physical body.

as departing (*i.e.* nothing leaving the body can be seen, smelled, tasted, felt, or heard). If life is something that actually leaves an organism at death, it would seem that it is not physical. Second, though life in the naturalistic worldview (naturalism) is thought to be nothing more than chemistry, there are few people who truly believe that their own life—or even the life of their pet cat or dog—is merely chemical reactions and nothing more. Third, biology textbooks written from a naturalistic worldview never define life. Sometimes the question "What is Life?" is never even discussed. In all cases, the 'characteristics' or 'common traits' of living things are listed but life is never defined. Life being something more than physical would explain why biologists have not been able to identify it after all these centuries of study.

To clarify, what is being suggested here is that *all* types of life (biblical life, nephesh life, *etc.*) are non-physical in nature. Even the biological life of bacteria and plants is non-physical in nature. In fact, the non-physical nature of all life might give us further insight into different types of organisms. Some organisms may differ only in non-physical ways. This would have to be true in the case of non-physical beings like God and spirit beings because they are only non-physical. But the non-physical nature of life itself makes it possible for two physical beings that are identical physically, might actually differ in non-physical ways. And, even between two organisms that differ physically, it is possible for the non-physical differences to exceed the physical differences. And, if life really is non-physical, we can neither directly perceive life, nor discern differences between different types of (non-physical) life. This in turn means that there may be many more types of (non-physical) life than we have so far recognized.

The non-physical nature of life also provides insight in the nature of biology. In the preceding chapter we observed that although biology is supposed to mean 'the study of life', biology cannot study all of life because biology cannot study living spirit beings. The further revelation that life is not physical suggests biology has yet another limitation—that it cannot even study life *in* physical beings. This creates the somewhat ironic situation that biology cannot study life itself! What biology does study, is organisms—the physical things that *possess life*.

2.5 | The Origin Of Life

The Law of Biogenesis

The smallest creatures of God's creation have been invisible to humans for most of human history. Most single-celled organisms (thousands of **species** of bacteria, protozoa, and algae) are too small to be seen without the aid of a microscope—something that was not invented until more than fifteen centuries after Christ. In fact, even for large organisms, egg cells and sperm cells are usually too small to be seen.

species—smallest recognizable group of similar organisms that persists naturally by the production of similar offspring; taxonomic level below genus

spontaneous generation—the natural (not created) origin of life from non-living things; once believed of all lower life forms, now only believed in naturalistic evolution for the first organism(s)

Consequently, until the last century and a half or so, the early stages of development were unknown for many organisms. Maggots, for example, seemed to appear on rotten meat from nothing other than the meat itself, and they seem to do this constantly and easily. Mice seemed to arise from dirty hay. Crocodiles seemed to be the products of logs rotting in water. For most of human history, humans have believed in **spontaneous generation**—the idea that lower life forms arose naturally from non-living material. For most of that time, few people ever considered the possibility that the idea was wrong. Since spontaneous generation is a claim about how the physical world works, spontaneous generation is an example of a 'scientific theory' of the past.

It was not until a series of famous experiments performed over the course of two hundred years (from the 1600s to the 1800s) that spontaneous generation came to be questioned, challenged, and ultimately rejected. According to spontaneous generation theory, for example, maggots spontaneously came into being on meat. This particular claim was challenged by an experiment performed by Francesco Redi (1626-1697) in 1668 (see Figure 2.3). Redi placed meat in each of three jars. One jar was sealed, one was covered with netting, and third jar was left open. Maggots appeared on the meat in the open jar, but not the other two.

If spontaneous generation were true, one would expect maggots to appear in all three jars—or at the very least in the jar exposed to the air but with netting over it. It was as if maggots appear only when

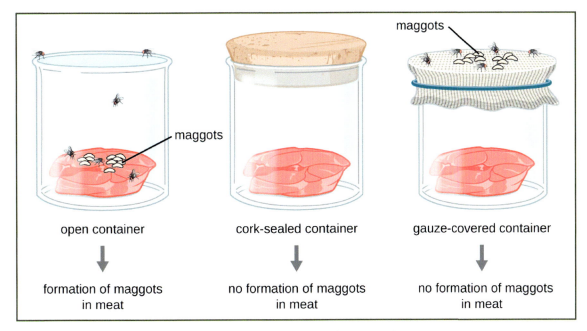

Figure 2.3
Francesco Redi's 1668 experiment. Image by CNX OpenStax via commons.wikimedia.org, used under CCby4.0.

something in the atmosphere leaves something on the meat that was too small to be seen. In the century to follow, microscopes were developed. With the use of microscopes, humans, for the first time, could actually see the tiny eggs that flies had laid, and from which the maggots were hatched. In a similar fashion, a variety of animals and plants were found to be generated from tiny eggs, spores, or seeds of previous animals and plants.

Ironically, however, the invention of the microscope that helped challenge spontaneous generation also encouraged its continued acceptance—in a slightly different form. Using the microscope, biologists not only discovered never-before-seen parts of known organisms, they also discovered never-before seen organisms—**microorganisms**. Such microorganisms in a broth would be killed if the broth were boiled but would reappear in the same broth after it was allowed to cool. These microorganisms seemed to have arisen by spontaneous generation. Even if larger, more complex organisms did not arise by spontaneous generation, these microscopic organisms must have. After all, it was thought, they were so small that they could not be that complex; therefore, it must be relatively easy for them to arise spontaneously.

This modified view of spontaneous generation—that microorganisms were spontaneously generated from nonliving matter—stood more or less unchallenged for nearly a century. In 1768,

microorganisms—organisms too small to be seen without magnification

Lazzaro Spallanzani (1729-1799) demonstrated that if the boiled broths were left sealed, microorganisms did not reappear. Once again, it would seem that the physical world was not consistent with the theory of spontaneous generation. The experiments continued and the issue was debated for another century.

Finally, the Paris Academy of Sciences offered a prize to anyone who could settle the debate once and for all. Enter Louis Pasteur (1822-1895), who designed an experiment in 1862 that put the issue to rest (see Figure 2.4). Pasteur boiled chemical broths in flasks with long necks. One neck was straight and the other was curved into an 's' shape and he exposed both of them to the air. The curved necked flasks trapped organisms, preventing them from reaching the broth. For over a year, the boiled broth produced no organisms in the curved necked flasks until the curved necks were removed. Not long after, living organisms began growing in the broth. In a sense, the physical world had finally spoken loud enough to convince just about everyone that spontaneous generation theory was wrong.[a]

If spontaneous generation was not true, what was true? It seemed instead that organisms came from other organisms. From Redi's experiments, maggots (baby flies) appeared on meat when flies were allowed access to the meat, suggesting that baby flies came from flies. So, just as humans come from humans and lions from lions, flies came from flies. In the case of microorganisms in broth, it would seem that microorganisms appeared in broth when some sort of 'eggs' or 'seeds' of those microorganisms floating in the air have access to the broth. So, just as oak trees come from oak trees and tulips come from tulips, microorganisms come from microorganisms.

As more and more specific claims were made, scientists realized that a general principle was being discovered. This justified describing a natural law called the **law of biogenesis**. The law of biogenesis claims that a given type of organism only arises from another organism of the same type. In its brief form the law of biogenesis states that life only

biogenesis, law of—a general observation that a given type of organism only arises from another organism of the same type, sometimes briefly summarized by 'life only comes from life'

---

[a] Note, however, to reinforce what we said about science in Chapter 1, spontaneous generation has never been *proven* wrong—and never can be. We have not tested—and we cannot test—every single individual in every single species to demonstrate that every single organism in every single species did not arise spontaneously. At best we can say that every known organism we have tested properly came into existence by spontaneous generation. This might be sufficient to convince most people that spontaneous generation is wrong, but (as is the case with science in general) we cannot know for sure.

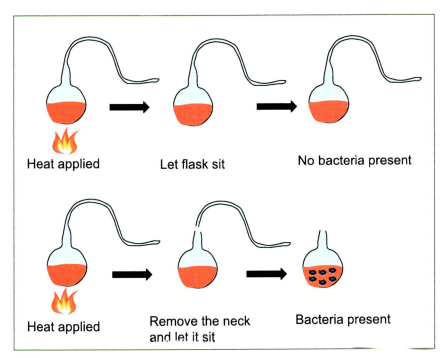

Figure 2.4
Louis Pasteur's 1862 experiment. Image by Kgerow16 via commons.wikimedia.org, used under CCby4.0.

comes from life. In the century and a half or so that this law has been tested, every organism we know about seems to follow the law[a].

How did Life Come to Be?

The law of biogenesis is true in the present. Was the law of biogenesis also true in the past? Has life always only come from life? *Biologists have never observed a contradiction to this law.* They have never observed a living organism arising spontaneously from nonliving things. With all their ingenuity and technology, and despite many attempts, humans have also not been able to create a living organism in the laboratory, even the simplest living organism.

Humans have not even been able to bring something back from the dead. God has resurrected a person that was dead for three days, but humans have done no such thing. People have been revived after they *seemed* dead for a few minutes, but humans have never brought back from the dead anything that had been truly dead—something where every cell of which it was composed was dead[b]. No human has

---

[a] Note, that, as is the case in science in general, the law of biogenesis has not been *proven* to be true, for it has not been tested on all organisms everywhere for all time, and it cannot be. The law of biogenesis has been shown to work in so many different cases that we may be pretty confident it is always true, but, as with any theory of science, we cannot know for sure.

[b] At any given moment in time, you have trillions of living cells in your body. Each living cell possesses biological life. Each cell does not have the nephesh or biblical or spiritual life that you enjoy, but each cell possesses its own biological life. Most of the cells of your body, for example, can be taken from your body and

accomplished what Dr. Frankenstein did in Mary Shelley's famous novel, nor have humans done anything approaching it. In fact, it seems that Christ purposely waited until Lazarus had been dead for several days, so that Jesus could demonstrate that Lazarus was not being revived from the *appearance* of death, but from death itself (John 11). The law of biogenesis would seem to be true not just of the present, but also of the past. But was it true *all* the way back to the origin of life itself?

Everyone seems to agree that there was a time in the past when organisms did not exist. Organisms had a beginning. According to the worldview of naturalism, the origin of the first organism must have also been the origin of life. In naturalism the first organism must have arisen from non-life. This process is called **abiogenesis** (Latin *genesis*, meaning 'beginning' + *bios*, meaning 'life' + *a*, meaning 'not'—the beginning or origin of life from non-life). Even though it contradicts the law of biogenesis and everything we know or have reason to believe about the origin of life, the naturalist believes that abiogenesis occurred at least once in the history of life.

Another difficulty with abiogenesis is that there is no reason to believe that the physical world can generate non-physical entities. Because "you can never know for sure" in science, abiogenesis has not been *proven* wrong, and *cannot* be proven wrong (because we can never prove that it is impossible in every situation, everywhere through all time), but it seems to be contrary to things that we are fairly confident of in science.

Scripture comes to the rescue with a better explanation for the origin of life. The Bible indicates that the living God created everything (Col. 1:16), including all that has any kind of life, whether it be angels or humans, or animals or plants, or any lower form of life. Anything alive is made alive by God (Deu. 32:39; I Sam. 2:6; II Kings 5:7; Job 33:4; Acts 17:25,28). Apparently, this is even the case for individual

---

abiogenesis—naturalistic theory for the origin of the first cell from non-living material

---

kept alive separately. In some sense, then, your body has 'a life of its own', or perhaps trillions of lives of its own. When a person's heart stops and blood stops bringing things to and from the cells that the cells need to survive, the cells of the body begin to die. How long it takes them to die depends upon many things (e.g. if the body is cold enough, cells in the body could remain alive for hours, days, or even years), but generally cells die in the matter of a few minutes. Under these circumstances, if a person is 'resuscitated' (i.e. the heart begins beating again and the lungs begin breathing again) before the cells of the body die, then the person's body can recover. In such a situation, when a person is resuscitated, the person has NOT been brought back from the dead, for that person's body had not yet died. Aside from the few cases mentioned in Scripture where God supernaturally raised a person from the dead, once the cells of a person's body have died there is no resuscitation. Those people who claim to have 'come back from the dead' WERE NEVER DEAD.

organisms, for the 'Spirit of God' is described as continually creating animals (Psa. 104:29). Even resurrection from the dead is something that *only* God can do.

God is also the source of spiritual life (John 6:54; 10:28; 17:2-3; 20:31; Rom. 6:23; I Cor. 15:22; II Cor. 3:6; 13:4; I Tim. 1:16; I John 5:11-13). All life comes from the ever-living God—the God of life. In the biblical worldview, the law of biogenesis is always true: life only comes from life. Furthermore, not only is an all-powerful God sufficient cause for life, but a non-physical God provides the kind of non-physical cause one might expect for the origin of non-physical life.

2.6 | Life: Our Responsibility

With the acquisition of knowledge comes responsibility—a responsibility to use that knowledge wisely. In the light of what we have learned about life, what responsibility falls upon our shoulders? As priests, we have a certain responsibility to God and the people around us. As kings and stewards we have a certain responsibility to the creation itself. Let us now briefly turn our attention to these responsibilities.

Our Responsibility to God

As priests of the creation (see Chapter 1) we have a responsibility to know God, worship God, make the creation a house of worship, and bring the creation into the worship of God. What we learn about

biological life can help us do that. A study of life gives us insight into God's very nature. The nature of life itself, for example, gives us at least two insights into the nature of God. First, we know from Scripture that God is spirit, but spirit, by its very nature, is not something we can sense with our physical senses. So, what is spirit, and what is it like? We get some insight into the very substance of God through our experience with life. Even though we cannot see life either (for it is non-physical), we nonetheless can see the *effect* of life on the physical and we can infer from that what is the nature of life itself.

We also gain insight into the nature of life as we experience our own life. Although we may be incapable of a full understanding of life, we grow in our understanding of it, and thereby grow in our understanding of the nature of other non-physical things, such as spirit—the very essence of God. Secondly, we know from Scripture that God is life. Life is one of the attributes of the invisible God that we cannot see. So that we understand His invisible attributes, God created the physical world so that we can understand them (Rom. 1:20). God has apparently created organisms with life so that we can get insight into what God's life is like.

Even though biological life is distinct from and different from God's life, we can infer (and Scripture confirms) that God's life has some of the same qualities (such as activity, vitality, intrinsic value). The more we learn about biological life, the more insight we are given into God's attribute of life.

The law of biogenesis provides at least three more insights into the nature of God. Since physical entities may not be capable of generating non-physical things, life itself—being non-physical—is likely to have a non-physical cause. God, being spirit and all-powerful, does seem capable of bringing life into being. Furthermore, the law of biogenesis suggests life only comes from life. God, being not only spirit and all-powerful, but also life, is not only a good candidate for the origin of all life, it seems that He is the only candidate. This makes God not only life, but the source of all life.

Just as Scripture indicates, we exist only because of Him ("For in Him we live and move and have our being": Acts 17:28). This not only provides insight into God's nature, but also reinforces how important it is that we abide in Him. The law of biogenesis also provides insight into God's activity. If life only comes from God, then each new life must come from God. The 'life' of one parent does not mate with the 'life' of the other parent and generate a 'new life'. God must be involved to create or infuse life into the offspring.

This means that God is continually active all over the world creating life in millions of organisms every single day ("You send out Your spirit and they are created": Psa. 104:30). Even though He cannot be seen, new lives that spring up about us all the time suggest that He is all about us, every moment of every day, actively working in His creation. We know that from His word, but biological life can keep that truth active in our thoughts. Thirdly, the law of biogenesis

suggests something about how interesting God is—how creative He is. Every human we meet is unique. Even when we encounter 'identical' twins—those that seem identical physically—they differ in personality and in character. They differ in the non-physical. Every human is a unique individual—different from every other. Even every animal is distinct in its personality—in its non-physical being.

Since God is the Creator of the non-physical in each individual organism, God is the author of astonishing variety. As creative as humans are, God is so much more creative. We would run out of different ideas long before we arrive at the millions of different organisms God must create in a single day! Such creativity makes God deeply exciting and fascinating, rich in character, and so very valuable to know.

Our access and comprehension of life provide at least two more insights into the nature of God. We can experience life and even contemplate the nature of life because God created biological life in such a way that we *could* better understand Him. This indicates that God desires to be known. Although God is so awesome as to deserve associating only with Himself—exclusively with His equal—God actually wishes to associate with us. Only the God of the Bible is One Who reveals Himself to us and desires to be known.

The spectrum of perfection of life provides yet another insight into the nature of God. From this spectrum we deduce that the life that God possesses is unlimited in nature. With unbounded life being part of the very character of God, death has no part of His character. When

God created, He created only life, not death. Humans were created with biological life, nephesh life, and spiritual life, with the purpose of living with God for eternity. Then man chose to rebel. Even when warned that on the day that he would eat of the tree of the knowledge of good and evil that man would die (Gen. 2:17), man cast away that life and disobeyed God. Man was immediately cut off from the source of spiritual life. Man's sin introduced the enemy of spiritual death into the experience of man and man died spiritually. And, as a result of man's sin, nephesh death was also introduced into the world (Rom. 5:12). Not only would man's body now die, but animals would also die.

This, of course, introduced a serious problem to humans. In rejecting the life God gave them, humans were now fated to both spiritual and physical death ("The wages of sin is death…": Rom. 6:23). Since the life they now needed could only come from God, humans were, and are, unable to rescue themselves. Yet, the God of life, Who gave man life in the first place, still desired man to know Him ("God so loved the world…": John 3:16a).

Consequently, the God of life who had no need of nephesh life, took on human nephesh life and became a man (John 1:14), the man Christ Jesus. Jesus not only lived as a man, He took upon Himself the sins of mankind, died a nephesh death ("I declare unto you …how that Christ died for our sins…": I Cor. 15:1-3), and received the judgment of a holy God for our sin. Jesus did this so that we might have life. And, being the source of life itself, Jesus had victory over death and rose

from the grave. He now offers spiritual life to us, only requiring that we believe upon Him ("Believe on the Lord Jesus Christ and you shall be saved...": Acts 16:32)—that we trust in what He has already done for us. And, as if providing eternal life for us was not enough, God even offers to believers life 'more abundantly' (John 10:10).

With all this insight into the very nature of God we ought to explode in worship of God. For the biological and nephesh life we were given, for the spiritual life which is offered to us, and for the abundant life He desires for us, and even more for the infinite life of God Himself, it ought to be that we cannot help but worship Him and bring others into that worship and praise. We ought to fill the creation with worship and make the creation a house of worship. And, when we consider the life that is about us, we ought to be reminded of the God of life, the source of life, the sustainer of life, and the fountain of abundant life.

The spectrum of perfection of life ought to remind us of the infinite grandeur of the life that God has. It ought to help us see beyond the physical to the spiritual, to recognize our own finiteness, and be awed by God Himself. For the remainder of our lives we should take the time to ponder the creation so as to think these thoughts and allow our hearts to be lifted towards God in praise and worship. We ought to bring others—in fact the entire creation—into that praise and join with those in heaven who "...worship Him Who lives forever and ever" (Rev. 4:10). In this way we fulfill our roles as priests to the creation.

Our Responsibility to the Creation

Preserving Life

In our responsibility as kings over the creation (see Chapter 1), we should serve God, and serve and protect the creation. As stewards who will have to give account for our actions, we should seek to preserve God's creation—to rule as God would rule. At the very least we should pass it on to the next generation (or to God Himself) in at least as good a condition as it was when we received it. God handed to man dominion over 'living creatures' (Gen. 1:26-28), expecting man to 'guard and keep' the life He created (Gen. 2:15). When it is in our power, we are to save life, and not take it.

For example, hunting merely for the sport of it, is wrong. We're not speaking here about hunting to survive or hunting to acquire

food or even hunting to control populations. We have been given the authority by God to kill other organisms, but it should only be done so as to sustain or promote or encourage that precious creation of God known as 'life'. Animals and plants can be killed as food to sustain humans, and so sustain the image of God. We can also harvest some organisms from a population to control that population so that it does not overrun or over-eat the organisms around it. This is especially true when we have killed off the organisms that God created to control that population or when we have introduced organisms into an area where God's designed controls do not exist.

In these cases, controlling a population can actually *in*crease the number of other organisms in that environment. This is actually a case where death promotes life itself. In contrast, the kind of hunting which is not appropriate, is hunting just for the fun of it—just to kill. An example would be the sport hunting of the American bison. The herds of millions of bison were almost completely exterminated from North America by hunting them for the sport of it.

We should also be careful not to kill so many of a given type of organism in a particular area that populations of that organism get too small to sustain themselves. This happened with the North American passenger pigeon. It was so heavily hunted—mostly for food—that too few passenger pigeons were left to find mates and lay eggs to grow into the next generation. As a result, the passenger pigeon went completely extinct.

Gluttony, wasting, and hoarding are three more sins that destroy life. When we eat more than we should we are eating more organisms than we should. We are killing organisms merely to satisfy our selfish consumption. When we waste paper by printing or copying more than we need and throwing the rest away, or when we waste food by taking more food from a buffet than we eat, and throwing the rest away, we are killing organisms without good reason. When we hoard more possessions than we ought, and build houses that are bigger than we need, we waste construction materials—many of which were constructed from organisms, such as wood and paper from trees.

We have even created substances designed to kill. We create 'antibiotics' which are designed to be 'anti' (against) 'bios' (life), insecticides which are designed to kill insects, herbicides designed to kill plants, fungicides designed to kill fungi, *etc*. Please do not misunderstand this. This is not a complete condemnation of the use of chemicals to kill. For when the killing of some is necessary to promote life in many more, chemicals designed to kill can actually promote life. When one microorganism threatens to kill a human being—the image of God—it may be a greater promotion of life to use an antibiotic and kill millions of microorganisms. Likewise, when a particular fungus or insect or weed threatens to decimate a food source for a starving human population, the use of a fungicide or insecticide or herbicide might actually preserve or promote or encourage life.

The pride of inconvenience is also an enemy of life. The killing of unborn babies when it is merely done because they are a burden is desperately wrong, and that would be true even if babies were not the image of God. The fact that babies *are* the image bearers of God makes this sin even more horrible. Killing the very old or the very ill merely because of their burden on society or the family is also wrong. Again, this would be true even if humans were not the image God. But, in fact, humans *are* the image of God, and they are from the moment of conception to the moment of death[a]. Abortion and euthanasia are both wrong.

Human life is to be more valued than animal life because humans are the image of God. Animal life is to be more valued than plant life because animals have such things as mind, will, and emotions that plants do not have[b]. Yet, all life is to be valued both because it is the creation of God and because life pictures something about the nature of God. When God calls us into account for what we have done to care for His creation we had better have been about the business of preserving and promoting life.

---

a   Allowing a disease to kill a person is one thing, but doing something that actually kills a person (*e.g.* withdrawing a feeding tube to starve a person to death, or giving a person an overdose of pain medication, or administering a lethal drug, or taking out a needed organ for transplantation) is murder.

b   See Chapter 7 for an elaboration of this distinction between plants and animals. The mind, will, and emotions of animals is part of what the Bible calls the *nephesh* or 'soul' of animals.

Enhancing Life

After creating the animals of the air and sea, God commanded them to 'reproduce and multiply and fill the earth' (Gen. 1:22). He gave the same command to man (Gen. 1:28). After the Flood He repeated this command to animals of the air and the land (Gen. 8:17) and to mankind (Gen. 9:1, 7). As the earth fills with life, a fuller picture of the living God emerges. God is glorified when life is abundant and flourishing on this planet. God made us rulers over His creation to do more than preserve life. As the servants in Christ's parable were commended for multiplying their talents, so we should *enhance* God's creation. We should seek to enhance life on earth—to make it fuller, more abundant.

Enhancing life is not just increasing population numbers. Notice that when God commanded organisms to multiply, He commanded them to do so *until* the earth is full. This suggests that the earth *can* be full. The nutrients and water in a particular area can only sustain a certain number of certain types of plants, which in turn, can only sustain a certain number of certain types of animals. When those particular plants and animals get to the correct population numbers, that area is full. That area is said to be at its **carrying capacity** and every species is at that environment's carrying capacity. The carrying capacity is the maximum population size for a particular species that the environment can sustain indefinitely. When population size is less than the carrying capacity, there is room for organisms to multiply, and with that increase, the area increases in vibrancy and fullness—in life.

However, when population size exceeds the carrying capacity, there are insufficient resources to support the population. Competition between organisms increases, stresses increase, diseases increase, death rates increase, all 'naturally' bringing the population back down towards the carrying capacity. Populations exceeding the carrying capacity are not as healthy or as vibrant or as 'alive' as populations at or just below the carrying capacity.

To maximize life, then, populations should be at or just below the carrying capacity for that environment. We should be striving to understand environments well enough to determine the carrying capacity for organisms, and then manage those organisms so as to attain and not exceed the carrying capacity for that environment. In agriculture, for example, we should be careful not to breed so much

---

carrying capacity—the number of organisms a particular environment can support over the long term

livestock as to overgraze the land, or to plant so densely as to sap all the nutrients from the soil. Unfortunately, in many locations across the planet we have already done that. We have a responsibility to God to restore the healthiness of such populations.

This reasoning applies to humans as well, for they were commanded to 'multiply and fill the earth' at the same time animals were. In every environment there is a carrying capacity for humans as well. We should manage human populations so as not to exceed the carrying capacity. In many areas, human population has exceeded the carrying capacity for that area. This is not only true in the desperate slums of the world's largest cities, it is even true in desert areas with low population where the water table is dropping because humans are consuming more water than the environment can supply.

What is different about managing human populations, of course, is that humans bear the image of God. Thus, whereas it may be appropriate to hunt animals or use chemicals to reduce animal or plant populations, it is NOT appropriate to kill humans so as to bring populations down to the carrying capacity of a particular environment. We may be able to redistribute people or resources, but we are not justified in any type of human eugenics.

In many cases it will be very difficult to enhance life. This is especially the case where humans have decimated natural populations through overharvesting or pollution or overpopulation. But this should be expected, for these decimations occurred because of human

sin. God paid a very heavy price to redeem us from our sin and the effects of our sin. Is it any wonder that we might have to work hard, or pay a heavy price—or even sacrifice—to even begin to redeem the creation that has been so stained by human sin? But we ought to pay that price so as to fulfill our role as kings of the creation to the glory of the Creator.

Summary of Chapter

A. Since life is one of the attributes of the invisible God, God illustrated His life by creating organisms (physical beings) with life.

B. The many different types of life probably differ in non-physical ways. Included among these different types of life are:

   a. biblical life (possessed by God, spirit creatures, humans, and animals and NOT by plants (and apparently not by) fungi, algae, protozoa, and bacteria)

   b. divine life (possessed only by God) is
      i. uncaused,
      ii. autonomous, and
      iii. infinite.

   c. creature life (possessed by spirit creatures, humans, animals, plants, fungi, algae, protozoa, bacteria) is
      i. created by God,
      ii. dependent upon God for its continued existence, and
      iii. finite in quality.

   d. life of spirit creatures (possessed by cherubim, and probably all spirit beings) is a type of creature life that apparently does not end, in a being that by nature does not have a physical body

   e. biological life (possessed by humans, animals, plants, fungi, algae, protozoa, and bacteria) is a type of creature life in a being that by nature has a physical body

   f. nephesh life (possessed only by humans and animals) is a type of creature life in a being that by nature has both a physical body and a 'soul'.

   g. human life (possessed only by humans) is a type of nephesh life that does not end at death.

   h. animal life (possessed only by animals) is a type of nephesh life that apparently ends at death.

   i. plant life, fungus life, alga life, protozoa life, and bacterium life are one or more type of biological life that apparently ends at death, in a being that has a physical body, but not a 'soul'. In the Bible, plants are most often referred to as 'green things' and as food for animals and humans: by what they *do* (photosynthesize and produce food), rather than for what they *are*.

C. God created a spectrum of perfection of life that leads us to conclude that something exists (God) that possesses infinite life:

  a. things without life
  b. viruses that might or might not be alive
  c. plants that have biological but not nephesh life
  d. animals that have biological and nephesh life but are not the image of God
  e. unregenerate humans that have biological and nephesh life and have the image of God
  f. believers that have what all humans have plus spiritual life
  g. mature believers that have what all believers have plus abundant life

D. Life is non-physical.

E. Evidence that life is non-physical:

  a. Biblical evidence that life is non-physical:
    i. God is life, but not physical;
    ii. cherubim are living, but not physical;
    iii. human life was added after his body was created;
    iv. the human soul and spirit and life are all said to leave when the physical body dies;
    v. an animal is called a *nephesh hayim* (living creature) just like Adam is called *nephesh hayim* only after he was given life
  b. Non-biblical evidence that life is non-physical:
    i. no physical quantity of an organism changes at the moment of death;
    ii. universal human intuition is that human life is something more than the physical; and
    iii. the definition of life has been elusive to naturalistic biologists.

F. Biology cannot study all things that are living (e.g. God and spirit beings) and cannot study life itself, so rather than being a 'study of life', it is more accurate to say that biology is the 'study of organisms that possess life'.

G. Life is a challenge to the naturalist worldview because naturalists do not believe in the non-physical

H. Creature life comes from God

I. In the present day, living creatures come from similar living creatures, and life itself may be a special creation of God for every organism. In the late nineteenth century, after two centuries of experiments, debates, and studies, the commonly believed theory of spontaneous generation (that organisms could originate spontaneously or naturally from non-living things) was rejected. In its place the law of biogenesis seems to be true (that organisms only arise from other organisms of the same type).

J. The first life was a special creation of God. Naturalism believes that abiogenesis (the origin of life from non-living material) must have occurred at least once in the past. However, abiogenesis appears to be false because

  a. there are no known contradictions to the law of biogenesis in the present,
  b. humans have never been able to create life in the lab, and
  c. humans have never been able to stimulate a truly dead organism back to life.
  d. Consequently, a more general form of the law of biogenesis seems to be true: Life comes only from life (with the living God being the only cause sufficient to explain the origin of life).

K. Studying life allows us to be better priests of the creation by

  a. giving us greater insight into Who God is (His spirit, His life, the infinite nature of His life, He as the source of life, His continuous activity, His creativity, His desire to be known) and giving us cause to
  b. worship God, and
  c. bring others and the creation into the worship of God.

L. As rulers and stewards of the creation we have a responsibility

   a. to sustain, promote and preserve life (i.e. not to kill for sport or gluttony or waste or hoarding or convenience),

   b. to enhance life for the glory of God by maintaining life at the earth's carrying capacity (the largest population that can be indefinitely sustained by the environment)—by bringing populations up to carrying capacity, by restoring damaged populations to carrying capacity, and by controlling populations prone to exceeding carrying capacity.

Advanced Discussion Topics

A. There is debate about whether 'life' is a real entity—like some sort of non-physical, spiritual 'body'—or whether it is merely epiphenomenal (consequences of some sort of activity of living things). Discuss the strength and weaknesses of these two perspectives as they relate to the nature of life as related in this chapter. Consider how each relates to biblical references to life, God's life, the life of cherubim, evidences for the non-physical nature of life, the difference between the life of multi-cellular organisms and the life of the cells of which the organism is made.

B. Discuss the difference between animals and plants and the significance of the difference. Consider such things as: different life/death references to animals and plants in the Bible; respective biblical references to plants/animals as to what they *do* versus what they *are* (*e.g.* plants as 'green things' vs. animals as 'living things', plants dead if non-productive vs. animals dead if not alive); respective purposes in original creation (*e.g.* plants as food & animals not); and different ethical obligations as per the differences between plants and animals.

C. Discuss the relationship between soul and body. Consider the following: humans created in the image of God, Christ taking on the form of a man, and (apparently comfortably) living in that form for eternity; humans, even unbelievers, being given (presumably comfortable) resurrected bodies for eternity; those in comas feeling emotion but being unable to express them; and the possibility that souls might guide physical development.

D. Could entities traditionally known as non-living things have some sort of life? Consider the following: the meaning of stones crying out (Luke 19:40), the meaning of trees praising God for not being cut down (Isa. 14:8), the detectability of life, the likelihood that Scripture mentions every different type of life that God created.

Test & Essay Questions

1. Define spontaneous generation / law of biogenesis / spectrum of perfection / spectrum of perfection of life / microorganism / abiogenesis / carrying capacity.

2. Compare and contrast biblical life and biological life / divine life and creature life / life of spirit creatures and biological life / human life and animal life / nephesh life and plant life / spontaneous generation and abiogenesis / spontaneous generation and law of biogenesis.

3. List three biblical reasons / three non-biblical reasons why one would argue that life is non-physical.

4. Short Essay: What insight does biological life give us into the nature of God?

Test & Essay Questions — The Living God

5. Short Essay: Why is it ironic that biology may not be able to study life? / Why is it ironic that biology may not be able to study life?

6. What kinds of things are biblically living / biologically living / biblically living but not biologically living / biologically living but not biblically living / both biblically and biologically living?

7. In what sense are plants living? In what sense are plants not living?

8. Short Essay: Why did God create the spectrum of perfection of life and what are we supposed to deduce from it?

9. Short Essay: Why did people once believe in spontaneous generation?

10. Short Essay: What caused people to reject spontaneous generation, and what did they conclude instead?

11. What is the law of biogenesis?

12. List two reasons why one might believe that the law of biogenesis was true in the past.

13. Short Essay: How is the biogenic law more completely true in a Christian worldview than a naturalistic worldview?

14. Short Essay: How does the Christian worldview provide a better explanation for the origin of life than the naturalistic worldview?

15. The observation of biological life should cause what sort of worship?

16. Short Essay: Is hunting for the sport of it ethical? Why or why not?

17. Short Essay: Why should we preserve life?

18. Short Essay: How does the principle of preservation of life apply to hunting / gluttony / insecticides / herbicides / abortion?

19. List two reasons why 'reproduce and multiply' is a good thing?

20. Short Essay: When should we decrease population size / keep population levels the same / increase population size?

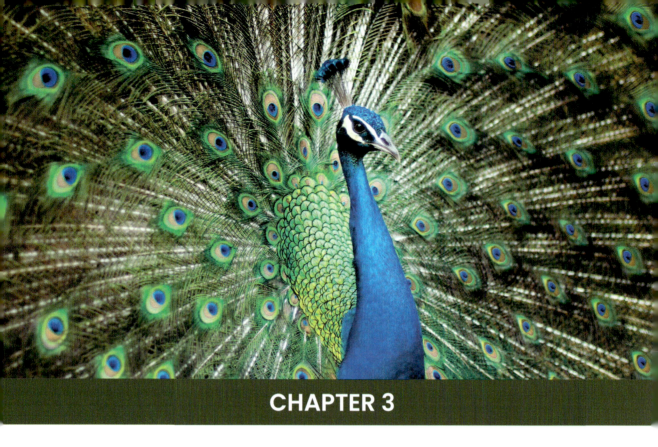

CHAPTER 3

THE GLORY OF GOD

Biological Beauty

"...worship the Lord in the beauty of holiness."
I Chr. 16:29, KJV

3.1 | The Beauty Of God

When the word is correctly applied, **beauty** is a holistic concept. Something is said to be beautiful when, upon considering it, one is struck by how all its characteristics fit together in a compelling way. It is not just that something is pretty—a type of one-dimensional aesthetic appeal—but also that it fits together well and that it is somehow good. We have trouble labeling a person as beautiful who is physically attractive but who is also artificial, cruel, miserable, devious, unclean, or lazy.

beauty—the attractive holistic fit of an entity's attributes; deep beauty is beauty at multiple scales

Beauty involves a sort of multi-dimensional aesthetic appeal. A beautiful person is attractive in their *whole* being—physically, emotionally, and morally. This is something of the sense in which Scripture refers to the beauty of God. Not only is any one of God's attributes (like mercy or love) attractive to us, but each of His other attributes like grace and patience and kindness and omniscience (and so on) are also attractive. And, these attributes are interwoven and fit together in an awesomely compelling way.

At the same time, we need to realize that God is not simply the addition and remarkable integration of a bunch of amazing attributes. No amount of addition can reach infinity. Each of God's attributes is infinite and a necessary part of God Himself. All of them are fully realized at the same time in a perfectly unified whole—so awesomely appealing (beautiful) as to compel us to worship.

In this way God's 'beauty of holiness'—the wholeness of the awesomeness of all His interwoven attributes—encourages worship and praise ('worship the Lord in the beauty of holiness': I Chr. 16:29 & Psa. 29:2 & 96:9; 'praise the beauty of holiness': II Chr. 20:21). When the psalmist writes "One thing have I desired of the Lord …to behold the beauty of the Lord…" (Psa. 27:4), he seeks the entirety of God's being—all of His attributes together. Consequently, the greatest blessing that can be bestowed upon a people is that God Himself is their 'crown of glory' or their 'diadem of beauty' (Isa. 28:5), for this means God dwells in, on, and among His people in all His fullness and being.

Because God is spirit, that awesome beauty of God cannot be seen by mortal eyes. But God desires that we know Him and that we know Him in His fullness, including His beauty. Therefore, God illustrates

that beauty that cannot be seen by physical beauty that we *can* see. For example, He creates physical beauty and 'clothes' Himself with it (Job 40:10).

From time to time He has demonstrated His presence with something the Bible refers to as 'the glory of God'—something so filled with manifestations of God's nature that it is overwhelming to man. It is described as so bright, brilliant, and shining as to 'fill' the area around it, such as the temple (Eze. 10:4; 43:5; 44:4)[a].

In fact, in eternity, the glory of God provides the light for all of heaven (Rev. 21:23)—a light that even seems to pass through objects and walls as if the glory of God cannot be hindered by anything created. Prior to our glorification, however, the glory of God is fearsome to sinful man (*e.g.* "…the glory of the Lord shone round about them, and they were sore afraid.": Luke 2:9[b]). This is because God's glory includes His holiness, and we fall so very short of that (Rom. 3:23). It is because of this, that after Moses spent just a short amount of time on Mount Sinai in the presence of God's glory, his face shone so brightly that even his brother was afraid to approach him (Ex. 34:29-30). Yet, as

---

a   And that glory will shine on even larger areas in future events (*e.g.* Hab. 2:14; 3:3-4; Rev. 18:1).

b   Other examples of human fear in response to the glory of God: (1) Moses being unable to enter the tabernacle (Exo. 40:34-35); (2) priests being unable to enter the temple (I Ki. 8:11; II Chr. 5:14; 7:1-3); (3) "…the sight of the glory of the Lord was like devouring fire… in the eyes of the children of Israel" (Exo. 24:17); (4) "Behold, the Lord our God has showed us His glory… Now therefore why should we die? For this great fire will consume us if we hear the voice of the Lord our God any more…": (Deu. 5:24-25).

intense as these physical manifestations of His glory were, they are still a dull image of God's true beauty.

When God chooses to reveal Himself, He often does so in beautiful form. For example, when Ezekiel sees Christ (Ez. 1:25-28), Ezekiel seems to struggle to describe what he sees. He refers to sapphire, amber, fire, and rainbow. Each of these items is visually stunning when considered alone; together they must have been beautiful indeed.

We see His glory illustrated in the beauty of His sanctuary (Psa. 96:6)—and (by illustration) by adornments of the places where He chooses to reveal Himself. Eden, for example, where God first dwelt with man, was a garden. Only locations of beauty are called gardens—much more so the 'garden of God'. In fact, God placed in Eden 'every tree that is pleasant to the sight and good for food' (Gen. 2:9; 3:6). He chose trees attractive to both the sense of sight and the sense of taste. At least one of the beings in the garden, the 'anointed cherub', was described as having 'beauty and brightness' and decked with precious gems and gold (Eze. 28:13-16).

Beautiful is also the description of the New Jerusalem (Psa. 48:2), the final, eternal home of man living with God. It is 'prepared as a bride for her husband' (Rev. 21:2). Its foundations are made of 12 gemstones, its walls are made of jasper, its gates of pearl, its streets of transparent gold, and it is lit with 'the glory of God' (Revelation 21). Any place God Himself creates for His own abode is a beautiful place[a]—a physical manifestation of the even greater beauty of God Himself.

When God instructed man to build a temporary abode for God, He gave meticulous detail to its beauty (Exodus 25-27). There were sweet aromas (spices for incense), luxurious textures (fine linen), and deep-grained wood (acacia) with beautiful carvings. There were shiny metals (gold, silver, and brass), translucent stones (onyx), and colorful dies (blue, purple, scarlet). The tabernacle was designed to stimulate the senses of sight, feel, and smell. Another chapter (Exodus 28) is devoted to the description of the garments of the high priest, God's

---

a The incarnation might be considered an 'exception' to this rule. Jesus, though not unattractive, was not of such compelling appearance as to draw people toward Himself because of His beauty (Isaiah 53:2). Instead, the Word (John 1:1-3, 14) condescended from His exalted state and took upon the humble form of a servant (Philippians 2:6-7) so as to touch man. In this case He is not seeking to illustrate His glory, rather He *chose* to taken on a less awesome form *so that* He could touch us, we could touch Him, and He could example how we are to live. It was done to connect with us. It is also true that even now the Holy Spirit chooses to reside in believers, even though those temples are less than beautiful. In those cases, it can be argued that that same Holy Spirit is perfecting those saints, so as to 'make them beautiful in His time' (Eccl. 3:11).

chosen representative among the Israelites. These garments involved linen, embroidery, lace, gemstones, and gold—all 'for glory and beauty' (Exodus 28:2, 40).

To orchestrate the construction of the tabernacle, such as its furnishings and priestly garments, Bezaleel and Aholiab were 'filled with the Spirit of God' (Ex. 35:30-36:2). God gifted these men to accomplish the necessary cutting, carving, engraving, molding, spinning, weaving, and needlework. And all of this, we were told, was done as a pattern of heaven itself, and the things in heaven (Heb. 9:23-24). In a similar manner, Solomon's temple was beautiful as well (I Kings 6-7; II Chr. 3-4). And when the temple was rebuilt in the days of Ezra, God put it in the king's heart 'to beautify the house of the Lord' (Ezra 7:27). God wishes His habitations to be beautiful as a testimony to the beauty of God Himself.

Yet, in spite of the beauty of things made by man, according to Jesus, the beauty of God is pictured even more powerfully in His creation—in fact, in His biological creation. Jesus claimed that Solomon in all his glory was not arrayed as beautifully as a single lily (Matt. 6:29).

3.2 | Biological Beauty

VIDEO 3.2

As an illustration of His awesome glory and beauty, God infused His entire creation—including the biological portion of His creation—with extraordinary beauty. Because that beauty is bestowed upon a finite creation, the beauty of the creation is limited, but it is nonetheless breathtaking. There is not just a lot of beauty, the beauty itself has an appealing array of qualities. There is, if you will, a beauty to the creation's beauty. What follows is an attempt to describe some of the amazing qualities of the beauty God placed in His biological creation.

Deep Beauty

God created beauty at every scale of the universe—infusing it into the smallest unit, the largest structure, and every entity between—including organisms. In contrast, the beauty of things made by humans is shallow—typically beautiful at only one scale[a]. An oil painting examined too closely amounts to unimpressive individual brush strokes; an oil painting far enough away is not seen as beautiful at all. A striking set of watch components examined too closely amounts to a set of ugly cogs. The most beautiful things made by humans lose their beauty when we get too close or back up too far.

In contrast, the creation's beauty is found at every scale. The beauty of a 'purple mountain' observed closer becomes a stunning mountain vista draped with a soft, green blanket. Closer yet it is a warm forest scene. Even closer, one can be struck by a bird's colorful feathers. At a smaller scale is the iridescence of a single feather, and smaller still is the repeating symmetry of microscopic interlocking barbules. Even closer

---

[a] The observation of the scale-independent beauty of God's creation in contrast to the single-scale beauty of human creations was first pointed out to me about 1990 by Dr. David Mention.

one finds the ordered arrangement of specially designed cells, and closer still are the intricate components of a single cell. The depth of beauty is not even limited by the bounds of biology. Stunning arrangements of molecules and atoms and subatomic particles are found at scales many times smaller than the smallest organisms. At the opposite end of the size spectrum, when all of biology gets too small to see, there is still beauty in the awe-inspiring scenes from satellites and spacecraft, and even beyond to stars and galaxies.

The beauty of God's creation is so much greater than the most beautiful things made by man in part because the beauty of God's creation is scale independent. How very deep indeed is the beauty of God. One can imagine that every brushstroke of His original creation must have been spectacularly beautiful. Not a single one was out of place and not a single one was unessential for the beauty of the whole. Even now, with the creation fallen and marred, God is interested in bringing out more glory to Himself by increasing its beauty. He seeks to make all things beautiful in His time (Eccl. 3:11), so that everything glorifies God to its fullest potential. This includes every imperfection of our lives—from a small, seemingly insignificant scratch to the very largest, seemingly insurmountable ugliness. Each one of us is an essential component of the beauty and the tapestry of the creation of God.

Ubiquitous Beauty

Beauty is found throughout His creation—not just at every scale, but in every location. This beauty is found in every corner of the universe. Even that portion of His beauty that He chooses to reveal through biology is found across the entirety of the earth. Organisms with beauty are found in the air, the land, and the seas.

Nearly every body of water and every landscape—no matter how harsh—is adorned with organisms. From the tops of the highest mountains, to the ice on the coldest pole, to the baked surface of the driest desert, biological beauty is found across all the continents. Organisms thrive from boiling hot springs, to seething cauldrons of acid, to springs at the bottom of the deepest oceans, and even between grains of sand under crushing pressures miles beneath the earth's surface. Biological beauty is ubiquitous across the planet.

Even when something makes His creation ugly, God begins a process that replaces that ugliness with a new beauty. In the 'natural' course of events, healing comes to landscapes devastated by pollution, war, or catastrophe. God built a remarkable facility into His creation that keeps its beauty in balance and restores beauty when beauty is lost, similar to how our bodies repair themselves after injury. Stinky, lifeless oozes become clear streams teaming with life. Stark landscapes dotted by fallen logs become vibrant forests. Drab underwater shipwrecks become colorful reefs. Land stripped of vegetation and deeply rutted

community succession—a sequence of communities in a particular area leading to the climax community in that area, each community altering the environment for the community to follow

community—(biological) system of plant, animal, and biomatrix species found in a particular location

climax community—the stable community for a particular area; the last in community succession

pioneer community/species—the first organisms (in community succession) to settle in an area

with erosion becomes reclaimed again by rain forest. Sometimes it takes weeks or months and sometimes it takes centuries or millennia. Sometimes it even requires a replacement of what was there with something new[a]. But, no matter how marred, God can make the fallen beautiful.

In the biological world this is accomplished in individual organisms by means of biological processes of repair and healing. On a larger scale, this is done in communities of many species of organisms by the process of **community succession**. In community succession, the species found in a particular area change over time as organisms at a given time prepare the area for another **community** of organisms to follow. Each such community alters the environment in such a way as to effectively phase itself out. Over the course of time, what results is a series of different communities, each replacing the previous one, finally resulting in a stable **climax community** that can persist more or less indefinitely at that location.

Community succession usually begins with the arrival of **pioneer species**. Pioneer species are specially designed for these situations. Photosynthesizers, like cyanobacteria, lichens, and plants, will usually be the first to arrive on a barren landscape because they can harvest their energy from the sun. Animals will come in later. If the environment is dry and without soil, the lichens will be the first because they can get

---

[a] Such as He does when He creates in believers a new spirit (*e.g.* Eze. 11:19; 36:26), and as He will in the future when He creates a new heaven and a new earth (Isa. 65:17; Rev. 21:1).

the water they need directly from the atmosphere, even in the driest of deserts. Lichens also erode rocks to produce soil for plants that follow after. The first plants that appear will usually be those that can fertilize the soil because they carry nitrogen-fixing bacteria.

Some plants are even designed to come in after a fire (*e.g.* lodgepole and Jack pine cones that open after being heated up in a fire). An example of community succession would be the development of communities following the retreat of Alaskan glaciers. Pioneer species, such as lichens, mosses, and nitrogen-fixing annuals build up a fertile soil. Other plants germinate in that soil and eventually a community dominated by nitrogen-fixing alders and willows and cottonwoods takes over. This community is replaced by one dominated by western hemlock and spruce, and finally that community is replaced by the climax community dominated by Sitka spruce.

God restores such communities for His own glory, to make sure His own beauty is always well pictured in the creation. If He is so interested in restoring the beauty of plants, animals, and the earth itself, how much more is He interested in restoring our own beauty and the beauty of our lives.

Profound Beauty

Beauty is not just everywhere at every scale, it is also intense. The beauty of the biological creation is awe-inspiring. Whether it is the bioluminescent sparkling of dinoflagellates on the waves of the ocean

at night, or the flurry of a flock of migrating butterflies, or the fiery brilliance of a maple tree in the fall, or the vibrant color of prairie flowers, or any of millions of other scenes, biological beauty has the power to stop us in our tracks. It strikes us with awe and convinces us again and again of the beauty of the Creator.

Biological beauty has inspired poets, painters, sculptors, composers and artists of all types and in all cultures. No one has matched it or fully captured it, though many have tried. Countless lifetimes of talented artists have been committed to sharing that beauty with others. As awesome as the creation's beauty is, how much greater is the beauty of the Creator.

Multifaceted Beauty

There are many different types of biological beauty in the world. Different sources of beauty can stimulate different senses. Our sense of touch reacts to a cool carpet of grass, a velvety cluster of mimosa leaves, the smooth scales of a boa, the sandpaper roughness of a shark, the soft fur of a chinchilla, and myriad other textures. God has designed organisms to stimulate our senses of smell, taste, sound, and sight in thousands and tens of thousands of different ways.

The same biological scene can not only be perceived with different senses, but also from different vantage points. We examine it from above and below, from the north and the south and the east and west. We can even examine it at different scales, with our own eyes and with aids from lenses and microscopes. And that same biology yields a distinct beauty in each case—from every perspective.

When God gifted humans to appreciate beauty, each person was graced with an appreciation of a unique type of beauty—giving rise to the phrase 'beauty is in the eye of the beholder'. Yet, God fashioned the biological creation in such a way as to stimulate each unique perspective. When a person shares the beauty he or she sees, others are given the privilege to appreciate a beauty they would otherwise have missed. When the perspectives of all people are combined biology is seen to be a magnificent tapestry of interwoven beauties. Like a well-cut diamond, the beauty of biology is multi-faceted. In a similar manner, God is uniquely beautiful in every situation and seeks to illustrate His beauty in all situations.

Sparkling Beauty

Finally, creation's beauty is not boring. It is diverse and it is ever-changing. The same highway turnout never manifests the same vista, for biology is always changing. Every stunning view changes because different organisms are active at different times of the day and different seasons of the year. Organisms also change shape and size as they grow, change color and form with the seasons. And movement characterizes most organisms—sometimes the movement of the whole organism, but always the movement of its components.

Creation's beauty is vibrant. It sparkles, if you wish, and that sparkle has its own beauty. The beauty of the creation is always surprising and always refreshing. God is an unchanging God—in the sense that He will always have an unbounded quantity of each of His attributes. He will always have unbounded holiness, power, beauty, and so on. For example, He will always and forever have unbounded love toward us, but every moment of every day we realize the impact of that love in a new and fresh way. It is as if His love 'sparkles'. Each of His attributes sparkles. God Himself—His nature and His beauty—is always surprising and always refreshing. God's beauty sparkles, and that sparkle has its own beauty.

The Origin Of Beauty

The ancient Greek philosophers were fascinated by beauty. They marveled at its splendor, speculated on its true nature, extrapolated to its perfection, and wondered about its origin. Millennia later, the origin of beauty is a puzzle to modern evolutionists because natural selection (thought to be the primary mechanism of change in naturalistic evolution) should select survival or efficiency over beauty. If organisms refocused the energy they currently use to 'be beautiful' in order to outcompete the organisms around them, they would be better survivors. They would be better players in the struggle for the survival game known as evolution.

If peacocks had shorter tail feathers they could fly better and more readily escape from predators. Charles Darwin (1809-1882) suggested that peacocks have beautiful tail feathers because peahens mated only with beautiful peacocks. But then, why did peahens do that? Their young would have survived better if they had chosen peacocks that could fly better[a]. In like manner, if butterflies did not stand out so well, fewer of them would be eaten by birds.

---

[a] Evolutionists have devised explanations (the most common of which is that beauty is an indication of health, so when peahens choose beauty they are implicitly choosing the desirable quality of health). Although many of these theories seem reasonable, none of them work well upon testing (*e.g.* there is little relationship between the 'beauty' of peacock feathers and the health of the peacock; peahens do not consistently choose either the most beautiful peacocks or the healthiest). Evolutionary theory continues to struggle to explain the origin and maintenance of biological beauty.

If natural selection was truly the source of biological variety, then we would expect there would be little to no beauty in the world. And, as long as natural selection has operated in the world (since the Fall of man), it has systematically taken beauty out of the world. Yet, there is still a profound amount of beauty in the world. Imagine how much beauty there must have been in the beginning! A God of beauty infusing the creation with beauty is the only reasonable explanation for the deep, profound, ubiquitous, multi-faceted, sparkling beauty we see in this world. Creation's beauty testifies of the Creator and stands as a substantial argument against naturalism and evolution.

Human Appreciation of Beauty

Humans appreciate beauty. Even though different humans appreciate different types of beauty, every human appreciates some sort of beauty, and that beauty is found in God's creation. Every human being is awed by biological beauty. The ability to appreciate beauty, and to be awed by it, are as God-given as beauty is itself. Freezing in awe at the beauty of His creation is not the best survival tactic in a savage world. Natural selection would tend to take out the appreciation of beauty. It would especially rid the world of the tendency to be awed by beauty. Consequently, the ubiquitous appreciation of beauty in humans and the tendency to be awed by beauty also testify of God's creation.

Consider, for example, how sensitive we are to beauty. If survival was our only reason for existence (which is more or less what is true in naturalistic evolution), we would not need our senses to be as sensitive as they are. Humans, for example, could get along quite well without seeing color. We could also get along with recognizing fewer tastes and smells. But God gave us these extreme sensitivities so that we could recognize beauty in biology. He desires for us to recognize biological beauty so that we can better understand the beauty of the unseen God. God has given us the capacity to see color—and to hear and taste and feel and smell to the extent that we can—as a blessing. They are a gift from the God Who desires us to perceive God's invisible nature from those things that were made.

Beauty: Our Responsibility

Our Responsibility to God

As priests of the creation (see Chapter 1) we have a responsibility to know God, worship God, make the creation a house of worship, and bring the creation into the worship of God. What we learn about biological life can help us do that. A study of biological beauty gives us insight into God's very nature. Even a casual acquaintance with the awe-inspiring beauty of the biological world ought to remind us—at the very least—that our God is a God of awesome beauty.

A closer look at biological beauty gives us even further insight. For example, God created a spectrum of perfection of beauty. God created some things that appear to lack beauty altogether. He garnished other

things with stunning beauty, and still others with various levels of beauty between these extremes. There is, in fact, a complete spectrum of things from the non-beautiful to the strikingly beautiful. In the sky, for example, such a spectrum exists: "one glory of the sun, and another glory of the moon, and another glory of the stars, for one star differs from another star in glory" (I Cor. 15:40-41).

Even among humans there is a spectrum of perfection of beauty. Some of us are not so attractive—in fact, some of us might be just downright ugly! Others are stunningly beautiful. Most humans, of course, find themselves somewhere on that spectrum between these two extremes. Beyond humans, as Christ indicated (Mat. 6:29), even a simple plant (such as a single lily) has been graced with more beauty than is found even among the most beautiful of humans and the most beautiful of human creations. Organisms, then, create another spectrum of perfection of beauty. There are organisms that appear quite ugly to us, and others that are ugly (or is it cute?) in a homely sort of way. Other organisms are tolerably pleasant to perceive and still other organisms are truly gorgeous. The millions—in fact, millions of millions—of organisms make up quite an impressive spectrum of perfection of beauty.

In our mind's eye, we can follow that spectrum from the most unattractive to the most beautiful, and project beyond, imagining even greater beauty than we observe in all the creation. We can extrapolate towards the existence of infinite beauty—God Himself. Even though our brains cannot fully grasp it, through this process we can get a bit of a glimpse of the boundless beauty of God. He is the source of all beauty, the sustainer of all beauty, and He Who makes all things beautiful in His time. His beauty is without bounds, infinitely greater than all the beauty we see about us. The awesome spectrum of perfection of biological beauty—whether found in landscapes, or butterflies, or birds, or flowers—ought to aid us in understanding a bit of His beauty and ought to cause our hearts to lift up in worship towards the Source of all beauty.

Scripture indicates that God is active, everywhere present, creative, and desirous of relationship with us. Biological beauty reinforces these claims and gives further insight into their meaning. The sparkling nature of biological beauty, for example, is consistent with God being an active, dynamic, living God, and provides insight into what it means

for God to be a living God. The ubiquity of that sparkle suggests that God is active over us, among us, and within us—even in those things we sometimes consider too mundane or lowly to concern the great God.

Beauty informs our understanding of the ever-present, on-going care by God of every aspect of our lives. The deep and ubiquitous nature of beauty is consistent with His omnipresence. God's presence at all places and all times is suggested by the beauty we see everywhere, at all times, at all scales—even when we are otherwise unaware of His presence. The multi-faceted nature of beauty impresses upon us the creativity of God. Providing each of us a beauty fitting our individual appreciation of beauty, reinforces the sense in which He interacts with us as individuals and cares about our individual needs.

The fact that we have been given the tools to appreciate the beauty of the creation when those are not 'necessary' for our survival is consistent with God's desire for a relationship with us. This reminds us of how He loves us so much that He steps down from His worthy greatness so that we can understand and know Him. Finally, the mechanisms He has put in place to make sure that biological beauty is restored when it is marred, reminds us of how He is able, willing, and even desirous of restoring the ugliness of our lives and 'make all things beautiful in His time'.

When we allow the Holy Spirit to reveal these aspects of God to us through the beauty of His creation, how can we not explode in worship? As you then acknowledge the origin of that beauty—in the person of

God—you glorify God. As you continue your study of the beauty of organisms and the greater beauty of His being, your worship increases, you then fill the creation with His worship. You cannot then help but share that wonder with others and bring them into His worship with you. In this way you fulfill your role as priests of the creation.

Our Responsibility to the Creation

Preserving Beauty. In our responsibility as kings over the creation (see Chapter 1), we should serve God, and serve and protect the creation. After God infused beauty into His vast creation, he handed it over to us to 'guard and keep it'. We have a responsibility to preserve the beauty of creation and pass that beauty on to the next generation. Some activities mar the beauty of the creation in an obvious and direct sense (*e.g.* clear cutting, strip mining, burying trash in landfills, polluting).

More generally, though, human development (such as the construction of new homes, highways, factories, office buildings, schools) often replaces the natural beauty of creation with something substantially less beautiful. It should be noted that many of these changes are *necessary*, because being in the image of God, the needs of humans have a higher priority than even preserving the beauty of the creation. We often *must* reduce the beauty of His creation so as to meet the needs of others.

However, it is often possible to replace the beauty that was there with another beauty that is at least as honoring to the God Who created it. In those locations where beauty has already been reduced, we should strive to re-infuse beauty so as to restore the glory that God receives for it. In those cases where we are planning to develop an area, we should include designs that generate at least as much beauty as we will have to destroy in the process of construction.

Enhancing Beauty. Although God created an enormous amount of beauty in the creation He created the *potential* for even more beauty. He also gave humans the ability to reveal that beauty—to 'create' beauty. He gave us the opportunity to bring God even more glory by bringing out more of the beauty than was evident in God's original creation.

One of the simplest ways to enhance creation's beauty is by rearranging the organisms of God's creation, thus creating more beautiful combinations of organisms. The 'Garden of Eden' was created by God (Gen. 2:8). God apparently selected a number of organisms, arranged them in a beautiful pattern, and placed them in the area He had set aside to be the 'Garden of Eden'. God then placed man into the Garden 'to dress and keep it' (Gen. 2:15).

It is likely that from that moment on throughout human history, in the pattern established by their Creator, humans have been active in what is sometimes called **'artificial selection'**[a]—selecting organisms and arranging them in beautiful patterns. Given that the organisms themselves were created by God, the beauty of such rearrangements is appropriately ascribed to God. Consequently, through gardening and landscaping we can glorify God by creating beautiful arrangements of organisms.

A second method of enhancing the beauty of God's creation is by **breeding**—where humans select what organisms are to be bred (mated or crossed) with what other organisms. Breeding can be used to make a desired characteristic more common or it can be used to combine characteristics of two organisms in a unique fashion. Changing the frequency of characteristics and creating new combinations of characteristics can produce beauty that was not recognized before.

Breeding can also reveal characteristics not seen in either parent, thus displaying beauty hidden in the organisms ('gifts of glory' placed there by God). Breeding is an ancient practice which may date back as far as the Garden of Eden. It is certain that it goes back at least to the time of Jacob, for God blessed Jacob by breeding livestock for him—and revealing to Jacob how He did it (Gen. 31:8-10). Breeding has produced hundreds of thousands of varieties, breeds, and cultivars of organisms, including hundreds of varieties of dogs, scores of varieties of pheasants, horses and cattle, roses, apples and peaches, and tens of thousands of varieties of orchids. Breeding new varieties of plants or animals that are cheaper to raise, or easier to keep healthy, or easier to get to market, or better-tasting or more attractive, *etc.*, is an enormous industry in our modern world.

Yet it is also clear that breeders are merely revealing designs that were already present in the parent organisms, something God probably placed within those organisms at their creation. Such breeding has increased the beauty that is actually manifested in the creation, thus directly bringing more glory to God. To whatever extent it is then

---

selection, artificial—human choice of some available organisms (e.g. for use or breeding)

breeding—human activity of controlling what sorts of organisms are produced in the next generation by selecting which parents are mated (or 'crossed')

---

a It is termed 'artificial' because it is done by humans and not by the 'natural' world.

acknowledged that the new breeds were not actually created by humans, but put there by God, God receives even more glory.

Finally, beauty can also be enhanced by creating more fundamental rearrangements of God's creation. Entire human professions have been created to enhance the beauty in God's creation in this very manner. Interior decorating, cosmetology, the culinary arts, the fine arts—these are all examples of entire disciplines devoted to *increasing* the beauty in the world. Properly executed, practitioners of these activities can glorify God in profound ways.

Summary of Chapter

A. Beauty is a holistic concept where multiple attractive characteristics are woven together in a compelling manner. Because the various attributes of God are attractive and woven together into a compelling whole, God is beautiful. And, given that the attributes of God are infinite, God's beauty is infinite.

B. Because God is invisible and He wishes us to know Him, He made Eden, the tabernacle, the temple, and heaven beautiful and He infused the creation with beauty (including the biological world).

C. God also created things with different measures or perfections of beauty (e.g. ugly things, less ugly things, mildly beautiful things, stunningly beautiful things). Humans have been created in such a way that they take this spectrum of perfection of beauty and extrapolate through the spectrum of beauty that we can see towards the infinite beauty of God Himself.

D. Biological beauty is deep (i.e. found at every size scale in the universe), ubiquitous (everywhere on earth), profound (intense, awesome), multi-faceted, and sparkling.

E. God created natural processes that restore biological beauty when it has been lost or damaged. These processes include healing processes in organisms and community succession in groups of species. Community succession (from a pioneer community to the climax community) is a series of communities of organisms, each modifying the environment to make it suitable for the community to follow.

F. The only reasonable explanation for the origin of biological beauty and the human ability to appreciate and be awed by beauty is that God infused beauty into the creation so that we could see illustrations of His infinite, but invisible, beauty.

G. Modern evolutionists believe natural selection is the process that makes evolution possible. Yet, in principle, natural selection selects efficiency over beauty—thus systematically reducing the beauty in the biological world. Natural selection would also tend to select against the human ability to appreciate beauty. Since evolution is the logical consequence of the naturalistic worldview, naturalism has no adequate explanation for how biological beauty came to be nor how human developed the ability to appreciate beauty, let alone how biological beauty could be so abundant, so deep, so ubiquitous, so profound, so multi-faceted, and so sparkling.

H. The abilities humans have to perceive beauty and to be awed by it are gifts of God given so that we could grow in our understanding of His infinite, but invisible beauty.

I. Studying biological beauty allows us to be better priests of the creation by

   a. giving us greater insight into Who God is (His beauty, the infinite nature of His beauty, He as the source of beauty, His continuous activity, His care for the details of our lives, His omnipresence, His creativity, His desire to be known, His desire to make our lives beautiful) and giving us cause to

   b. worship God, and

   c. bring others and the creation into the worship of God.

J. Our kingly responsibility to biological beauty is

   a. to preserve the beauty of the creation (e.g. prevent pollution, replant logged areas, design parks and landscaping in new construction),

   b. to restore creation's beauty wherever it has been marred by human sin (e.g. clean up pollution), and

   c. to enhance creation's beauty to the glory of God. We can enhance creation's beauty

      i. by rearranging organisms of the creation into beautiful new communities of organisms (e.g. gardens),

      ii. by breeding organisms to generate plants and animals with new combinations of characteristics, and

      iii. by breeding organisms to reveal beauty that God hid inside organisms.

Advanced Discussion Topic

Some argue that beauty is an illusion because 'beauty is in the eye of the beholder'. Develop a response to this argument. Consider the following: the number of people who believe beauty does not exist at all; the existence of things that everyone finds beautiful; and the difference between the existence of beauty and types of beauty.

Test & Essay Questions

1. Define deep beauty / ubiquitous beauty / community succession / climax community / pioneer species / profound beauty / multifaceted beauty / sparkling beauty / spectrum of perfection of beauty / artificial selection / breeding.

2. Compare and contrast deep beauty and the beauty of human creations, and then the origin of biological beauty in the naturalistic and Christian worldviews.

3. Short Essay: In what sense can an invisible God be beautiful? (Not how does he *show* His beauty, rather how *is* He beautiful when He cannot be seen?)

4. Why did God make the biological world / New Jerusalem beautiful? // Why did God specify that the tabernacle / temple be beautiful?

5. Short Essay: What does it mean for biological beauty to be deep / ubiquitous / profound / multifaceted / ever-changing and what does this suggest about God's nature?

6. Short Essay: What is the spectrum of perfection of beauty, and what are we supposed to deduce from it?

7. Short Essay: How is the origin of biological beauty / human appreciation of beauty better explained in the Christian worldview than in the naturalistic worldview?

8. Short Essay: Why should we preserve / enhance biological beauty?

9. Short Essay: Why did God hide potential for beauty in the creation?

10. Short Essay: How can interior decorating / landscaping / cosmetology / the culinary arts / the fine arts be used to glorify God?

CHAPTER 4

GOD IS DISTINCT

Created Kinds and Biological Discontinuity

"...You are great, O Lord God. For there is none like You, nor is there any God besides You..."
II Sam. 7:22, NKJV

4.1 | The Uniqueness Of God

God is unique and distinct. He is like none other (*e.g.* Exo. 8:10 and Mark 12:32[a]). In fact, God is not just greater in *quantity* than everything else (bigger, stronger, wiser, *etc.*), He is markedly different in *quality* (substance). God is so great that He had the power—and every right—to create a world and universe that could not know Him.

However, His desire is to be known, so He endowed the creation with finite illustrations of His attributes and created us capable of knowing Him. That is not to say that we can *fully* know God or even that we will *ever* fully know Him at some time in the future, but we have been created in such a way as to know *about* Him and even *know* Him personally. We have been created to recognize God, to recognize His attributes, and to develop a relationship with Him.

As much as God has allowed Himself to be knowable, His uniqueness is still one of His attributes, and He wishes us to understand that. God has chosen to illustrate His uniqueness and distinctness in the biological creation with biological **discontinuity**—observable gaps or differences between organisms. These gaps seem to indicate that groups of organisms exist that are different from other organisms, and perhaps always have been. The absence of organisms within those gaps suggests that the gaps are not traversable. The Bible hints at the existence of such biological discontinuity when it refers to the 'biblical **kind**'.

discontinuity—gap in form between or among organisms. Different types of discontinuity: very shallow discontinuity is bridgeable, persists only with breeding, and separates breeds and varieties and cultivars; shallow discontinuity is bridgeable, persists naturally, and separates species; deep discontinuity is unbridgeable and separates baramins

kind, biblical—recognizable group of similar organisms surrounded by deep discontinuity that persists by members interbreeding and producing similar offspring (aka baramin)

The Biblical Kind

In the creation account of Genesis 1, a Hebrew phrase is used when describing the creation of plants and animals that is not used in describing anything else in His creation. The phrase occurs ten times in Genesis 1 and translates into English as 'according to their kind'[b]. God commanded the earth to produce fruit trees 'according to their kind' and the earth brought forth herbs and fruit trees 'according to their kind' (Gen. 1:11-12). God created every living creature in the water and in the air 'according to their kind' (Gen. 1:21). God commanded the earth to bring forth living creatures of the land 'according to

---

a Other references that indicate there is none like God include Ex. 9:14, 15:11, Deu. 33:26, II Sam. 7:22, I Ki. 8:23, I Chr. 17:20, Psa. 86:8, 89:6, Isa. 40:18, 46:9, and Jer. 10:6-7.

b 'According to' can be replaced with 'after'. 'Their' can be replaced with 'its' or 'his'.

their kind', and God made the beast of the earth, the cattle, and every creeping thing 'according to their kind' (Gen. 1:24-25).

Later in the biblical account, God commands Noah to take birds, cattle and every creeping thing onto the ark 'according to their kind' (Gen. 6:19-20), and every bird, every beast, all cattle, and every creeping thing entered the ark 'according to their kind' (Gen. 7:13-14). Finally, at the end of the Flood, every beast, every creeping thing, and every bird exited the ark 'according to their kind' (Gen. 8:18-19). Later yet, humans were told to eat and not eat specific organisms 'according to their kind' (Lev. 11:14-16, 19, 22, 29; Deu. 14:13-15, 18).

Linguistically, considered alone outside of its larger context, it is possible that the Hebrew phrase merely refers to a diversity of things. In other words, 'God created birds according to their kind' might be translated 'God created a number of different birds' or 'God created a variety of birds' or 'God created a diversity of birds'. However, the specific way in which the phrase is used in Scripture suggests there is more to it than a mere statement of diversity.

First, since the biblical phrase 'according to their kind' is only used in reference to organisms, 'kind' appears to be something biological. Second, the fact that Noah is to take two of every 'kind' onto the ark (Gen. 6:19-20) suggests that a 'kind' is a group with more than one member. Putting these two deductions together, a 'kind' appears to be a group of organisms. Third, Genesis 2:19-20 indicates that Adam named all the 'beasts of the field' and the 'fowl of the air', and this occurs between the creation of Adam (Gen. 2:7) and the creation of Eve (Gen. 2:22).

Since the creation of both Adam and Eve occurred on the sixth day of creation (Gen. 1:23-31), Adam must have named all these animals on a single day. Getting all this naming done in a portion of one day suggests that Adam named animals rapidly. Considering also that Adam's names were accepted without question by humans following him (Gen. 2:19), suggests that Adam named these animals in some sort of natural or intuitive manner.

All this suggests that God created organisms in groups ('kinds') and created humans in such a way that they can recognize those 'kinds'. That humans have the intuition necessary to recognize 'kinds' would also explain how Noah could follow God's command to permit pairs of each 'kind' onto the ark (Gen. 6:19-20), and how after the Flood, the Israelites could choose the right organisms to eat and not eat 'according to their kind' (Lev. 11:14-16, 19, 22, 29; Deu. 14:13-15, 18).

From the third deduction about 'kinds'—their recognizability—flow two other deductions (a fourth and a fifth). For a 'kind' to be recognizable there must be similarity within the 'kind' (the fourth deduction) and dissimilarity between the 'kinds' (the fifth deduction). To correctly identify all members of a 'kind' they must all be similar to each other—or at least have in common one or more noteworthy features.

At the same time, to make sure that no member of another 'kind' was mistakenly identified as part of any given 'kind' there needs to be substantial difference or differences between each 'kind' and every other 'kind'. Described another way, the 'kinds' are separated by gaps. The fact that each of Adam's names 'became the name' thereafter (Gen. 2:19) leads to a sixth deduction—that each 'kind' remained the same for generations to follow Adam. And, given that most of the pre-Flood patriarchs lived into their tenth century (Genesis 5), it follows, then, that 'kinds' must persist for at least thousands of years.

A seventh deduction is derived from the creation of organisms 'according to their kinds' on three different days of the Creation Week: each biblical 'kind' was separately created. Combined with the distinction and persistence of 'kinds', it seems reasonable to conclude that the 'kinds' remained distinct from each other up until the Flood. At the time of the Flood, the animal 'kinds' were taken 'two by two' onto the ark so as 'to keep them alive' (Gen. 6:19-20), 'multiply on the earth' (Gen. 8:17) and persist for 'perpetual generations' (Gen. (9:12).

From this we derive our final (eighth and ninth) deductions. The eighth is that each biblical 'kind' was created to remain distinct for *all* of earth history, so that the gaps between 'kinds' are too broad to cross—or, at the very least, have never been crossed. Every 'kind' is therefore surrounded by a discontinuity that cannot be (or has not been) crossed—what might be called a deep discontinuity. The last (ninth) deduction is that members of a 'kind' have the capacity to breed with one another and produce other members of the 'kind' for successive generations, but probably lack the capacity to breed with other 'kinds'. Although interbreeding may be possible among some or many members of a 'kind', it may not be possible between members of two *different* 'kinds'.

Combining all nine of our deductions, Scripture's use of 'kind' suggests that organisms should be classifiable into 'created kinds'—with each 'kind' being a *recognizable group of similar organisms which is surrounded by deep discontinuity and persists by members interbreeding and producing similar offspring*.

Adam's names were probably consistently used as long as there was a single human language. After the confusion of tongues at Babel, however, it is likely that each language ended up with different names for the same organisms. Even within the various language groups,

individual tribes and families over the millennia invented their own names for organisms.

By the middle of the eighteenth-century A.D. many, many thousand names existed for organisms. In fact, *too many* names existed. In many cases a particular organism had lots of names, making the study and understanding of the organisms of the world a bit challenging. Carl Nilsson Linnaeus (1707-1778) determined to standardize the naming of organisms. Since he believed in the concept of the biblical 'created kind', Linnaeus sought to assign each 'kind' a single, standard name that would be used by biologists all over the world. He created names in Latin, the standard language of the former Roman Empire (and a substantial fraction of the known world).

Linnaeus recognized that different people consistently united organisms into the same groups, or taxa (sing. *taxon*). The fact that such taxa were consistently identified by very different people suggested that the groups were recognizable by humans—just as was expected of biblical 'kinds'. Members of these taxa were also observed to be 'interbreeding and producing similar offspring'—a second thing expected of biblical 'kinds'. Finally, in Latin, these taxa were called *species*, and *species* was the Latin translation for the Hebrew word *mîn* in the Vulgate (the Latin Bible)!

This suggested that species really were the biblical 'kinds'. Fulfilling as they did, many of the requirements of biblical 'kinds', Linnaeus began his naming task with the conviction that species were

inter-specific hybrids—offspring generated by parents from two different species

hybrid zone—geographic region located between adjacent ranges of different organisms, which region contains intermediates between the different organisms

stratomorphic intermediate—a fossil having attributes appearing intermediate between two other organisms, and which is also located in strata (rock layers) located between the same two organisms; a stratomorphic series is a sequence of stratomorphic intermediates

inter-generic hybrids—hybrids between different genera

to be equated with 'kinds'. Because he thought biblical 'kinds' could not change and could not interbreed, Linnaeus began his career believing that *species* could not change, and that members of two different species could not possibly interbreed.

As Linnaeus began to name hundreds, and then thousands of species[a], he discovered organisms he believed to be hybrids between two different species (**inter-specific hybrids**). As a result, Linnaeus revised his hypothesis that species were biblical 'kinds' and began to believe that the 'kind' was a larger grouping than the species. At first Linnaeus theorized that the 'kind' was to be equated with the 'genus', a biological taxon which contains one or more species.

In the nearly quarter of a millennium that has elapsed since Linnaeus, Linnaeus' suspicion has been verified in three different ways. First, many of the species of the world have been shown to produce inter-specific hybrids[b]. Second, some pairs of species that occupy adjacent tracts of land have a **hybrid zone** between them—a zone at the boundary where members of the two species naturally interbreed and produce intermediates between the two species. Third, a small percentage[c] of pairs of fossil species have 'populations'[d] of **stratomorphic intermediates** between them. These are groups of fossils that are younger than one species and older than the other[e] and have a body shape intermediate between the two species[f]. Stratomorphic intermediate 'populations' between two species appear to record the transformation of the older species into the younger one. It seems, then, that the biblical 'kind' must be a larger grouping than the species, just as Linnaeus surmised nearly two hundred and fifty years ago.

As Linnaeus continued to name organisms, however, he eventually discovered organisms he believed to be **inter-generic hybrids**

---

a   Plant species, as they were Linnaeus' specialty.

b   Some of these hybrids are formed under natural conditions. Others are formed 'artificially', meaning either they produce offspring by natural means, when members of one or both species are taken from where they live naturally and placed together, or humans mix the genetic material of the two species.

c   Something less than 1% of fossil species.

d   Technically, populations are groups of interbreeding organisms. Technically, since it is impossible to determine if fossil organisms could actually interbreed, fossil populations are impossible to demonstrate—thus the quote marks around 'population' when referring to fossils.

e   *i.e.* stratigraphic intermediates—in other words, located in intermediate or in-between strata (where 'strata' are rock layers)

f   *i.e.* morphological intermediates—in other words, having an intermediate morphology (where 'morphology' refers to the 'morph' or form of an organism) (hybrids between different genera). This led Linnaeus to theorize that the biblical 'kind' was an even larger group than the genus. Since the time of Linnaeus this inference has also been confirmed. Though not as common as inter-specific hybrids, a large number of inter-generic hybrids are known. Many orchids used in corsages, for example, are inter-generic hybrids[a].

However, whereas hybrid zones are known between different species, no hybrid zones are known between different genera. And, whereas 'populations' of stratomorphic intermediates are known between fossil species, 'populations' of stratomorphic intermediates are not known between fossil genera. Yet, fossil genera themselves can be arranged into a **stratomorphic series**. This is a sequence (or series) of fossil genera found in successive layers that step through a sequence (or series) of forms, rather analogous to footprints in a trackway through time. For example, successive horse genera in Tertiary and Quaternary rocks of North America step through a sequence of increasing size, decreasing number of toes, and increasing tooth height.

stratomorphic series— a sequence of fossil genera found in successive layers that step through a sequence of forms

There seems to be considerable evidence, then, that the biblical 'kind' is a larger grouping than the genus, just as Linnaeus surmised in the latter part of his life.

---

a   Some of the most popular corsage orchids are the Rhyncholaeliocattleya orchids, produced by crossing the genera *Rhyncholaelia*, *Laelia*, and *Cattleya*. The Gladysyeeara orchid is another type of orchid hybrid produced by crossing 8 genera (*Brassavola*, *Broughtonia*, *Cattleya*, *Cattleyopsis*, *Caularthron*, *Epidendrum*, *Laelia*, and *Sophronitis*)!

family, taxonomic—taxonomic level above genus and below order

In the last part of his life Linnaeus believed that the created 'kind' was equated with the taxonomic[a] **'family'**—a biological taxon which contains at least one genus. Linnaeus died believing that taxonomic families do not change and members of one taxonomic family cannot interbreed with members of any other taxonomic families. In the last quarter millennium, evidence has accumulated to confirm Linnaeus' last inference as well. No hybrids have been discovered between members of different taxonomic families, no hybrid zones are known between living taxonomic families, no stratomorphic intermediate 'populations' are known between fossil taxonomic families, and no stratomorphic series are known made up of fossil taxonomic families.

Because many taxonomic families seem to be separated by the sort of non-traversable gaps (deep discontinuity) expected between biblical 'kinds', some creationists have suggested that the biblical 'kind' should be equated with the taxonomic family, just as Linnaeus died believing. However, there are some cases where deep discontinuity seems to exist *below* the level of the taxonomic family and other cases where deep discontinuity cannot be found until the group is larger than the taxonomic family. More study is needed to clarify exactly how the biblical 'kind' relates to the groupings that biologists have created over the centuries.

---

a   So as not to confuse the word 'family' as it is used here with the more familiar meanings of family (e.g. human parents and children), the somewhat awkward term 'taxonomic family' will be used in the text. Taxonomy is the science of naming organisms—the very thing Linnaeus was doing. The 'family' to which we are referring is one of the groupings of organisms that taxonomy identifies and names—thus why it is called a 'taxonomic' family.

In the meantime, creation biologists think it is best to identify the created 'kinds' independently of the biological groupings created by Linnaeus and other taxonomists. Furthermore, to avoid the confusion that might come from using the word 'kind'—a term already having a variety of common meanings—many creationists refer to the 'biblical kind' as a **baramin** (Hebrew: *bâra*, 'to create' + *mîn*, 'kind')[a].

4.2 | Baraminology

The science attempting to identify the baramins of the world is called **baraminology**[b]. Baraminologists search for **holobaramins**, groups of organisms surrounded by deep discontinuity. They suspect that these holobaramins correspond to the 'biblical kinds' or baramins of Scripture. Preliminary baraminology research suggests that, in contrast to the nearly two million fossil and living species that have been described so far, the number of extinct and living baramins probably totals only a few thousand. The relatively small number of baramins in the world may resolve a couple challenges that have been leveled against the biblical account over the centuries.

At various times, critics of the Bible have claimed that there are too many species on this planet for Adam to have named all the animals in a single day, or for the ark to have carried all the land animals through the Flood. However, since the biblical account repeatedly refers to animals 'according to their kind'—and specifically does so in both the creation and the Flood account—it is likely that Adam was naming baramins, not species, and the ark was carrying baramins, not species. The biblical account indicates Adam named all the 'beasts of the field' and 'fowl of the air'. The 'beasts' do not include the creeping things (as seen in the lists of Gen. 1:24-25), and the 'beasts of the field' probably refer only to the animals closely associated with humans (as opposed to the more general term 'beasts of the earth' of Gen. 1:24-25).

Therefore, even if Adam named the field and flying animals of the entire world—as opposed to naming all the field and flying animals of Eden—the number of baramins involved is probably less than 300 (versus thousands of species). Adam could have named this number of

VIDEO 4.2

baramin— recognizable group of similar organisms surrounded by deep discontinuity that persists by members interbreeding and producing similar offspring (aka biblical kind)

baraminology— the science of the discovery and study of baramins; a creationist biosystematics method

holobaramins— groups of known organisms surrounded by deep discontinuity hypothesized to correspond to baramins

---

a 'Baramin' was coined by Frank Lewis Marsh (1899-1992) in 1941.

b Baraminology is a creationist biosystematics method (method of classifying and naming organisms), created and named (Hebrew *bâra*, 'to create' + Hebrew *mîn*, 'kind' + Greek *logos*, 'discussion' or, perhaps, 'study of') by Kurt Wise in 1990.

animals in a few hours or less. In the case of the ark, since its purpose was to save land animals through the Flood, only land animals had to be included. The number of land animal baramins is probably about one or two thousand, so at most only a few thousand animals had to be placed on the ark (versus tens or hundreds of thousands of species). The size of the ark described in the biblical account (Gen. 6:15) could easily accommodate this many.

4.3 | Biological Discontinuity

Deep discontinuity is not the only sort of discontinuity that has been discovered. In fact, there seems to be a spectrum of perfection of discontinuity in the biological world, with the deep discontinuity occupying only one position in that spectrum. What follows is a summary of that spectrum as it is now recognized and understood, beginning with the shallowest discontinuity and moving through to the deepest.

Extremely Shallow Discontinuity Within Species

species—smallest recognizable group of similar organisms that persists naturally by the production of similar offspring; taxonomic level below genus

A **species** can be described as the smallest recognizable group of similar organisms that persists naturally[a] by the production of similar offspring. *Within* a species the dominant pattern is the strong similarity found among individuals of the species AND between parents and their offspring. The young in these species mature to look just as similar to one another and their parents as their parents appear to each other. In species where mating occurs, each member of the species seems readily able to mate with any other member of the species. So, in both mating and similarity, there appears to be no discontinuity at all within a species.

However, individuals in the species have distinct bodies. Furthermore, closer examination indicates that individuals can be distinguished by differences in DNA, behavior, personality, and/or body form. Since this discontinuity is accompanied by no apparent barrier to mating, we describe as extremely shallow that discontinuity found between individuals within a species. Asexual organisms, like bacteria, seem to show the shallowest of within-species discontinuity.

---

a The word 'naturally' in this definition is to stress that the similarity between offspring and parents persists when the mating within the species is *not* being determined by humans—i.e. it is natural. This natural persistence of form is characteristic of species, but not human-maintained varieties, cultivars, and breeds. (See the section on varieties, cultivars, and breeds.)

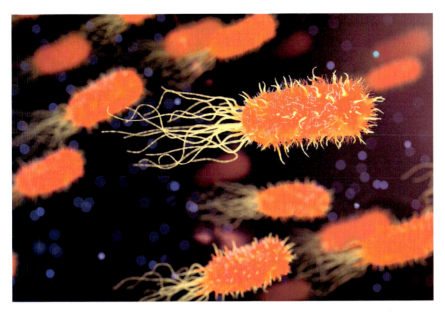

The young of asexual organisms are clones of their parents and appear to be nearly identical—even as far as having identical DNA.

Very Shallow Discontinuity Between Varieties, Cultivars, and Breeds

Breeding[a] is the human activity of controlling what sorts of organisms are produced in the next generation by selecting which parents are mated (or 'crossed'). Humans have been breeding plants and animals for thousands of years—most likely from the time of the Garden of Eden. Breeding has generated many thousands of organisms that have an appearance different from any organisms found among the natural or wild species of the world. The organisms produced by this breeding that have more distinctive, recognizable forms are even given their own names.

Thousands of animals have been named as **breeds**, and thousands of plants have been named as **varieties** or **cultivars**. For example, there is the Beagle dog breed, *Canis familiaris* 'Beagle', the Aspiran grape variety, *Vitus vinifera* var. Aspiran, and the Jonathan apple cultivar, *Malus domestica* 'Jonathan'. Such varieties, cultivars, and breeds can be just as different from one another as species are, but they are not persistent. When allowed to mate as they wish, they mate just as readily with their own as they do with other varieties, cultivars, or breeds of

breeds— distinguishable group of animals that remains distinct only with breeding

varieties— distinguishable group of plants that remain distinct only with breeding (aka cultivars)

cultivars— distinguishable group of plants that remain distinct only with breeding (aka varieties)

---

a Also known as 'artificial selection', opposed to 'natural selection'. It is considered 'artificial' because it is done by humans and it is selection in the sense that there is some sort of selection of what organisms will produce young in the next generation.

the same species, so the gaps between and among them are readily bridged by mating. Because of the unbridled breeding in the natural world, varieties, cultivars, and breeds quickly disappear when humans are not controlling the breeding.

Consequently, since mating between varieties, cultivars, and breeds is as free as it is between other individuals of the same species, but there is more difference in appearance, varieties, cultivars, and breeds are said to be separated by very shallow discontinuity (rather than extremely shallow discontinuity between individuals in natural species populations).

Shallow Discontinuity Between Species

The smallest recognizable group of similar organisms which is surrounded by some sort of *persistent* discontinuity is the species. Unlike varieties, cultivars, and breeds, species in the wild produce offspring that look the way their parents did in the previous generation. This allows species to remain unchanged for successive generations. In this way, species we first recognized in our childhood remain recognizable throughout our entire life.

Species have even persisted throughout the history of human culture (*i.e.* throughout most of the period following the Flood). Aside from species that have gone extinct, descriptions of species in ancient cultures and depictions of them in ancient art remain identifiable and identical with species we recognize in the present. Even in the fossil

record, species typically remain recognizable and unchanged from the oldest fossil through to the present (or to the youngest fossil, in the case of extinct species)—something called **species stasis**[a].

The discontinuity surrounding species is further confirmed by the abrupt appearance of species in the fossil record[b]. This refers to the fact that for a vast percentage (>99%) of species in the fossil record, the oldest fossils are not preceded by a 'population' of stratomorphic intermediates that would connect that species with any other[c]. Considering all these observations at the same time—the recognizability of species, the persistence and stasis of living and fossil species, and the **abrupt appearance** of species in the fossil record—the case for discontinuity surrounding species seems rather strong.

However, there is also evidence that the gaps among and between species are crossable. First, among some pairs of species that live in

stasis—status of a fossil taxon when it shows no change through successive layers of the fossil record

abrupt appearance—status of a fossil organism when it is not preceded in the fossil record by intermediates to any other fossil organism

---

a   As will be addressed in Chapter 15, most of the fossil record preserves the organisms living at and after the time of the Flood. The stasis observed in that portion of the fossil record formed in the Flood is NOT evidence of the persistence of species, for this stasis is merely due to fossils being sampled at different layers in the Flood from the SAME population. It is only the stasis observed in the fossil record from *post*-Flood times that is actually evidencing the persistence of species.

b   Similar to the comments about the stasis of fossil species, the abrupt appearance of fossil species has a different interpretation in the portion of the fossil record deposited in the Flood as it has in that portion deposited after the Flood. Abrupt appearance of species in Flood sediments is evidence of a discontinuity surrounding species at the time of the Flood. In post-Flood sediments, abrupt appearance of species is evidence of no intermediates being preserved *as species came into being*.

c   And, notably, the only fossil species that have been argued to be preceded by intermediates are those in post-Flood sediments. These are the only sediments that might be documenting how a species came to be. As might be expected in a young-age creationist interpretation of the fossil record, *no* species in Flood sediments are preceded by intermediates.

adjacent regions, the boundary areas show a continuous transition between the two species (a hybrid zone). In many of these hybrid zones every conceivable intermediate can be seen between the species on either side of the hybrid zone. Second, among some pairs of species at different levels in the fossil record, the rock layers between the two species contain 'populations' of stratomorphic intermediates that show a continuous transition between the species below and the species above. Although fewer than 1% of all species show these zones of transition—both in the fossil record and in the living world—they still suggest that the discontinuity between species is crossable.

A third line of evidence is inter-specific hybridization. This happens when members of two different species, such as a grizzly bear and a polar bear, mate and have young. Typically, somewhere around three quarters of the species in any given biological family can not only be mated but can also produce healthy offspring that can in turn produce offspring of their own. In families with many species, an astonishing number of inter-specific hybrids can be produced. For example, since 1940, at least 50,000 inter-specific hybrid orchid species[a] have been produced from an estimated 25,000 natural orchid species. This ability to hybridize suggests that at least within families it is possible to overcome the discontinuity that seems to exist between species.

---

a   Even though these organisms have been produced by breeding, they are not referred to as cultivars or varieties, because when allowed to mate without human control, these plant forms persist. They cross only with others nearly identical to them and only produce offspring just like them—i.e. they function just like wild or natural species.

On the one hand, the discontinuity *between* species is real, and much greater than the discontinuity between individuals *within* the species. On the other hand, there is considerable evidence that the gap between species can be crossed. Therefore, the discontinuity between species should be described as a shallow discontinuity.

Semi-Deep Discontinuity Between Genera

Biologists group species together into larger groups known as 'genera' (sing. *genus*). Genera seem to be surrounded by discontinuities that are deeper than those surrounding species. First of all, genera are more readily distinguishable than species, suggesting the gap between genera is wider than the gap between species. Secondly, the hybrid zones and 'populations' of stratomorphic intermediates occasionally found between species are *not* found between living or fossil genera.

At the same time, there is evidence that the discontinuities between genera can be crossed—though not as commonly as the discontinuities between species. For instance, some genera have been successfully crossed with other genera. Examples of such inter-generic hybrids include the some of the most popular showy orchids (resulting from the cross of three or more different orchid genera[a]) and the 'cama' (a cross between a camel and a llama). Furthermore, although continuous transitions between genera are absent in the fossil record, a few sets of genera are found in series or sequences analogous to footprints in a trackway through time (for example, horse genera through Tertiary and Quaternary rocks).

Consequently, the discontinuity between genera is deeper than the discontinuity between species, but still bridgeable. We might then describe the discontinuity between genera as semi-deep discontinuity.

Deep Discontinuity

Holobaramins are the smallest recognizable groups of similar organisms completely surrounded by large, apparently non-traversable gaps (*i.e.* separated by deep discontinuity). Holobaramins are identified as groups of similar organisms surrounded by a substantial gap of form from all other organisms which lacks any evidence of crossability. In other words, no hybrids or hybrid zones are known between living

---

a   The result of an inter-generic hybrid (generated from two different genera) being crossed with another inter-generic hybrid (generated from a third and fourth genus).

members of different holobaramins, no 'populations' of stratomorphic intermediates are known between different fossil holobaramins, and no sequences of fossil holobaramins are known. Baraminology's identification of a number of holobaramins suggests biblical 'kinds' are real. Deep discontinuity seems to divide life into several thousand baramins.

Deeper Discontinuity in Higher Groups

Careful baraminology studies are also confirming what has been suspected intuitively by biologists for a very long time—that the size of discontinuities between groups gets larger with larger, more inclusive, groups of classification. Discontinuities between genera tend to be smaller than the discontinuities between families, discontinuities between families tend to be smaller than the discontinuities between orders, discontinuities between classes are even larger, discontinuities between phyla are larger still, and discontinuities between kingdoms are even larger than those between phyla.

In other words, discontinuity becomes deeper with larger and larger (more and more inclusive) groups of organisms. This pattern of discontinuity is not only true for living organisms; it also seems to characterize the fossil record. Every biological group (genera, families, orders, *etc.*) remains distinct and unchanging in the fossil record, and the size of the discontinuity between fossil groups tends to increase with more and more inclusive groupings. Discontinuity in both living

and fossil groups gets deeper when more and more inclusive groups of organisms are considered.

Extremely Deep Discontinuity Surrounding Life

The very deepest discontinuity in biology seems to separate organisms from everything else. As full as the spectrum of perfection of life is—from viruses, bacteria, algae, plants, animals, and man—there still seems to be an unbridgeable gap between the non-living creation and biological life. Looked at in one way, viruses seem to blur the line between non-life and biological life. Yet they have to use cells of biologically living organisms to perform any of the traditionally understood characteristics of life (growth, reproduction, metabolism, *etc.*).

Viruses may actually be better described as extensions, or tools, or external organs of biological living things, rather than living things themselves. This leaves a substantial gap between the nonliving world and biological life. First there is a gap in complexity. Even the simplest biological organisms are much more complex than the most complex of non-living things. Secondly, the non-living world on its own does not seem capable of generating even the simplest of living things. There are no known examples of any biologically living thing arising naturally from non-living things. In fact, the non-living world does not even seem capable of producing the biological molecules (the macromolecules) that make up organisms, such as RNA, DNA, and

proteins. There even appears to be a number of theoretical reasons why many of these molecules cannot be generated under any natural conditions outside an organism. Thirdly, even humans are not capable of generating biological life. Because the worldview of naturalism believes life did arise from non-life, many scientists have attempted to create life from non-living components in order to demonstrate that it can be done and to determine how it happened.

Even when all the ingenuity of very intelligent humans is applied to the task, however, a living thing has never been created from non-living materials. Finally, the non-physical components of life suggest that life involves more than just greater complexity. Life seems to be composed of a substantially different substance. There is both a quantitative and qualitative difference between non-life and life. The gap between the non-living and living appears to be larger and more unbridgeable than any gap between organisms. An extremely deep discontinuity exists between life and non-life.

4.4 | The Origin Of Discontinuity

According to naturalism, all living organisms are related by the process of evolution, so no true discontinuity should exist. Organisms are descended and modified from parent organisms, and parent organisms were descended and modified from organisms before them. And so it goes, all the way back to the first organism, the ancestor to all the organisms that have ever lived. In the naturalistic worldview, all organisms are connected by a family tree—by a continuous lineage. Even the first organism is thought to have developed from non-living matter.

If the naturalistic worldview is true, there must be some path of continuity connecting all organisms. The **deep discontinuities** (large gaps that seem unbridgeable) that divide life into thousands of baramins suggest that naturalistic evolution of all life is impossible and, in fact, never occurred. Evolution is even more impossible considering the even deeper discontinuities that divide baramins into groups—that group them (successively) into orders, classes, phyla, kingdoms, and domains. Surrounding life itself, there is an even deeper discontinuity that suggests that life could not possibly have arisen from non-life. The deep, deeper, and deepest discontinuities of biology are much more

deep discontinuities— large gaps that seem unbridgable that divide life into thousands of baramins

consistent with the separate creation of organisms by God as described in the Bible than they are with the naturalistic origin of organisms by evolution.

Deep discontinuities are not the only challenges to the naturalistic worldview. The shallow discontinuities are also a challenge. Naturalistic evolution suggests that there should be continuity between species which are in the process of separating. If naturalistic evolution is true, at least some **sister species** (those recently descended from the same ancestor) should be difficult to distinguish from one another. In fact, sister species tend to be easily distinguished from each other. This distinction between species is even maintained when interbreeding between the two species is known to occur, such as when a hybrid zone is present.

This is even the case when the hybrid zone has been there for a long time[a]. And, when a hybrid zone is lacking, geographically adjacent species remain distinct even when breeding can generate a nearly continuous spectrum of intermediates between two species (such as between grizzly and polar bears). Some species can even live in the same area and remain distinct, even though crosses between them are known to produce intermediates (such as between coyotes and dogs).

sister species—species recently descended from the same ancestor

---

[a] Such as a hybrid zone between two snail species which Stephen Jay Gould argued had been there since the low-stand of sea level in the Pleistocene, more than 12,000 radiocarbon years ago (interpreted by naturalists as 12,000 actual years ago; interpreted by young-age creationists as about the time of Babel, about 4000 years ago).

Although shallow discontinuity (distinction of species even when they can readily interbreed) can be explained if the discontinuity was programmed into the organisms by God, naturalistic evolution is hard-pressed to provide an explanation. In fact, finding the same pattern of discontinuity throughout the creation—*even in places where breeding should eliminate it*—would seem to be powerful evidence of a God purposely infusing a discontinuity pattern into the creation.

But God created even more evidence in the form of the very shallow discontinuities among varieties, cultivars, and breeds. In naturalistic evolution, the only way known to introduce new things into the world of biology is by the process of mutation. However, if any of the mutation's contributions are helpful, they are an extremely small percentage. Most mutations are harmful to organisms. This is why natural selection is so important in naturalism, for natural selection is thought to get rid of the harmful mutations and select only the beneficial ones.

However, natural selection can only operate on what an organism looks like or on what an organism actually does. If the organism has a *hidden* trait, natural selection cannot act on it. Natural selection cannot eliminate hidden mutations and it cannot select hidden breeds or varieties, even if they are good designs.

In the case of varieties, cultivars, and breeds, most of them were completely hidden within organisms before humans revealed them. The hundreds of dog breeds, for example, are thought by naturalists to have been generated from the wolf. However, the vast majority of dog breeds have physical and behavioral characteristics that are completely unknown in any wild population of wolves. A careful examination of breeding reveals also that humans did not actually *create* dog breeds. Through many generations of selecting certain traits that they saw in dogs (or wolves), they eventually 'discovered' the breeds we have today. It seems that these dog breeds already existed, somehow programmed into dogs (or wolves), but hidden. Humans merely found ways to reveal these hidden breeds.

But how did the breeds get there? Natural selection cannot act on hidden things, so natural selection cannot explain how breeds were selected from among the bewildering horror that mutation must have produced. Even if the breeds did reveal themselves sometime in the past, the typically brief survival of breeds would not allow them to last long enough for natural selection to select them! If naturalistic evolution were true, we would have found thousands of horrible mutants for every functioning variety, cultivar, or breed that we find. That is not what breeders find. Instead they find a host of truly amazing and wonderful designs. Although they find mutants, the mutants were extremely rare—no more common than mutants are in natural dog or wolf populations.

In general, varieties, cultivars, and breeds seem to show the same pattern of design and discontinuity that is found in the wild, except that under natural conditions the discontinuity disappears. The existence of discontinuity when it cannot exist under natural conditions seems to be strong evidence of an invisible God filling the biological creation—even the unseen biological creation—with discontinuity as an illustration of His uniqueness.

Inter-specific hybridization presents yet another difficulty for naturalistic evolution. Evolution is thought to be a slow process, requiring hundreds to thousands of generations to generate new species and millions to tens of millions of years for a family of organisms to be populated with a range of species. In this amount of time, species should lose their ability to interbreed.

If inter-specific hybridization occurs, it might only be expected between sister species—the ones most recently separated from each other—and not be at all common in the world. In fact, inter-specific hybridization is not only common, but individual species in many families can interbreed successfully with a *majority* of the other species in the family. This suggests that even if these species are related, they are related very recently—at most thousands of years ago, not millions. Inter-specific hybridization is not only more consistent with creation than evolution, but it is more consistent with the time scale of Scripture than the time scale of naturalism.

Discontinuity: Our Responsibility

Our Responsibility to God

As priests of the creation we have a responsibility to know God, worship God, make the creation a house of worship, and bring the creation into the worship of God. What we learn about biological discontinuity can help us do this. First of all, the pattern of discontinuity among organisms gives us some understanding of what it means for God to be distinct or unique. Secondly, since both shallow and deep discontinuities challenge the naturalistic worldview, the pattern of discontinuity points to God as Creator of organisms. Thirdly, the spectrum of perfection of discontinuity provides insight into the uniqueness and transcendence of God.

On one end of the spectrum, organisms exist that are identical in every way that can be measured. On the other end of the spectrum is an enormous unbridgeable gap between life and non-life. Between these extremes, there are organisms that are very similar and freely interbreed, other similar organisms that are capable of interbreeding but tend not to, and others separated by various depths of discontinuity.

This spectrum leads the human mind to extrapolate that spectrum to infinity—to the existence of something surrounded by perfect discontinuity. This spectrum points to the God of Scripture—one God infinitely greater than everything He created. The spectrum of perfection of discontinuity helps us understand God's transcendence. He is so much greater than we are that we do not deserve to even know *about* God, let alone know Him. At the same time, the fact that He created discontinuity as an illustration and created both the spectrum of discontinuity and our ability to recognize it as a teaching aid, suggests that God desires us to know Him.

The pattern of biological discontinuity that we currently recognize provides substantial insight into the awesome nature of God. But we still have an incomplete picture of that pattern. Not only have we not identified all the species of the world, we do not even know how many exist. We have named perhaps less than one in every ten. We have identified an even lower percentage of baramins.

Knowing such a small percentage of the pattern, there is little doubt that we have much to learn—both about the pattern and the character of God. It is our responsibility as priests of the creation to continue the task of naming organisms (**taxonomy**) that God assigned to Adam on the first day of his existence. We should continue to identify and name the species, genera, families, baramins, *etc.* of the world. As we do, we will continue to gain insight into the unique, transcendent, creator God Who desires to know us.

taxonomy—
the science of naming organisms

This contemplation of biological discontinuity ought to lift our hearts in worship of the God Who, though infinitely greater than we, has seen fit to allow us to know Him, and know Him intimately. As we engage in this worship, we can fill the creation with the worship

of God and bring others into that worship. In this way we fulfill our responsibility as priests of the creation.

Our Responsibility to the Creation

Preserve Discontinuity. In our responsibility as kings over the creation, we should serve God, and serve and protect the creation. After God created the desired pattern of discontinuity into His vast creation, he handed it over to us to 'guard and keep it'. We have a responsibility to preserve the pattern of discontinuity and pass that pattern on to the next generation. We should preserve the biological groupings that now exist in our world. If it is within our power we should not allow the **extinction** (complete disappearance of every member) of any species, or genus, baramin, or any other biological taxon.

extinction—death of every member of a taxon

That having been said, of course, all biological taxa are not equal. No animal or plant taxon, for example, is as valuable as the image of God, so human lives should not be sacrificed to save any non-human taxon. Furthermore, as we shall see in Chapter 14, it is likely that every member of a baramin contains a large amount of hidden information for developing a wide range of species—the very thing that also explains inter-specific hybridization and hidden varieties, cultivars, and breeds. If that is so, then lost species might be recoverable from hidden information in other species of the same baramin. Consequently, if we are forced to choose which organisms to save and which to let go, greater priority should be placed on preserving baramins than

preserving species (or any other group of organisms found *within* baramins).

Enhanced Discontinuity. In our first role as kings we are to preserve the God-glorifying pattern of discontinuity. In our highest calling as kings, we are to enhance that pattern. And we can do so because God hid within His creation even more evidence of discontinuity and gave humans the ability to find it and reveal it. Breeding of a variety of organisms (such as dogs and cats, roses and orchids, pheasants and pigeons, hostas and azaleas) has revealed distinct, well-designed varieties, cultivars, and breeds that were completely unseen and unexpected in the organisms from which they were bred. The discovery and revelation of these hidden designs increases the amount of discontinuity in the world and fills in more of the spectrum of perfection of discontinuity.

Continuing that breeding in organisms that have already generated varieties, cultivars, and breeds, and beginning such breeding in organisms where none are known, allows us to fulfill our kingly role of enhancing biological discontinuity. And, as we admit that each new form was not actually created by us, but revealed from among the variety God originally created, then we can increase the glory of God in the process.

Summary Of Chapter

A. baramins

   a. Biblically, *mîn*, the Hebrew word translated 'kind' seems to refer to a separately created, recognizable group of similar organisms surrounded by deep discontinuity, that persist by members interbreeding and producing similar offspring (called a baramin).

   b. Three centuries ago baramins were thought identical to the species (the Latin translation of *mîn*), but inter-specific hybrids and fossil intermediates suggest the baramin is larger than both the species and the genus, and perhaps often close to the 'family' of biological taxonomy.

B. Biologically, deep discontinuities (non-traversable gaps) seem to exist, dividing organisms in thousands of baramins.

C. If Adam named baramins rather than species, he would have had sufficient time to name the animals.

D. If Noah took baramins onto the ark rather than species, there would have been enough room on the ark to save the animals.

E. Baraminology is the study, identification, naming, and classifying of holobaramins (groups of similar organisms surrounded by deep discontinuity; groups of known organisms thought to be approximations of biblical baramins).

F. To illustrate His (invisible) uniqueness, God created a spectrum of perfection of discontinuity across the entire biological creation:

 a. extremely shallow discontinuity between different individuals within a species
   i. on the one hand being unique, distinct individuals (suggesting discontinuity)
   ii. on the other hand, readily interbreeding with one another to produce offspring (suggesting the discontinuity is extremely shallow)

 b. very shallow discontinuity between different varieties, cultivars, and breeds (recognizable groups of similar organisms that do not persist naturally) within a species
   i. on the one hand being distinguishable from other varieties, cultivars, and breeds (suggesting discontinuity)
   ii. on the other hand, under natural conditions, so readily interbreeding with other varieties, cultivars, and breeds in the same species that varieties, cultivars, and breeds do not persist from one generation to the next (suggesting the discontinuity is very shallow)

 c. shallow discontinuity between species (the smallest recognizable group of similar organisms that persists naturally [by the production of similar offspring])
   i. on the one hand (suggesting of discontinuity):
     ◦ distinguishable from other species
     ◦ persisting unchanged through human cultural history persisting unchanged in the fossil record (stasis)
     ◦ (about 99% of the time) abrupt appearance (appearing in the fossil record without showing transition from another species)
   ii. on the other hand (suggesting the discontinuity is shallow):
     ◦ producing inter-specific hybrids with other species
     ◦ (<1% of the time) maintaining a hybrid zone with an adjacent species
     ◦ (<1% of the time) different fossil species in two different layers separated by a layer containing intermediate fossils

 d. semi-deep discontinuity between genera (groups of one or more species)
   i. on the one hand (suggesting discontinuity):
     ◦ more readily distinguishable than species
     ◦ always stasis in the fossil record
     ◦ never maintaining a zone of hybrids with a neighboring genus
     ◦ always abrupt appearance in the fossil record
   ii. on the other hand (suggesting the discontinuity is semi-deep)
     ◦ producing inter-generic hybrids with other genera (though less commonly than between species)
     ◦ occasionally, a sequence of fossil genera in successive rock layers

 e. deep discontinuity between holobaramins (the smallest group of similar organisms completely surrounded by deep discontinuity), at approximately the level of the biological family (which appears to corresponds to the biblical baramin)
   i. hybrids or hybrid zones with other organisms are never observed
   ii. intermediates with other fossil groups are never observed
   iii. holobaramins never observed in a fossil sequence

 f. increasingly deeper discontinuity between successively larger groups of organisms above the level of the biological family (bigger gaps between orders than between families, bigger gaps between classes than between orders, bigger gaps between phyla than between classes, and bigger gaps between kingdoms than between phyla)

g. extremely deep discontinuity around all living things

   i. living things are vastly more complex than non-living things

   ii. the non-living world is unable to generate life, or any living thing, or any of the macromolecules of life

   iii. even with all their remarkable intelligence and manipulative ability humans can create the macromolecules of life, but have not been able to create life

   iv. life seems to be composed of a qualitatively different (non-physical) substance from non-life

G. God creating biological discontinuity to illustrate His (invisible) uniqueness is a better explanation than naturalism for

   a. the very shallow discontinuity between varieties, cultivars, and breeds (because natural selection cannot choose, maintain, or take out mutations from among things that are hidden and not expressed)

   b. the shallow discontinuity between species (because evolution expects less discontinuity between sister species, no inter-specific hybrids between other species, and no distinction between species with persistent hybrid zones)

   c. the deep discontinuity between holobaramins and the deeper discontinuity between larger groups than the taxonomic family (because evolution requires some sort of continuity must have connected all living things)

   d. inter-specific hybrids in 70-80% of the species in a taxonomic family (because the long chronology of evolution expects inter-specific hybrids only between sister species)

      i. The commonness of inter-specific hybridization (often 70-80% of species in a family) is more readily explained by the thousands of years of earth history presented in the Bible than it is by the billions of years of earth history postulated in naturalism.

H. Studying biological discontinuity allows us to be better priests of the creation by

   a. giving us greater insight into Who God is (His distinctness, He as Creator, His transcendence) and giving us cause to

   b. worship God, and

   c. bring others and the creation into the worship of God. Continuing Adam's job of taxonomy expands our understanding of biological discontinuity, so further fulfills our responsibility as priests of the creation.

I. As rulers of the creation we should

   a. preserve biological discontinuity (by preventing extinction whenever possible and preserving baramins over species when we are forced to make a choice), and

   b. enhance biological discontinuity (by breeding organisms to reveal the varieties, cultivars, and breeds that God hid in His creation).

Test & Essay Questions

1. Explain how God demonstrates His uniqueness in the biological creation.

2. List two/three/four reasons for believing that there is an unbridgeable gap between the non-living creation and biological life.

3. Explain why some people would argue that viruses are evidence that life evolved from non-living things and explain how others would understand viruses differently.

4. What does it mean to say that an 'extremely deep discontinuity' surrounds life?

5. Where does the phrase 'according to their kind' come from, and what does it seem to mean?

6. Why did Carl Linnaeus refer to readily distinguishable groups of nearly identical-looking organisms as 'species'?

7. What does it mean for something to show 'stasis and abrupt appearance'?

8. List two/three reasons for believing that the biblical 'kind' is a larger group of organisms than the species.

9. Compare and contrast 'shallow discontinuity' and 'deep discontinuity'.

10. Define 'baramin' / 'baraminology'.

11. Skeptics claim that one of the errors of the Bible is its story about Adam naming the animals, for it was impossible for anyone to name all the animal species in a single day. How would you respond to this challenge?

12. Skeptics claim that one of the errors of the Bible is its story about Noah's ark, for it was impossible for anyone to fit all the animal species in a single boat. How would you respond to this challenge?

13. How is shallow discontinuity explained in creation theory?

14. Explain what the spectrum of perfection of discontinuity is and what it suggests about God.

15. Why is deep / shallow / life-surrounding discontinuity a challenge to naturalism?

16. Why is inter-specific hybridization a challenge to naturalism?

17. Why should a baramin be preserved before a species?

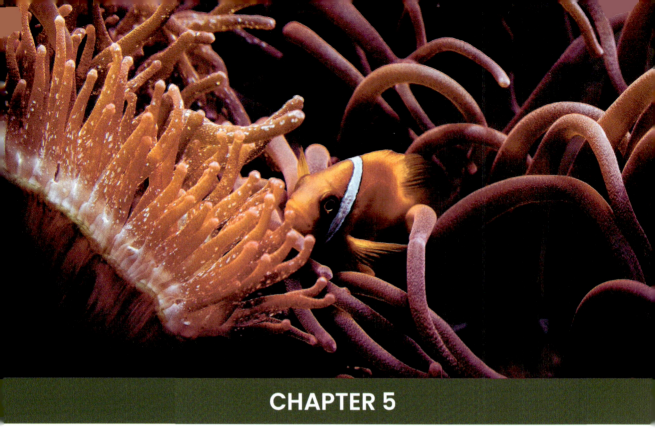

CHAPTER 5

GOD IS GOOD

Biological Cooperation and Biological Evil

"O give thanks to the Lord, for He is good; for His steadfast love endures for ever!"
Psa. 107:1, RSV

5.1 | God is Good

God is not only good (*e.g.* Psa. 34:8 and Luke 18:19[a]), He *does* good by doing things for the benefit of others (*e.g.* I Chr. 17:26 and James 1:17[b]). Part of that goodness is providing what is needed by those things that are made (something we will examine more closely in Chapters 7 and 8), but God's goodness goes much further than that. He also provides 'abundance in all things' (Deu. 28:47; Job 36:26-31) *beyond* what is needed. He does not just provide life, for example, He provides life 'more abundantly' (John 10:10). In fact, He wishes to overwhelm us with His abundant heart (*e.g.* Mal. 3:10) and is able to do 'exceedingly abundantly' beyond anything we can imagine (Eph. 3:20).

Biological Cooperation

God could have provided everything organisms need to survive through a type of balanced war. If all organisms fought for survival and were apportioned with a perfect balance of power, it is theoretically possible that the system could operate indefinitely. This is something like the balance of opposing powers that the founding fathers established for the U.S. government, hoping this balance would allow the United States to persist indefinitely.

This same balanced war concept is how naturalism understands the biological world to work—all organisms fighting for survival against all other organisms. However, a good creator Who wishes to illustrate His goodness might well design the biological world in a state of mutual cooperation rather than conflict. In fact, it seems that the original creation was created just that way.

Symbiosis and Mutualism

Organisms interact in a variety of ways. Some interactions are fleeting and have little impact on either organism. Other interactions are long-term. Historically, a long-term interaction between two or more organisms is called a **symbiosis** (pl. *symbioses*). One type of symbiosis is called **commensalism**, where one organism is benefited, and the other is neither harmed nor helped. Moss, for example, grows on the bark of trees. The moss is benefited by having access to sunlight

symbiosis (pl. symbioses)—long-term interaction between two or more organisms

commensalism—symbiosis where only one organism benefits; the other(s) neither harmed nor helped

---

a   Further references that God is good include Ex. 33:19, Ex. 34:6, Psa. 100:5, Psa. 106:1, Psa. 107:1, and Psa. 119:68.

b   Further references that God does good include Psa. 33:5, Psa. 84:11, Psa. 107:8-9,15,21,31, Psa. 119:68, Psa. 145:7 & 9, Zech. 9:17, Matt. 7:11, Acts 14:17, and Rom. 8:28.

without being covered by leaf litter on the forest floor. The tree, however, seems to be neither hurt nor helped in this process.

When long term interaction brings a benefit to both organisms, the symbiosis is called **mutualism**. Today researchers are discovering that many symbioses are mutualisms. A symbiosis is a **facultative mutualism** when each organism is benefited by being with the other, but the two organisms can survive apart from each other. For example, sea anemones and anemone fish are in facultative mutualism. The anemone fish eats invertebrates that could potentially harm the sea anemone and the anemone feeds on the feces of the fish, but either species could survive without the other.

A mutualism is an **obligate mutualism** when each organism not only benefits from the relationship, but also cannot exist without the other. Flowering plants produce colorful, aromatic flowers, each one containing a small amount of nectar. Pollinators (*e.g.* bees, butterflies, hummingbirds) are attracted to the flowers and go from one flower to another, collecting nectar. In the process, pollinators carry pollen from one flower to another, thereby pollinating the flowers. Many of the pollinators would not be able to survive without the nectar they collect from flowers, and many flowers would be unable to reproduce without the pollinators. Examples of obligate mutualisms between species pairs would include yucca plants and yucca moths, figs and fig wasps, and acacias and acacia ants.

---

mutualism— symbiosis where two (or more) organisms benefit. Types of mutualism: facultative mutualism organisms that can survive apart; obligate mutualism organisms cannot survive apart

faculative mutualism— a symbiosis when each organism is benefited by being with the other but can survive apart from each other

obligate mutualism— a mutualism when each organism not only benefits from the relationship, but also cannot exist without the other

In the case of animals interacting with the plants they eat, it first appears that these relationships are only benefiting the animals. Upon closer inspection, both can be benefited—at least in some way. Most fruits are designed with good taste and bright colors in order to encourage animals to eat them. Fruits have seeds designed to survive the digestive system of the animal. In this manner, seeds are carried away from the plant and even deposited in nutrient-rich feces.

Most nuts are designed with a hard shell on the outside to allow for long-term storage and even require burial to soften them enough to eat. This encourages animals to bury the nuts for later consumption. If, in the process, not all the nuts are consumed, the plants have effectively used the animals to 'plant' their seeds! In grasslands, most of the plants store most of their food supplies below ground, protecting the plants from the grass fires that are so frequent on the dry grasslands. Although cropping (being cut or chewed off above ground) damages the plant, it is a relatively small percentage of the plant, *and* the cropping actually stimulates the plant to grow. Furthermore, most grasses are benefited by the feces and ground disturbances caused by grazing animals. In general, producers survive because they create and store enough food for themselves and for the animals that eat them. Producers overproduce for the benefit of the entire community.

Perhaps the most impressive examples of mutualism are those where the two species are in such a close mutualism that they seem to function as a single organism and look unlike either of the species of which they are composed. A lichen, for example, usually consists of two very different organisms, in close relationship—a fungus which has a remarkable capacity for absorbing water from the environment but not able to produce its own food, and a photosynthetic bacterium or alga which can produce its own food from sunlight but must live in a watery environment. The fungus surrounds the bacterium or alga, protecting it and providing the water it needs to thrive, and the bacterium or alga generates food from sunlight to feed both itself and the fungus. The two species produce such unique forms when combined that a lichen often looks different from both the fungus and the alga or bacterium from which it is formed. They look so different, in fact, that thousands of species of lichens are classified separately from fungi, photosynthetic bacteria, and algae.

In contrast, other examples of symbiosis involve a readily recognized species interacting with a much smaller, even microscopic, species. In most of these cases the larger species was classified and studied for a long time as if it were the only species. Only more recently have we come to recognize mutualistic relationships with much smaller species.

For example, in many cases a mutualism exists between plants and microscopic **mycorrhizal fungi**. The mycorrhizal fungus is located in the soil and interacts with plant roots. The plant is able to deliver nutrients to the fungus from photosynthesis that occurs in its leaves, and the fungus provides the plant with more water and soil nutrients than the plant could absorb by itself. We suspect that many of these

mycorrhizal fungi—fungi in mutualism with plant roots that provide water and nutrients to the plants

special fungi may also protect the plant from disease and dangerous chemicals.

Another type of mutualism between plants and microscopic organisms is found in legumes (such as beans and peas). Nitrogen-fixing bacteria enter into mutualism with legume roots. As in the case of mycorrhizal fungi, the plant 'feeds' the microorganism. In return, the nitrogen-fixing bacteria take nitrogen from the air and convert it into a nitrogen-rich fertilizer for the plants.

Further examples of mutualisms are found between animals and algae. For example, many corals are in mutualism with dinoflagellates (a type of photosynthetic algae). The animals protect the dinoflagellates, and the dinoflagellates provide the animals with food when food is lacking in the environment. This relationship is why coral reefs tend to grow upwards towards the light at the ocean surface even though animals make up most of the structure.

Another interesting example is the mutualism between an alga and one particular freshwater jellyfish. The jellyfish protects the alga and carries it towards sunlight. The alga provides food to the jellyfish by photosynthesis. Yet another example is the mutualism between an alga and spotted salamanders. The symbiosis between the two has been known for a long time, but recently the alga has been discovered within the cells of the salamander embryo. The evidence suggests that the algae are receiving carbon dioxide from the respiration of the embryo and the embryos are being provided with food from photosynthesis of the alga.

A common mutualism involves bacteria in the digestive systems of animals (commonly called **gut flora**). These bacteria thrive in the oxygen-free environment inside an animal and profit from a continual supply of food eaten by the animal. In return the animal is benefited because the bacteria help to break down food that the animal cannot digest. In many cases, bacteria also produce other nutrients the animal needs.

No animal, for example, can break down cellulose, the molecule that makes up most of the **cell walls** of plant cells and makes foods like lettuce and celery stiff and crunchy. It is intestinal and stomach bacteria that break down cellulose so that cows can digest grass, termites can get nutrients out of wood, and humans can digest the cellulose of vegetables. In fact, in the intestines of humans there is a diverse

gut flora—biomatrix microorganisms aiding digestion by living in mutualism in animal intestines

cell wall—structure outside the cell membrane strengthening cells of bacteria, algae, plants, and fungi

community of bacteria helping us with digestion. Some bacteria in the small intestine even produce vitamin B_{12} while other bacteria living in the large intestine produce vitamin K.

The more we learn about the biological world, the more tiny organisms we find inside the bodies of larger organisms. The human body, for example, has scores—probably hundreds—of species of microorganisms inhabiting it (many of them *not* in our intestines). The more we learn about those microorganisms the more we learn about the essential and beneficial roles they play within our bodies. It may well be that no large organism is just one species at all, but rather a community of organisms in complex mutualistic relationships. At the very least, mutualism is found everywhere in the biological world. Every community has multiple examples of these relationships, and it is quite probable that every organism on the planet is in mutualistic relationship with at least one other organism.

The Origin of Mutualism

Mutualism is a challenge to naturalism. The closer the relationship between the two organisms and the more obligate is their mutualism, the more difficult it is to conceive of how the two organisms could have come into such a relationship in the first place. Even more problematic is the commonness and depth of mutualism. In the naturalistic worldview, the primary mechanism of biological change is thought to be natural selection. Natural selection involves the elimination of organisms in a struggle for survival. A world generated by natural selection is a world at war—every organism struggling for survival against other organisms. Every other organism is a competitor, a foe, an enemy. Interactions that provide benefits to both organisms should be relatively rare in such a world.

But such is not the case. The more we learn about the world, the more mutualisms we discover. They are common and everywhere in every kind of environment and every kind of organism. For each disease-causing bacterium and virus we discover, there are hundreds more that are in cooperative relationships. Virtually every plant may be in mutualism with mycorrhizal fungi. Virtually every animal may be in a symbiotic relationship with gut bacteria. Mutualism is also deep in the creation.

At the largest biological scale, oxygen rejected by plants is consumed by animals, and the carbon dioxide rejected by animals is consumed by plants. Every biological system is a mutualism formed by a host of organisms or cells, and biological systems are found at every biological scale. Mutualisms exist between multicellular organisms of similar size and every multicellular organism might actually be a community of mutualistic relationships between the multicellular creature and a host of microscopic organisms. This is not at all what would be expected in a naturalistic worldview. But it might well be expected of a world created by a good God.

5.2 | Biological Evil

The physical world does *not* seem to be inherently evil, such as the ancient Greeks believed. Nor is the world in such conflict as naturalistic evolution would expect. Rather, the common and deep mutualisms suggest that the creation is basically good, perhaps not far removed from the label of 'very good' that the Creator originally gave it (Gen. 1:31). However, there is still something wrong with the biological world. Disease, death, and especially suffering do not seem to be expectations from the good God described in Scripture, so what is the origin of these things?

The Origin of Biological Evil

The immediate effect of Adam's sin was his spiritual death (the dying referred to in Gen. 2:16-17). And, given Adam's role as ruler, God responded to man's sin by cursing man's entire dominion. Wives became rebellious and husbands became abusive (Gen. 3:16b), human diet changed (Gen. 3:18), human childbirth became more painful (Gen. 3:16a), the human body became mortal (Gen. 3:19; Rom. 5:12), work became burdensome (Gen. 3:19), animals were cursed (Gen. 3:14[a]), plants were changed (Gen. 3:18), and the ground was cursed (Gen. 3:17).

The strong tie between human sin and animal death in the Bible's sacrificial system suggests that human sin is not only the cause of human death, but also the cause of animal death. This would mean that before man's sin, animals did not eat each other, which, in turn, is consistent with the original animals eating only plants (Gen. 1:30). And, since animals and humans are the only organisms that are biblically alive, it follows that *all* death is due to man's sin. The original creation would have had no death of biblical life. Being 'very good' (Gen. 1:31), then, apparently not only means there was no death, but that the creation lacked suffering (excessive pain) and anything that would cause suffering. There would have been life without death, eating without predation, health without disease, maturation without deterioration. And, as the original creation was, the new heaven and new earth will be (Isa. 11:6-9; 65:17-25; Rev. 21:1-4).

In short, in the original creation there was no evil at all. There was neither moral evil (sin) nor natural evil (anything in the physical world that causes suffering). Consequently, there was no **biological evil** (anything in the biological world that causes suffering). There was no disease, and upon reaching maturity, humans and animals did not degenerate, nor did they die. All natural evil is part of God's **curse** in response to man's moral evil. This is consistent with the claims of Romans 8, that all created things are prone to aging, decay, and death (Rom. 8:20-21) and the *entire creation* will suffer (Rom. 8:22) until the curse is lifted with the final glorification of humans (Rom. 8:19; Rev. 21:3-4; 22:3).

---

biological evil—anything in the biological world that causes suffering of humans or animals

curse—natural evil allowed into the creation by God in response to Adam's disobedience. Types of changes: negative effects: truly negative changes, probably imperfections in cycles and repair processes; evil-minimizing effects: changes introduced to minimize natural evil in a cursed creation

---

[a] According to Gen. 3:14 the serpent was cursed *more* than the beasts of the field.

Sin, in its essence, is due to a desire to be independent of God—to live outside of a relationship with God. When Adam sinned, he was choosing something other than a perfect relationship with God. God's response was to curse all the relationships of the ruler's dominion. Rather than each individual providing benefit to others, individuals began seeking their own benefit at the expense of others. Man seeks his own benefit rather than rely upon God. Wives try to rule over their husbands and husbands try to put down their wives. Men deal selfishly with other men, very soon after the Fall resulting in a man murdering his own brother (Gen. 4:4-8) and deteriorating thereafter to an earth filled with violence (Gen. 6:11).

The mutualistic relationship between the shepherd king and the animals of the world would change to include fear, subjection, and killing (Gen. 9:2-5). Relationships among animals changed as animals began eating other animals and deteriorating to a time when 'all flesh had corrupted itself' and the earth was 'filled with violence' (Gen. 6:12-13). Even the relationships between plants and the organisms that ate them changed (Gen. 3:18). Some of the changes occurred immediately at the curse; other changes developed over the years and centuries that followed.

5.3 | Negative Effects of the Curse

Before the curse humans and animals were designed to live forever. If this was due to the design of the physical bodies of organisms, the

processes that maintain physical bodies would have had to have been highly effective. All parts of all organisms would have to be kept in good, if not perfect, repair. Damaged, dying, or dead components would have to be detected and replaced quickly so that no tissue, organ, or organ system would fail or even degenerate.

This repair process would have to continue without diminishing. There would even have to be effective mechanisms for detecting and repairing damage to the information stored in DNA. Many of the negative effects of the curse may have been the result of God altering the efficiency of these repair processes. Just the slightest inefficiency would lead to the kinds of biological evils we see today.

Degenerative Aging. Once biological repair systems were unable to completely restore an organism, each component of the organism would become less and less efficient with time. Damage (*e.g.* mutations, chromosomal aberrations, injuries) would persist unrepaired. Each part would gradually wear out, and one organ after another would falter or fail. This is the cause of **degenerative aging**. Note, that this is not developmental aging, where an organism 'grows up' or matures. Developmental aging would have been a normal part of the world before the fall of man. What came as the result of the curse was degenerative aging, the type of aging that weakens an organism towards its death (that which leads to descriptions such as 'over the hill' and 'growing old').

aging, degenerative—gradual wearying of an organism—a negative effect of the curse

Disease. Some of the unrepaired damage would be pathological (*i.e.* resulting in suffering, the word being derived from the Greek word *pathos*, meaning 'suffering'). The breakdown in biological repair systems is likely to be the biological cause of most, if not all, **pathogenesis**—the 'origin' (Gr. *genesis*) of 'suffering' (Gr. *pathos*). Some of the damage causes harm (**pathology** or **disease**) directly to the altered organism (*e.g.* the mutation that causes sickle-cell anemia or the chromosomal aberration that causes Down's syndrome).

Other damage does not so much hurt the altered organism directly, as much as it alters its relationships. In some of those relationships the organism becomes a **pathogen**—an organism that causes suffering in another organism. Examples would include the pathological varieties of the normally harmless anthrax bacteria found in soil, and the normally helpful *E. coli* found in our intestines. The fact that in most situations helpful varieties of microorganisms (such as *E. coli*) will outcompete the harmful varieties of the same species, suggests that pathology is due to deterioration of the design of the original biological system.

In fact, the breakdown of complex biological systems is likely to be how *all* diseases came to be, not only at the curse, but ever since then to the present day. For example, new mutations for sickle-cell anemia are thought to be occurring regularly in the human population.

Parasitism. Some of the unrepaired damage in organisms is likely to have introduced **parasitism** into the world. Parasitism is when one organism (the **parasite**) is benefited by harming another organism

pathogenesis—the cause of a disease

pathology—degeneration of an organism due to genetic error(s) and/or harmful organism(s); a negative effect of the curse (aka disease)

disease—degeneration of an organism due to genetic error(s) and/or harmful organism(s); a negative effect of the curse (aka pathology)

pathogen—a disease-causing organism

parasitism—symbiosis where one organism benefits and one or more organism is harmed; a negative effect of the curse

parasite—the organism in parasitism that benefits, thus harming the host organism

host—the organism in parasitism that is harmed (by the parasite organism)

(the **host**). Examples include plants that parasitize other plants (*e.g.* mistletoe, Indian pipe, dodder), ticks that suck the blood of mammals, and lampreys that attach to and feed off the bodies of other fish.

Some of the relationships between parasites and hosts are rather sophisticated. For example, ticks find mammal hosts by being attracted to higher temperatures and higher carbon dioxide levels. Many have pain killing chemicals in their saliva to avoid detection and an anticoagulant to prevent blood from clotting (*e.g.* leeches). Such sophisticated designs were probably part of an originally mutualistic relationship that became parasitic following the curse.

Evil-Minimizing Effects of the Curse

The breakdown in efficiency of biological repair systems would be expected to gradually introduce more and more suffering into the biological world. God, however, did not wish suffering to be unlimited in this life. He wanted the horrible effects of sin to be felt, but He still desired that we would see His nature in those things He made (Rom. 8:20), including His mercy (*e.g.* Psa. 103:17), graciousness (*e.g.* Jonah 4:2), kindness (*e.g.* Psa. 36:7) and goodness (*e.g.* Psa. 119:68).[a] Consequently, at the time of the curse God introduced things into the biological world that would put limits on how much biological evil would enter in the world.

---

a   Other references include Exo. 34:6-7, I Chr. 16:34, Ezra 3:11, and Eph. 4:2 for God's mercy; Psa. 103:8 & 145:8, Joel 2:13, and Eph. 2:5-9 for God's graciousness; Neh. 9:17, Isa. 54:8, Joel 2:13, and Jonah 4:2 for God's kindness; and Psa. 34:8, 100:5, 106:1, and 107:1 for God's goodness.

Death. After the Fall, God drove Adam and Eve out of the Garden of Eden 'lest they live forever' (Gen. 3:22-24). Most likely, God prevented humans from living forever for our *benefit*. If degenerative aging had been introduced into the world, but animals and humans had not been permitted to die, then animals and humans would become feebler and feebler for eternity. If animals and humans were immortal in a world where suffering was increasing, then every individual would suffer more and more every day without limit. If humans remained physically immortal after dying spiritually, humans would remain separated from God forever, and increase in suffering without limit.

In the world *before* man sinned, death of man and animals would have been a bad thing. *After* man's sin, death is a good thing. Death is a consequence of sin and is the last enemy to be destroyed for the follower of Christ (1 Corinthians 15:26). However, in the cursed creation, God introduced physical death to nephesh life as an act of mercy. In a cursed world, the deaths of animals and humans places limits on the suffering and decay that occurs as a consequence of the curse. Also, human death permits people to be resurrected with perfect bodies. Death of nephesh life places limits on biological evil in a cursed creation, so its introduction at the time of the Fall was a good thing.

Competition and Natural Selection. As good as death might have been to put limits on suffering, the introduction of death created another problem. Before the Fall, every animal born would have survived to adulthood. To fulfill God's command to 'fill the earth', populations would have increased until the earth was 'full' (*i.e.* carrying capacity had been reached). There were probably designs in place to cease reproduction in organisms when they reached the earth's carrying capacity. Remnants of some of these controls on reproduction are seen in creatures whose reproductive rates are controlled by environmental factors (*e.g.* not enough food in the **habitat** or severe weather conditions may cause females not to produce young that year).

habitat—space in a biological community in which species can thrive

We do not know how fast God originally designed the world to be filled, but before man's Fall it did not have to be quickly. After all, they had an eternity of time to do it! Consequently, animals before the Fall probably produced young at very low rates. Under such conditions today, the spread of a disease would run the risk of killing off animals faster than they could be replaced, thus bringing about the extinction of species. Even if an organism has already filled the earth and does not

overproduction—organisms producing more offspring than will survive to reproduce; leads to competition and natural selection, and introduced by God after the curse to replace dead organisms

competition—the struggle of organisms to survive when resources are limited; caused by overproduction, leads to natural selection, and designed as an evil-minimizing effect of the curse

selection, natural—natural process resulting in death of (the less capable) part of a population; caused by overproduction and designed and introduced by God as an evil-minimizing effect of the curse

need to increase its numbers, it must produce more young to replace those that are dying.

Consequently, God probably increased reproductive rates of all organisms over pre-Fall levels. If the reproductive rates are raised high enough there are not enough resources available for every organism to survive—something called biological **overproduction**. Biological overproduction brings about **competition**, as organisms compete against other organisms to get the resources for themselves. Organisms must struggle to survive, and some must die or move away for the population to survive. Biological overproduction sounds bad, and it would be in a world without death.

However, in a world with death, the weaker organisms will tend to be eliminated and the stronger ones will survive. Diseased individuals will tend to be taken out of the population, effectively slowing and in some cases preventing the spread of disease. If there are individuals in the population that are stronger or more disease-resistant and can genetically pass that information on to the next generation, those individuals will be 'selected', and the population will be improved in the process (something called **natural selection**).

In finch populations on the Galápagos Islands, for example, natural selection changes the size of finch beaks to match the size of plant seeds as climate changes alter the size of plant seeds. Competition and natural selection tend to reduce the total amount of disease in the world. Biological overproduction seems to minimize the total amount of biological evil in the world. Since biological overproduction is only a good thing in a world with death, it would have been a bad thing in the world before the curse. In a cursed world, however, biological overproduction places limits on biological evil, so its introduction as part of the curse was a good thing.

Predation. Another control on biological evil is **predation**. Predation is where some animals (**predators**) eat other animals (**prey**). As in the case of competition and natural selection, predators tend to take out the weaker individuals in a prey population. Consequently, predators tend to remove diseased organisms, slowing, stopping, or even eliminating disease in the prey population.

Studies have shown, for example, that when humans kill off wolves and other predators in a region, the deer populations in that region suffer from more disease than deer populations in areas with predators. To make sure that predation is most effective, predation designs were given to predators[a]. They were given behaviors to attack prey organisms, teeth capable of tearing meat, and digestive systems capable of digesting meat. A variety of special predation designs were scattered

predation—the eating of animals by other animals (aka carnivory), an evil-minimizing effect of the curse

predator—an animal that eats another animal (aka carnivore)

prey—an animal eaten by another animal

---

[a] Whether these designs were given to organisms at the Fall or whether they were designs hidden in the organisms at creation, is not specified in Scripture. Since many other designs seem to have been hidden in organisms at the creation and God terminated his creation activities at the end of the Creation Week (Gen. 2:2), it is most likely that God hid these designs in organisms at the original creation with the intention of revealing them when they were needed.

among predators, such as exceptional eyesight (*e.g.* hawks and eagles), camouflage (*e.g.* scorpionfish and mantises), silence and stealth (*e.g.* big cats and owls), agility and speed (*e.g.* weasels and cheetahs), and cooperative group behavior (*e.g.* wolves and lions).

To make sure that predators did not wipe out entire populations of prey, **defenses** were given to prey[a]. were given to prey. These include passive defenses like camouflage (*e.g.* most moths, snakes, and female birds), spines (*e.g.* sea urchins, echidnas, and porcupines), hard shells (*e.g.* mollusks, turtles, and armadillos), distasteful toxins (*e.g.* many frogs, monarch butterflies, ladybugs), warning coloration (*e.g.* many bees, wasps, and newts), and mimicry of undesirable prey (*e.g.* syrphid flies looking like wasps).

They also include designs intended to scare off predators, like producing obnoxious smells (*e.g.* skunks and stink bugs), spewing distasteful chemicals (*e.g.* bombardier beetles and blister beetles), flashing large eye spots (*e.g.* Cecropia moths), and loud noises (*e.g.* hissing sounds in some butterflies and rattles on rattlesnakes). They also include designs that allow for defense with direct contact with predators, such as detachable body parts (*e.g.* lizard and salamander tails), stingers (*e.g.* bees and hornets), strong kicks (*e.g.* horses and deer), sharp horns/antlers (*e.g.* antelope, and deer), sharp teeth (*e.g.* **carnivores** and rodents), and clustering (*e.g.* crows against hawks, elephants surrounding young).

deep discontinuities—large gaps that seem unbridgable that divide life into thousands of baramins

carnivore—an animal that eats another animal (aka predator)

a See the previous footnote.

Plant Defenses. Although animals and humans did not die before the Fall of man, other organisms did. And, because bacteria, algae, protozoa, fungi, and plants were consumed in the original creation, they must have been designed from the very beginning to overproduce in order to maintain their populations and feed others. Given the optimal design of the original creation, this was probably so perfectly balanced that competition and natural selection did not exist even among these organisms. However, when the curse occurred and overproduction was introduced into the animal world, the reproductive rates of *all* organisms had to be increased, including those of the biomatrix (bacteria, algae, protozoa, fungi) and plants.

At the same time, the decay of biological repair systems would have also introduced degenerative aging and disease into the biomatrix and plants. Consequently, even among the plants and organisms of the biomatrix, competition and natural selection would help minimize the biological evil they experienced. Furthermore, consumers were already filling the roles of predators in the animal world, further minimizing the biological evil they experienced. It only remained for God to provide plants with defense mechanisms so that consumers would not over-eat them. Such plant defenses include spines and thorns (*e.g.* thistles, locust trees, and holly leaf points), tough leaves (*e.g.* cellulose and silicon dioxide in plants), irritants (*e.g.* stinging nettles and poison ivy) and distasteful toxins (*e.g.* nicotine in tobacco leaves and caffeine in coffee beans).

5.4 | Goodness of the Creation

In the worldview of naturalism, naturalistic evolution is the best theory available for the origin of life's diversity. If this *is* how life's diversity arose, then every organism came to be by natural selection and every organism continues to exist because of natural selection. Natural selection, however, involves a struggle for survival, and survival means other organisms must die. In the naturalistic worldview the world is at war, every organism is struggling against every other, and every organism is an enemy to every other. According to evolution, differences in organisms arise from mutations, changes in the sequence of DNA nucleotides, from one generation to another. So far, the data show that most mutations are harmful to the organism. If organisms have arisen by evolution, they should be full of mutations, and pathology ought to be abundant in the world.

Unfortunately for evolutionary theory, the world does not correspond to the expectations of evolution. Although there are pathologies in the world, they are not as common as evolution expects. New viruses and bacteria are being discovered all the time, but only a small percentage of them seem to be pathological. The total number of disease-causing organisms (pathogens) in the world seems to be in the hundreds, or at most thousands. Yet there are millions of organisms in the world. It seems that something less than 0.1% of organisms are pathological.

As for mutations, if the **genetic load** (the number of harmful mutations carried in an organism, also known as the **mutational load**) gets too high, an organism cannot survive. Natural selection slows down how *fast* the number of mutations increase, but it cannot eliminate them, nor can it stop the increase. There is no known mechanism to prevent mutations from continuing to accumulate and ultimately bring about the eventual destruction of life. Research suggests that on average, at any particular location on the human DNA, a mutation will occur about once in a million generations.

Consequently, in a million generations every single nucleotide on the DNA will have undergone a mutation. There seems to be no way life could have been doing this for even close to a million generations, but evolution requires humans to be the most recent portion of a family tree that is billions of generations long!

---

genetic load—the number of harmful mutations carried by an organism or species (aka mutational load)

mutational load—the number of harmful mutations carried by an organism or species (aka genetic load)

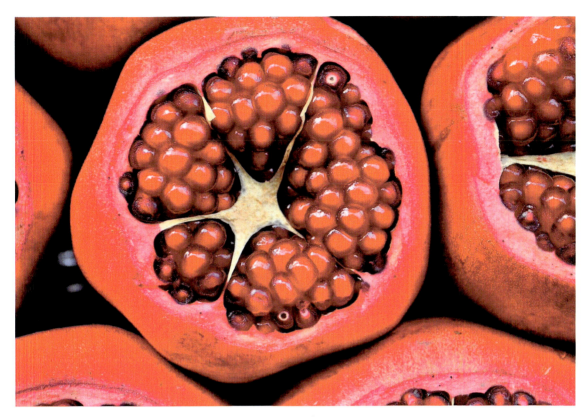

Not only do organisms exist, but they also do not seem to carry anything close to the mutational load that would exterminate life. Among the millions of long-term relationships that exist between organisms, parasitism seems to be rather rare (less than 0.1% of relationships). Commensalism does not even seem to be the dominant type of relationship in the world. Most long-term relationships among organisms seem to be mutualisms, where each organism benefits the other in some way. Evolution would expect an abundance of pathology and mutation, and a rarity of cooperative symbiotic relationships.

The account of creation in the Bible is far more consistent with the frequency of mutualism, mutation, and pathology that we actually see in the biological world. If biological life was created without any pathology and mutation, the mutational load and pathologies we see in the present have accumulated since creation. If that creation was only a few thousand years ago, it would be easy to explain why the mutational load is as low as it is, and why pathologies are as rare as they are. If a good God created the biological creation desiring us to see His goodness in creation, then the abundance of mutualism would also be explained.

Mutualistic Symbiosis: Our Responsibility

Our Responsibility to God

There is a spectrum of perfection of mutualism that points to the perfect relationship that exists among members of the Godhead. Outside the biological world there are entities that do not interact at all. Others interact in a very limited, programmed manner to serve a larger whole, providing no real benefit to either organism. Other organisms impact one another but do so for short periods of time. Other relationships are long-term and commensal, providing benefit to one organism and no impact on the other.

Then there are symbioses (mutualisms) where each organism is benefited. Some of those mutualisms are facultative and others are essential for the survival of all organisms involved. Some involve organisms in immediate proximity. Others involve organisms living *within* other organisms. Some of these relationships are so intimate that the organisms behave as if they are a single organism.

There is a spectrum of perfection of mutual benefit, pointing to an unseen relationship where each member of the relationship receives and provides infinite good. This spectrum helps us understand what the Bible means when it describes the three persons of the Godhead not only united in perfect unity but being infinitely good as well.

The abundance and ubiquity of cooperative symbiosis should also lift our eyes and hearts in worship of the God of goodness. God filled the

original creation with goodness, even providing beneficial relationships to every organism He made. Mutualism so abounded in that creation that even after thousands of years of decay and deterioration, mutualistic relationships still far outnumber pathological ones. Organisms are far closer to perfection than they are to self-destruction. How much more does God desire to pour out blessings on those whom He loves?

The biological evil in the world (suffering, disease, predation, death) ought to grieve us. After all, it was man's desire to be independent of God (sin) which led to this evil in the first place. The sin of Satan did not bring biological evil into the world. It was the sin of creation's king. It was our sin. The horrors of biological evil ought to remind us of the seriousness and horror of sin in the eyes of God, something we are too quick to forget or too anxious to ignore.

Yet even an examination of biological evil should stimulate worship. Even while cursing His creation in response to man's sin, God put limits on that curse. He introduced death to reduce the total amount of suffering and to allow a permanent escape from the presence and effects of sin. He introduced predation and natural selection, not as creative forces, but as evils that minimize the total biological evil and maximize the health of organisms in a cursed world. He provided defenses for organisms so that they would survive an evil-tainted world. Finally, He became man to live among men as an example and be killed in order to provide us with salvation. He permitted the evil He was willing to suffer for, so that we could see His awesome goodness. As grievous as

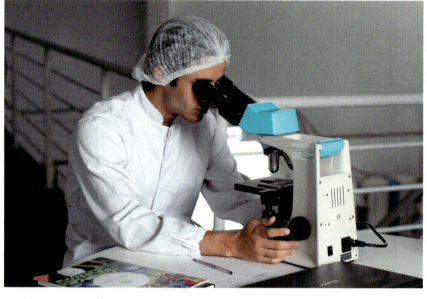

man's sin is, and as tragic as its consequences are, God's goodness is infinitely greater and worthy of our praise.

One of the reasons God created biological mutualism was to give us a visible and finite illustration of the invisible and infinite goodness of God. Because He did, study of biological mutualism can improve our personal understanding of God and draw us closer to Him. To help us fulfill our priestly role, God not only created us to respond to illustrations of His nature, He also gave us the Holy Spirit to lead us into truth. Because of this, contemplation of biological mutualism can stimulate us to worship God, glorify God, and bring others into our worship of God.

Our Responsibility to the Creation

Preserving Goodness—Reverse Negative Effects. When God handed over His creation to us so we could rule it, it was without fault. There was no spiritual death, no pathology, no predation, no degenerative aging, and no death of nephesh life. All these biological evils were introduced in direct or indirect response to the sin of man.

As rulers over the creation, we have the responsibility to guard and keep the creation. If it were the best of all possible worlds, we should strive to return the world in at least as good a shape as we received it, but the curse makes this impossible. The best we can hope for is to reverse as many of the negative effects of the curse as possible. We can spread the gospel to provide spiritual life and we can disciple as a cure for spiritual pathology. We can provide comfort to those suffering emotionally, and we can minimize the pain of those who are suffering physically. We can also cure pathologies and reverse harmful mutations.

Our responsibility to reverse the negative effects of the curse, in fact, provides justification for most of the field of medicine. At the same time, while we are reducing the natural evil in the world, we should be careful to make sure we are reversing the truly negative effects of the curse—those things that result in true suffering. Remembering that God loves diversity and the spectra of perfection have been created to glorify God, we are not to rid the world of the variety of body shapes, skin colors, intelligences, manipulative abilities, voices, *etc.*, that grace the human population.

Fighting Pathogens. Historically, most human medicine has focused on killing pathogens (disease-causing organisms). Sometimes this involves killing the pathogen directly. In centuries past this ranged from killing ticks and tapeworms to burning the bodies and possessions of plague victims. In the mid 1800's Louis Pasteur invented **pasteurization**—a process of heating or boiling—to kill off microorganisms to delay the spoiling of milk and wine. From that time on, a variety of methods have been developed for killing microorganisms that might cause disease, some using heat (*e.g.* boiling water or flame), some using chemicals (*e.g.* alcohol, hydrogen peroxide, bleach, antibiotics), and others using radiation.

Pathogens can also be killed *in*directly by helping the **immune** system kill the pathogens. God designed the immune system to locate an identification tag[a] on something that does not belong to the body and produce antibodies for that ID tag. If the invader enters the body again, the antibodies for that invader allow the body to respond quickly to destroy the invader, making the person immune to that invader. To help the human immune system kill a pathogen, a person can be injected (**vaccinated**) with a vaccine that contains the identification tag for that pathogen, making the person immune to the pathogen. The first vaccine was developed by Edward Jenner in the late 1700s by injecting people with the cowpox disease. Those injected developed the very mild symptoms of cowpox and developed immunity both to later invasions of cowpox and to the much deadlier disease of smallpox—

pasteurization— process of heating or boiling that kills microorganisms which might spoil beverages

immune— resistant to an infection or toxin

vaccinated— when a person is injected with a vaccine that contains an identification tag for a pathogen, making the person immune to the pathogen

---

a   Such an 'identification tag' is some molecule or part of a molecule on the surface of the invader that is not found on any of the cells of the body.

apparently because both viruses had the identification tag found on the smallpox virus.

Since that time, vaccines have been developed for many pathogens[a]. One of the challenges of all these pathogen-killing approaches is that God has designed most organisms to detect when something is attempting to kill it and modify itself in order to survive. This has resulted in what are called 'superbugs'—organisms that have found ways to survive our attempts to kill them.

An alternative approach to medicine would be to *cure* pathogens rather than kill them. Not only were there no pathogens in the original creation, but all organisms were in mutualistic relationships. Most of the pathogens are probably degenerated from some of the mutualistic organisms of the original creation. Since the creation was only about 6000 years ago, most pathogens are the result of only slight changes from their harmless—even beneficial—ancestors[b]. If we focused on restoring these pathogens to their mutualistic relationships, we would not only cure the disease, but we would be curing the pathogen. Curing, rather than killing, pathogens would seem to better fulfill our obligation to care for the creation.

Preserve Evil-Minimizing Effects. Although we are justified in reversing the negative effects of the curse, we are not justified in reversing the evil-minimizing effects of the curse. For example, although we are justified in minimizing the suffering that comes with degenerative aging, we are not justified in eliminating death itself. Although death is not a good thing in a perfect world, it is a good in a cursed one. It would seem that increased deterioration and suffering, without escape, is not a good thing. Likewise, we are not justified in eliminating natural selection or predation.

---

a Some vaccines involve injecting a harmless organism similar enough to the pathogen to contain the identification tag of the pathogen (as in the case of Jenner's vaccine). Other vaccines contain dead or weakened pathogens. Still other vaccines contain harmless viruses that have been modified by humans to contain the identification tag from the pathogen. Still other vaccines contain DNA that codes for the pathogen's identification tag, hoping cells in the body will pick up the DNA and produce the identification tag and thereby alert the body's immune system to make the appropriate antibodies.

b This conclusion is confirmed by such evidences as: 1) Although people only think of *E. coli* as a pathogen, *Escherichia coli* is actually a helpful bacterium that resides in the human intestines. Only some rare strains are harmful to humans. 2) Although the 2001 anthrax scare alerted many to the disease for the first time, the disease-causing organism, *Bacillus anthracis* is normally a harmless bacterium of the soil. The bacterium is pathological only when its DNA contains a particular very small segment of DNA known as a 'pathogenicity island'.

Again, although these things are not good in a perfect world, they make a cursed world as good as it can be. For example, it is not uncommon for humans to kill off the predators in a particular region. Even if people think that they are protecting themselves, killing off predators actually increases the biological evil in the world. If wolves and other predators of white-tailed deer are eliminated, for example, diseased deer that would have been eaten by wolves survive and the deer population rises. Consequently, in places where there are no predators, the deer population is larger and greater percentages of the deer are sick, than is the case in places where predators exist. When tough times come, such as a hard winter or a dry summer, and food becomes scarce, the deer die in large numbers (**catastrophic die-off**).

catastrophic die-off—the death, due to harsh conditions, of a large percentage of a population—usually because that population exceeds the carrying capacity of the environment

In situations where the predators have already been exterminated, it becomes our responsibility to either restore the predators or to assume the role of predator. For example, a population of deer and its environment can be studied and monitored to determine the ideal carrying capacity of the deer population. Then the proper number of hunting licenses can be issued to maintain the healthiest possible size for deer herds.

Enhancing Biological Relationship. Although we have a responsibility to preserve and protect the creation, we also have a responsibility to increase cooperative symbioses in the world—to bring more glory and honor to God. Almost certainly, God placed more potential for mutualism in His creation than has yet been revealed. He has given us the ability to reveal it. Numerous medicines for human illnesses have been discovered in plants, fungi, and bacteria. It could be that God placed a treatment for every disease in one or more organisms on this planet.

Furthermore, God 'hid' the information for building a number of organisms inside other organisms. Among the unrevealed organisms are probably several cooperative relationships yet to be discovered.

Summary of Chapter

A. As a physical illustration of His invisible goodness, God created organisms in such a way that the dominant type of long-term relationship among organisms across the biological world is mutualism. Even when harmful relationships entered the world following man's Fall, God placed limitations upon them, so that mutualistic relationships still dominate.

B. God created a spectrum of perfection of organism-organism relationships:

   a. from organisms that do not interact, to organisms that interact briefly, to organisms that interact over the long-term (symbioses)

   b. from a symbiosis that provides benefit to one organism but neither harms nor hurts the other organism (commensalism), to a symbiosis that benefits both organisms (mutualism)

   c. from a facultative mutualism where the two organisms can live apart from one another, to an obligate mutualism where the two organisms cannot live apart from one another

   d. from an obligate mutualism between distinct equivalent-sized organisms (e.g. flowering plants and pollinating insects and birds; oxygen-producing plants and carbon dioxide-producing animals; fruit-producing plants and nut-distributing animals)

e. to an obligate mutualism between a multicellular organism and (unseen) microscopic organisms (e.g. animals and gut flora; plants and mycorrhizal fungi; legumes and nitrogen-fixing bacteria; corals and dinoflagellate algae)
  f. to an obligate mutualism so interwoven that the two organisms function as one organism (e.g. lichens)
C. We can begin to understand the Trinity (three persons of the Godhead in perfect mutualistic relationship).
D. The original creation was without biological evil: There was no death (of nephesh life [i.e. of animals and man]), there was no disease, and there was no suffering. When man sinned, man died spiritually and God cursed the physical creation over which man had rule. This curse was the origin of biological evil.
  a. Biological evils that are negative effects of the curse (probably due to less efficient biological repair systems):
    i. degenerative aging, or 'growing old' (NOT developmental aging, or 'growing up');
    ii. disease (or pathology), both
      ◦ degeneration of the organism, and
      ◦ harm from a pathogen (another organism that is probably degenerated);
    iii. parasitism—probably a once mutualistic relationship that changes when one organism (the parasite) gets more benefit from the relationship than the other organism (the host)
  b. Biological evils that are evil-minimizing effects of the curse
    i. introduced by God to limit the biological evil of the curse so that God's attributes are still seen in the creation;
    ii. NOT good in an absolute sense (i.e. not good in an uncursed world), but good in a cursed world:
      ◦ death, which mercifully puts an end to suffering and permits resurrection with perfect bodies
      ◦ overproduction (the production of more young than there are resources to support), which results in competition and natural selection (the death of weaker and sicker organisms), thus minimizing the suffering that comes with disease and degenerative aging.
      ◦ predation: predation designs allow predators to consume prey, especially those that are weaker and sicker, thus minimizing in prey the suffering that comes with disease and degenerative aging.
      ◦ defenses are given to plants and animals to protect species from extinction.
E. According to the naturalistic worldview, conflict (not mutualism) should dominate relationships among organisms (because organisms came to be and are continually acted upon by natural selection—a process of conflict). However, mutualism is actually
  a. common (the dominant form of symbiosis—even after the Fall of man)
  b. ubiquitous (found everywhere; probably involve every organism)
  c. deep (at every level of biological organization)
  d. Long-term biological relationships are better explained by God creating a physical illustration of His invisible goodness.
F. According to the naturalistic worldview, biological evil has been around for billions of years (as long as organisms have been around). Organisms should have a very high mutational load (the number of harmful mutations carried by an organism) and pathogens (disease-causing organisms) and parasitism should be abundant. However,
  a. overall, organisms seem to be quite healthy (i.e. they seem to have low mutational loads)

b. <0.1% of organisms are pathogenic

c. parasitism accounts for much less than 0.1% of all biological relationships

G. The biological world is good, as would be expected in a world that was created 'very good' and which has been degenerating under the effects of biological evil for only thousands of years.

H. In our role as priests, the goodness of the creation should cause us to worship God and bring others into worship of God.

   a. How awesome the goodness of the original creation must have been (and thus the goodness of its Creator) for the creation to be as good as it is after thousands of years of deterioration under the curse.

   b. Even as God cursed the creation in holy response to man's sin, He displayed His mercy and compassion by setting limits on that curse.

I. In our role as kings of the creation, the creation was cursed because of our sin.

   a. We have a responsibility to reverse the negative effects of the curse (those things that cause physical suffering [i.e. prolonged physical pain])
      i. correct mutations (copying mistakes in DNA)
      ii. restore pathogens to mutualistic relationships
      iii. minimize pain

J. We are justified to engage in improving the health of organisms (i.e. human and veterinary medicine).

   a. We should NOT reverse the evil-minimizing effects of the curse (death, natural selection, predation). If we eliminate a predator we have a responsibility to function as a predator for the prey populations.

Advanced Discussion Topics

A. What is a biblical justification for medicine? Things to consider: (1) Does the Bible specifically condemn or justify medicine? (2) Are there instances in the Bible when going to doctors is condemned? Why? (3) Are there instances in the Bible when medicine is encouraged? (4) Luke is a doctor; (5) What should be the purpose of medicine (*i.e.* are there types of medicine that may not be justifiable)?

B. Why does modern medicine kill pathogens rather than cure them as is suggested by young-age creationism? Things to consider: (1) What worldview dominated as modern medicine was being developed? (2) What worldview dominates the organizations that fund medical research? (3) What sort of things must happen and how much does it cost to get a new treatment for a disease approved? (4) What worldview dominates in medical research facilities?

Test & Essay Questions

1. Define mutualism / biological evil / pathogen / pathology and give two / three examples.

2. Compare and contrast parasitism, commensalism, and mutualism / facultative and obligate mutualism / negative and evil-minimizing effects of the curse / degenerative and developmental aging / host and parasite / predator and prey.

3. How does God illustrate His goodness in the biological creation?

4. Describe a lichen and what it is made of and how it works.

5. What is a mycorrhizal fungus and what does it do?

6. Describe what the spectrum of perfection of mutualism is and what we are to conclude from it.

7. How is mutualism a challenge to the naturalistic worldview?

8. For each of the following, indicate whether, in a young-age creationist worldview, it originated for nephesh organisms at or before the Fall: biological evil / pathology / death of life / natural selection / overproduction / struggle for survival / predation / predation designs?

9. List two / three different negative effects of the curse.

10. How is death / natural selection / predation / both a good thing and a biological evil?

11. How is mutualism / mutational load / pathology frequency a challenge to the naturalistic worldview / better explained by the young-age creationist worldview?

12. What responsibility do we have regarding the negative effects of the curse? / regarding the evil-minimizing effects of the curse?

13. How is the elimination of death / predation a bad thing?

CHAPTER 6

GOD IS PERSON

Biological Personality

"Our God is in heaven and does whatever He pleases. Their idols... have mouths, but cannot speak, eyes, but cannot see. They have ears, but cannot hear, noses, but cannot smell. They have hands, but cannot feel, feet, but cannot walk; They cannot make a sound with their throats. ...The Lord remembers and will bless..."
Psa. 115:3-12, HCSB

6.1 | God is a Person

So, what is a person? First of all, a person has no exact copy; every person is unique and has a different combination of traits from any other. Secondly, a person is dynamic and active. Thirdly, a person is discerning, understanding, and knowing. Fourthly, a person is not an automaton, but has a will, making decisions, choosing options, and being at least somewhat unpredictable (unless, of course, they are completely known). Fifthly, a person has a personality, with emotions, expressions of emotion, and even humor. Sixthly, a person is self-aware. Finally, a person can enter into a personal relationship with another personal being.

God is person. First, *God is unique.* He is an individual like none other (*e.g.* Exo. 8:10; I Ki. 8:23; Mark 12:32[a]). God is qualitatively different; He is hugely different than anything in His creation[b].

Second, *God is active.* He sees, hears, smells, and tastes. He speaks, eats, breathes, moves, walks, sits, and He rises. He drives out and He shuts in. He creates, and He rests. He plants, He places, He marks, He forms, and He divides[c]. Even if some of these descriptions are examples of giving human characteristics to God (anthropomorphisms) to help us understand something beyond our comprehension, it is clear that God is an active God. In fact, God's activities are unlimited (Matt. 19:26; Luke 1:37; Eph. 3:20). God is not just mighty (*e.g.* Isa. 9:6; Jer. 32:18-19), He is *all*-mighty (Gen. 17:1; 35:11; Psa. 91:1; Eze. 10:5; Rev. 1:8).

Third, *God is intelligent.* He remembers (*e.g.* Psa. 105:8[d]), He knows (Gen. 18:21), He has thoughts (*e.g.* Psa. 33:11; 139:17; Isa. 55:9),

---

a   Further references indicating that there is none like God include Exo. 9:14; 15:11; Deu. 33:26; II Sam. 7:22; I Chr. 17:20; Psa. 86:8; 89:6; Isa. 40:18; 46:9; and Jer. 10:6-7.

b   It should be remembered that when dealing with the uniqueness of God we are dealing with a qualitatively different being than the uniqueness of individual humans in relation to one another. The uniqueness in creation is only a dim illustration of the uniqueness of God in relation to His creation.

c   The references for God's activity include passages that indicate God sees (*e.g.* Gen. 1:31; 6:12; Exo. 2:25; Psa. 33:13-14; Lam. 3:50), hears (*e.g.* Exo. 3:9; II Chr. 7:14; Psa. 94:9; Isa. 38:5; I John 5:14-15), smells (Gen. 8:21; Eph. 5:2; Philip. 4:18), tastes (Lev. 1:9; 3:5; Psa. 11:5), speaks (*e.g.* Gen. 1:3-30; 6:7), eats (Gen. 18:8), breathes (*e.g.* Gen. 2:7; Job 33:4; 37:10; Psa. 33:6; Isa. 30:33), moves (Gen. 1:3), walks (Gen. 3:8; Lev. 26:12), sits (*e.g.* I Ki. 22:19; Isa. 6:1; Dan. 7:9; Mat. 25:31; Rev. 20:11), rises (Psa. 68:1), drives out (*e.g.* Gen. 3:23; Exo. 23:29-30; Deu. 9:4-5; Josh. 3:10; II Chr. 20:7), shuts in (Gen. 7:16), creates (*e.g.* Gen. 1:1; Psa. 104:30; Isa. 45:18; 65:17; Col. 1:16), rests (Gen. 2:2-3), plants (Gen. 2:8), places (Gen. 1:17; 2:8, 15; 3:24), marks (Gen. 4:15), forms (Gen. 2:7, 19), and divides (Gen. 1:4, 7).

d   Further references that indicate that God remembers include Gen. 8:1; 9:15-16; 19:29; 30:22; Ex. 2:24; 6:5; Lev. 26:42; I Sam. 1:19; 15:2; Psa. 98:3; 105:42; Jer. 2:2; and Luke 1:72.

including thoughts about us (*e.g.* Psa. 40:5, 17), and He understands the thoughts of others (*e.g.* I Chr. 28:9 and I Cor. 3:20[a]). He is wise (*e.g.* Job 12:12-13 and Rev. 7:12) and He is the source of wisdom (*e.g.* Psa. 111:10 & James 1:5)[b]. He has understanding (*e.g.* Psa. 147:5 & Isa. 40:28) and He is the source of understanding (*e.g.* Job 32:8 & Pr. 2:6[c]). He has knowledge (*e.g.* I Sam. 2:3 and I Jn. 3:20) and is the source of knowledge (*e.g.* Pr. 1:7)[d].

Fourth, *God has a will* (*e.g.* I Cor. 12:11; Eph. 1:11; 5:17) and He has a will regarding us (I John 5:14). God changes his mind in response to human obedience (Jer. 18:8-10[e]). In fact, God has complete freedom to do whatever He wishes (Psa. 115:3)[f].

Fifth, *God feels*. He has the emotions of grief, anger, vengeance, pity, favor, compassion, pleasure, joy, and love[g]. In fact, God's grief is described as residing in God's heart (Gen. 6:6) and soul (Jud. 10:16)—both considered sources of emotion and personality.

Sixth, *God is self-aware*. God refers to Himself in the first person (*e.g.* Gen. 1:26; 3:22; 26:2-5; Exo. 3:6-4:23; 15:25-26). Perhaps most obviously, God has a name—such as the 'I am' in Exo. 3:13-15, and 'The Lord' in Ex. 15:3, 'The Lord thy God' in Deu. 28:58, 'Jehovah' in

---

[a] Further references God knowing the thoughts of men include Psa. 44:21; 94:11; 139:2; Jer. 17:10; Rom. 8:27; and I Cor. 2:10.

[b] Further references that God is wise include Job 9:4; 12:16; 36:5; Psa. 104:24; 136:5; Pr. 3:19; Jer. 10:7, 12; 51:15; Dan. 2:20; Rom. 11:33; I Cor. 1:24-25; 12:8; Eph. 1:17; 3:10; Col. 2:3; I Tim. 1:17; Jude 25; and Rev. 5:12. God was the source of wisdom for Bezaleel & Aholiab (Ex. 31:3, 6; 35:31, 35; 36:1-2), for Solomon (I Ki. 4:29; 5:12; 10:24; I Chr. 22:12; II Chr. 1:12; 9:23), and Daniel (Dan. 1:17 and 2:23) and is the source of wisdom for all (Ezra 7:25; Job 11:6-7; 28:28; Psa. 19:7; 15:33; Pr. 2:6; 9:10; Eccl. 2:26; Isa. 11:2; Dan. 2:21; Luke 21:15; Acts 7:10; I Cor. 1:30; 4:10; Eph. 1:8; James 3:17).

[c] Further references that God has understanding include Job 12:12-13; 26:12; 34:21; Pr. 3:19; Isa. 11:2; and Jer. 51:15 and further references that He is the source of understanding include Ex. 31:3; 35:31; 36:1; I Ki. 4:29; 7:14; I Chr. 22:12; 28:19; Job 32:8; Isa. 11:2; Luke 24:45; and II Tim. 2:7.

[d] Further references that God has knowledge include Job 36:4; 37:16; Pr. 2:5; 3:20; Isa. 11:2; Rom. 11:33; Col. 2:3; and Heb. 4:13, and that He is the source of knowledge include Ex. 31:3; 35:31; I Sam. 2:3; II Chr. 1:12; Psa. 94:10; Pr. 2:6; 9:10; Eccl. 2:26; Isa. 11:2; Jer. 11:18; Dan. 1:17; 2:21; and I Cor. 1:5.

[e] Examples of God changing his will based on human response include Gen. 6:6-7 & 8:21; Jud. 2:17-18; II Sam. 24:16; and Jonah 3:4-10.

[f] Note that this says God can do whatever He *wishes* to do; it does not say that God can do anything. God cannot sin, He cannot lie, He cannot deny Himself, *etc.* God cannot be something contrary to Himself or do something contrary to His nature. But then, He would never wish to do anything contrary to His nature. Thus, God has the complete freedom to do whatever He wishes.

[g] A few references for each of the emotions of God listed in the text: for grief, Psa. 78:40, Isa. 63:10, Mark 3:5, and Eph. 4:30; for anger, Exo. 22:24, II Sam. 6:7, I Ki. 15:30, Ezra 10:14, and Zeph. 2:2-3; for vengeance, Psa. 94:1, Mic. 5:15, Nah. 1:2, Rom. 12:19, and Heb. 10:30; for pity, Psa. 103:13, Joel 2:18, and James 5:11; for favor, Gen. 4:4-5, Exo. 2:25, and II Ki. 13:23; for compassion, Deu. 30:3, I Ki. 8:50, II Ki. 13:23, Psa. 86:15, and Mic. 7:19; for pleasure, I Chr. 29:17, Mat. 3:17, Philip. 4:18, Heb. 11:5-6, and Rev. 4:11; for joy, Deu. 28:63, Psa. 104:31, Isa. 62:5, Zeph. 3:17, and Heb. 12:2; and for love, Isa. 54:8, Psa. 103:17, John 3:16, and 17:24.

Psa. 83:18 and 'The God of hosts' in Amos 4:13). In fact, throughout Scripture the name of God is interchangeable with the glory of God, God Himself, and the fullness of God's glory.

Seventh, *God enters into personal relationships.* He participates in activities characteristic of relationships. He responds, helps, and heals. He comforts, blesses, curses, and chooses to destroy or not destroy. He wipes away tears, counsels, teaches, guides, and leads. He witnesses, redeems, intercedes, advocates, and does not lie. He does good and He judges with justice. He bestows mercy and grace. He is kind, gentle, and patient. He is faithful and He is faithful with a steadfast love. In fact, God actually engages in relationship with humans. He dwells among men, speaks personally to man, walks with humans, renames humans, and knows humans by name. He covenants with humans, communes with man, develops friendships with man, and believers can call him *'Papa'*[a].

---

[a] A few references for each of the inter-personal actions of God listed in the text: God responds (*e.g.* Num. 21:3; Josh. 10:14; Jud. 13:9; II Ki. 13:4; II Chr. 30:20), helps (*e.g.* Deu. 33:29; Psa. 54:4; Isa. 41:10; 50:9; Heb. 13:6), heals (Gen. 20:17-18; Exo. 15:26; Deu. 32:39; II Ki. 20:5; II Chr. 30:20), comforts (*e.g.* Isa. 51:3; Jer. 31:13; Zech. 1:17; John 14: 26; II Cor. 1:3-4), blesses (*e.g.* Gen. 1:28; 9:1; 26:12; I Chr. 13:14; 26:5), curses (*e.g.* Gen. 3:14-19; 4:11-12), destroys (*e.g.* Gen. 7:23; Num. 21:3; Deu. 2:21-23; I Sam. 5:11; Amos 9:8) or not destroys (*e.g.* Gen. 9:11; 18:23-32; Deu. 4:31; II Chr. 21:7; Amos 9:8), wipes away tears (Isa. 25:8; Rev. 7:17; 21:4), counsels (Isa. 9:6; John 14:16, 26; 15:26), teaches (*e.g.* Ex. 4:15; Deu. 4:36; Psa. 25:12; Isa. 2:3; Micah 4:2; John 14:26), guides (John 16:13-14), leads (*e.g.* Deu. 8:2; Psa. 27:11; 77:20; 139:9-10; Isa. 63:14), witnesses (John 15:26; Rom. 8:16), redeems (*e.g.* Job 19:25; Psa. 78:35; Pr. 23:11; Isa. 47:4; Jer. 50:34), intercedes (Isa. 53:12; Rom. 8:26-27, 34; Heb. 7:25), advocates (I John 2:1), does not lie (I Sam. 15:29; Titus 1:2; Heb. 6:18), does good (*e.g.* Job 84:11; 145:9; Matt. 7:11; Acts 14:17; James 1:17), judges with justice (*e.g.* Deu. 32:4; I Sam. 2:10; Psa. 94:2; II Cor. 5:10; Rev. 20:13, Pr. 16:11; Isa. 45:21; Jer. 31:23; Acts 22:14; Rev. 15:3), judges with mercy (*e.g.* Exo. 34:6-7; I Chr. 16:34; Ezra 3:11; Psa. 103:17; Eph. 2:4) and judges with grace

Finally, when Scripture refers to God by simile or metaphor, it often alludes to human persons, such as a man of war (Exo. 15:3), a farmer (John 15:1; I Cor. 3:9), a builder (I Cor. 3:9-10; Heb. 11:10-11), a potter (Isa. 64:8; Jer. 18:1-6), a father (e.g. Psa. 103:13; Prov. 3:12), a husband (Isa. 54:5; Jer. 31:32; II Cor. 11:), a bridegroom (Isa. 61:10; 62:5), and a shepherd (e.g. Ps. 23:1; Jer. 31:10; John 10:2-18; Heb. 13:20; I Pet. 2:25; 5:4). Not only is God a person, each of the three members of the Godhead is a person. God the Father, God the Son, and God the Holy Spirit are each unique, expressive, active, dynamic, willful, self-aware, and all are in personal relationship with one another.

Personality in the Biological World

The personhood of God is an important characteristic of God and He wants us to recognize, know, and love that quality about Him. God's personhood, however, is invisible to us. Therefore, God illustrated it in the physical creation. He placed a spectrum of perfection of each of the personhood characteristics across the biological world.

(e.g. Psa. 103:8; 145:8; Joel 2:13; Jon. 4:2; Eph. 2:5-9), is kind (e.g. Neh. 9:17; Psa. 36:7, Isa. 54:8; Joel 2:13; Jonah 4:2), is gentle (II Sam. 22:36; Psa. 18:35; Isa. 40:11; 42:3), is patient (e.g. Exo. 34:6; Num. 14:18; Jonah 4:2; Rom. 15:5; II Pe. 3:9), is faithful (e.g. Deu. 7:9; Psa. 36:5; 119:90; Isa. 49:7; Rev. 19:11), and is faithful with a steadfast love (Deu. 7:9)., dwells among men (e.g. Exo. 25:8; I Ki. 6:13; Eze. 43:9; Zech. 2:10-11; Rev. 21:3), speaks personally to man (e.g. 3:9-22; 6:13-21; 28:13-15; Exo. 6:1-8; 19:20-24:3), walks with humans (e.g. Gen. 5:22-24; 6:9; Lev. 26:12; II Sam. 7:7; I Chr. 17:6), renames humans (e.g. Gen. 17:5, 15; 35:10) and knows humans by name (Deu. 34:10; Jer. 1:5; Hos. 13:5; John 10:14, 27; Gen. 3:9; Exo. 3:4; 31:2; 33:17; Isa. 43:1), covenants with humans (e.g. Gen. 9:9-17; 15:9-21; Exo. 2:24; Psa. 89:3; Heb. 10:16), communes with man (e.g. Gen. 18:33; Ex. 25:22; 31:18; Num. 12:8; Deu. 34:10), develops friendship with man (Ex. 33:11; II Chr. 20:7; Isa. 41:8; James 2:23), and we can call him *'Papa'* (Mark 14:36; Rom. 8:15; Gal. 4:6).

Biological Uniqueness

A 'person' is someone who is unique and relates to others in unique and personal ways, so an essential characteristic of personhood is individuality. God illustrated His own uniqueness by infusing the creation with biological uniqueness and individuality. This uniqueness, for example, is seen in the ways organisms differ from the non-living world. Organisms are many times more complex than anything else in the physical world. Living things also develop along what appear to be programmed pathways. As an organism develops, it seems to be drawn towards some final and mature form, unique to that type of organism.

Although non-living things can and do change (*e.g.* the sun burns, rocks erode, crystals grow), and they often change in predictable ways, they do not change as if they are drawn towards some final form. Even when non-living things follow very strict rules of change (*e.g.* in crystal growth), obstacles or variations early in the process produce very different outcomes in the end. In contrast, very often when an obstacle is encountered in the growth of an organism, development adjusts in an effort to go around the obstacle. If the obstacle is not too large it returns to its normal developmental path when the obstacle is passed. It is as if the creature 'knows' where it is going, or something is drawing it toward its mature form.

As a consequence of such strongly programmed development, individuals of a given species might seem much more similar to each other than non-living things are to each other. In actual fact, organisms are much more likely to be unique in the macromolecules that make them up. Because of tiny variations that can occur in a host of molecules, it is likely that each organism differs from all other organisms. This is probably even true for every **clone** (an organism having the same genetic information as another organism).

clone—an organism with identical DNA as another organism

Even more uniqueness is found in sexual organisms (organisms having a combination of genetic information from both parents). Even the sex cells (sperm cells and egg cells) of sexual organisms are unique. It is even likely that each of the millions of sperm and egg cells are different from each other. DNA is different enough among sexual organisms for all but identical twins to be readily distinguished by DNA fingerprinting. Even identical twins have distinguishable personalities.

Among complex organisms, each individual of a species can be distinguished by differences in form and behavior. Distinctive songs, for example, allow the distinction of individuals among both porpoises and song birds[1]. Many organisms can distinguish individual members of their own species. Elephants, for example, have no difficulty distinguishing more than 200 individual elephants. In the case of vervet monkeys, playing the distress call of an infant not only causes the mother to look at the source of the sound but causes unrelated females in the troop to look at the infant's mother[2]! Organisms can even distinguish individuals of other species. For example, an octopus can become accustomed to a particular human diver, and crows can identify one specific human who raids their nests[3].

Biological Activity

For a 'person' to exhibit the many characteristics of personhood, he or she must be capable of a considerable diversity of activities. God illustrated His dynamic and active nature by granting organisms a host of different kinds of activity not found in the non-living world.

Metabolism. Organisms are always changing. Organisms are many times more complex than anything in the environment around them. As per the second law of thermodynamics, the complexity of an organism is continually being compromised. Since organic molecules are always breaking down, organisms must be continually active in replacing them. Aside from some structures used for support (like wood, shells, bones) nearly every molecule in any given organism is replaced every few weeks, months, or years. This self-maintenance is called **metabolism**.

Even for the simplest organisms, metabolism involves a bewildering array of complex processes—some breaking down complex molecules into **monomers** (molecular building blocks), some assembling monomers into the needed polymers (larger compounds made from the monomers), and others storing or releasing the energy needed to run all these processes. All organisms metabolize without stop from the beginning of their life to the end of their life.

Growth. All organisms grow, in the sense of increasing in size. Each individual cell starts out as a portion of another cell or as a growth on the periphery of another cell. To 'grow' in this case means that the cell must increase in size in order to reach the proper size for that particular

---

metabolism—cell processes storing & releasing energy and building & breaking down macromolecules

monomer—molecular building block of (biological) macromolecule

kind of cell. Organisms made of more than one cell usually begin their development as one cell. For a multicellular organism to 'grow', one cell must produce other cells. Most of the growth you observe in multicellular organisms occurs by increasing the number of cells, rather than by increasing the size of cells.

Activity and Rest. Many organisms have different levels of activity at different times of their lives. Most of the time, this is merely the direct consequence of changes in the environment. Since most chemical reactions speed up with increased temperature, much biological activity slows down in cold weather. Reactions that require light, such as photosynthesis, slow down or even cease in the dark. Reptiles sun themselves to warm up their bodies and raise their metabolism. Some organisms behave differently according to the season or time of day (*e.g.* jellyfish with algae in their bodies move upwards to the light during the day and downwards to safety at night, raccoons sleep during the day and forage at night, turkeys and deer sleep at night and forage during the day, grizzly bears and some bats hibernate for the winter, various bats and birds migrate to eat or mate).

Development. Development is the process of an organism maturing from how it starts life to its final, adult form. For some organisms constructed of only one cell, maturation merely involves growth. All the cell and its components must do is simply grow in size.

For most organisms, however, development is more complex. In the simpler cases, the organism might start out with all the components

found in an adult, but not as many of them, so the components have to be multiplied. In most cases, however, the organism starts out lacking a number of components found in the adult, so these things must be developed. In organisms composed of multiple cells but one tissue, development mainly involves multiplying cells. In organisms with multiple tissues and organs and organ systems, development is extremely complex, including one cell type generating different cell types.

Some organisms even develop through life cycles that involve more than one very different looking multi-celled body. Sometimes these different life stages even have different amounts of genetic material (*e.g.* mosses have two different body forms with the most familiar one having only one version of each gene, fungi have three different body forms with the most familiar one having four versions of each gene[a]). Although different animals begin from very similar starting points (single, fertilized eggs), and maintain the same amount of genetic material, they can develop along very different looking paths.

direct development—animals developing directly into a miniature version of the adult and grow in size from there

Some animals undergo **direct development**, meaning they develop directly into a miniature version of the adult and grow in size (cell number) from there (*e.g.* immature grasshoppers and snakes look like miniatures of their final forms and molt their outer coverings as they grow, humans develop into adults from adult-similar infants and toddlers and adolescents). In contrast, some animals undergo

[a] The most familiar body form of most animals has two versions of each gene.

indirect development, meaning they develop through more than one radically different stage (*e.g.* flies develop from maggots, butterflies develop from caterpillars, shelled oysters attached to underwater surfaces develop from free-swimming worm-like larvae).

6.2 | Biological Activity Continued

Reproduction. In sexual organisms the genetic material of two parents is combined. This requires some sort of reproductive activity. Genetic material from the two parents must be brought together. Stationary organisms either grow towards each other (*e.g.* the hyphae of most fungi) or develop special structures that send out mobile sex cells (*e.g.* mosses produce sperm that swim to the eggs, flowering plants produce pollen that releases sperm when it lands on the female organ of the flower).

In the case of most organisms that can move, one or more of the parents move towards the other in response to chemicals (*e.g.* hyphae of fungi growing towards one another, female butterflies producing a chemical to attract a male), sound (*e.g.* crickets, bullfrogs, songbirds, elk), light (*e.g.* female fireflies flashing to attract a male[4]), visual displays (*e.g.* tails of peacocks, courtship dances in spiders and butterflies and fish and chameleons and songbirds). Some organisms choose mates (*e.g.* a female fiddler crab chooses a mate according to claw size and a satisfactory burrow, a female sage grouse chooses a mate on the basis of a male's dance, a female bowerbird chooses a mate based on the

> **indirect development**—animals developing through more than one radically different stage

VIDEO 6.2

arrangement and colors of the structure (bower) that he builds for mating purposes).

Additional behavior is required to unite the sex cells of the two organisms (*e.g.* several male honey bees giving up both their sperm-containing internal organs and their lives so a queen can fertilize eggs for years to come, females picking up sack of spore in some pseudoscorpions, male horseshoe crabs spreading sperm on clusters of eggs laid by the females, male mammals depositing egg-seeking sperm inside the bodies of females). Many animals engage in further types of birthing (*e.g.* eggs brooded in the mouth of *Tilapia* fish, female seahorses laying eggs in a pouch on the 'belly' of the male where the eggs hatch and develop, fetal chickens developing inside eggs laid by and kept warm by mothers, fetal marsupials crawling into and developing in a pouch of the mother, fetal placental mammals developing inside the body of the mother). Some animals have complex behaviors in raising their young after hatching or birth (*e.g.* songbirds bringing food to the young in their nests, raccoons and lions and octopi training their young to forage, female mammals nursing their young with milk from mammary glands).

Resource Acquisition. Since an organism's metabolism must be constantly active, organisms must also have ways to get to the energy sources and the molecules that they need. Some of that motion involves growth towards water, food, or light. Cells detect gravity or light or chemicals and produce more cells in that direction. This allows

algae and plants to grow towards light, plant roots to grow downward towards water, and fungi to grow towards food. But growth is a slow process. Bacteria, protozoa, and animals can respond more rapidly by moving their whole body in the direction of desired water, food, light, or air.

Some of these behaviors are simple, involving organisms moving in the direction of higher concentrations of desired molecules (*e.g.* bacteria moving towards higher concentrations of sugars, sharks moving towards higher concentrations of blood) or greater light (*e.g.* submerged organisms rising to the surface). Some behaviors involve becoming invisible until the food wanders along (*e.g.* octopi and squids and chameleons changing color to match the background[5]).

Other behaviors are more complex, trapping or luring food (*e.g.* fungi growing in loops to lasso roundworms, antlions digging pits to capture ants[6], predatory fireflies imitating the flashing behavior of the species they wish to eat[7], alligator snapping turtles wiggling worm-shaped tongues to draw in fish, frog-eating bats imitating frog mating calls to draw in frogs).

Still other behaviors involve stealthy approach to prey (*e.g.* leaf-mimicking praying mantises, color-changing chameleons, stalking lions). Still other behaviors involve some level of reasoning power, such as the use of tools (*e.g. Aphaenogaster* ants use crude sponges to carry fruit pulp or body fluids[8], archerfish spit droplets of water to knock insects out of the air and into the water, various birds use twigs to get insects out of small holes[9]), or cooperation (*e.g.* honeyguide birds drawing humans to bee nests so that the birds can eat the honeycomb after humans disperse the hive[10], flocks of pelicans corralling fish[11], lions and wolves hunting in packs[12]).

Once the organism reaches a resource, other behaviors allow the organism to take the resource in. Different animals, for example, drink by siphoning (*e.g.* elephants), sucking, lapping, or licking. Different organisms eat by absorbing (*e.g.* fungi), sucking (*e.g.* aphids, ticks, leeches), swallowing (*e.g.* sharks, alligators, birds), or chewing (*e.g.* mammals). Although most animals eat only what food they contact, other organisms harvest their food (*e.g.* leaf-cutter ants eat fungi raised on leaves), and some organisms store their food (*e.g.* chickadees, jays, crows, ravens[13], squirrels, the red fox)

Defense. Some organisms are prey (food sources for other organisms). Such prey need to protect themselves against being eaten. Some of these defenses involve hiding, others involve looking dangerous (*e.g.* non-poisonous scarlet kingsnakes looking like the venomous eastern coral snake).

Other defenses involve scaring off potential predators (*e.g. Cecropia* moths having eye designs on their wings to scare away birds[14], bombardier beetles spraying chemicals at frogs, frogs swelling up to look menacing or unswallowable, skunks spraying musk, rattlesnakes rattling). Others still require simple escape (*e.g.* amoebas moving away from poisons, octopi squeezing through remarkably small slits[15],

squirrels scurrying from hawk shadows, deer running in response to quick movements). Others protect themselves by seeking shelter, carrying protection (*e.g.* hermit crabs carrying shells, *Difflugia* amoebas studding themselves with sand grains), constructing shells (*e.g.* many mollusks) or burrowing. Others involve fooling the predator (*e.g.* opossums playing dead) or fighting the predator (*e.g.* bees stinging, birds pecking, rodents biting, cats scratching). Others use tools (*e.g.* crabs that hold anemones in their claws to ward off attackers[16], a wild capuchin monkey that attacked a venomous snake with a club[17]). Still others defend themselves by using cooperative behavior (*e.g.* caterpillars eating together and rearing up and regurgitating in response to a predator, birds sending out alarm calls, crows ganging up on hawks, prairie dogs setting up lookouts, adult elephants gathering around young when attacked).

Locomotion and Orientation. Different organisms move in a wide variety of ways, such as by whip-like flagella (*e.g.* some bacteria, sperm), oar-like cilia (*e.g.* some protozoa), pseudopods (*e.g.* amoebas), undulation (*e.g.* some worms), jet propulsion (*e.g.* octopuses and squids), climbing (*e.g.* monkeys), gliding (*e.g.* flying fish, flying squirrels), flying (*e.g.* pterodactyls, bats, birds, flying insects), digging (*e.g.* many insects, moles), swimming (*e.g.* fish), and walking.

Many organisms travel extremely long distances at one or more points in their lifetime. Some migrate according to the seasons (*e.g.* the gray whale migrates 10,000 kilometers a year from summer

feeding grounds in the arctic and winter breeding grounds off Mexico; successive generations of the monarch butterfly migrate a total of 4000 kilometers a year between Canada and Mexico; many bat and bird species nest in the temperate regions and winter in tropical or subtropical regions). Other species migrate long distances at one or a few stages of their lifetime (*e.g.* Coho salmon grow up in a stream, migrate downstream to live most of their lives in the ocean, and then return to the same stream to mate, lay eggs and die). Still other organisms travel shorter distances but do so with great precision (*e.g.* honey bees going to and from particular flowers to store pollen and nectar in their hive).

To accomplish these feats organisms must be able to orient themselves. Different organisms determine orientation in different ways. Up and down can be determined by detecting the greater brightness usually associated with the sky (in many animals of the sea), or horizontal streaks on retinas (in some desert and marine birds), or by gyroscope designs (in flies), or by tiny solid objects rolling around in cells (in some corals, mollusks, echinoderms, and crustaceans). Compass direction can be determined by the rotation of stars near the North Star (*e.g.* indigo buntings[18]), the sun (honeybees and various species of spiders, crustaceans, fish, amphibians, reptiles, birds, and mammals[19]), the moon (in many moth species), the polarized light (some insects, crustaceans, mollusks, and fish[20]), and tiny crystals of iron orienting themselves in the earth's magnetic field (*e.g.* some species of bacteria, honey bees, Chinook salmon, porpoises, various bird species[21]).

Organisms move in the direction of other organisms by means of body heat (*e.g.* ticks seeking hosts, pit vipers and rattlesnakes finding prey), chemicals (*e.g.* butterflies and moths finding mates), sound (*e.g.* echolocation in bats and porpoises), vibrations (*e.g.* spiders finding prey on webs, whirligig beetles and water striders detecting organisms on the surface of water), and electrical fields (various fish, newts, and the duck-billed platypus locating prey; electric catfish, electric eels, and knife fish using electrolocation).

6.3 | Biological Intelligence

For a 'person' to engage in personal relationships, he or she must be able to distinguish and identify others, remember them, and respond to each of them in unique ways. Personhood requires more than programmed behavior. A person needs to be able to discern subtle differences, to be able to change behavior to respond to new experiences, and to infer the best response to whatever happens. Personhood requires memory and **reason**; personhood requires intelligence. God illustrates his own intelligence with a spectrum of biological intelligence—examples of learning, memory, and reason among organisms of the creation.

Instinct and Learning. Organisms not only illustrate degrees of activity in animal behavior, they also illustrate a spectrum of intelligence. Some behavior, for example, appears to be fully programmed into the organism, and the organism performs it perfectly the very first time it is tried (*e.g.* newly hatched honey bees properly caring for larvae, mother birds feeding any open mouth, mother dogs birthing and cleaning newborn puppies). This kind of behavior, is called **instinctive behavior**. Behavior which the organism does not do perfectly the first time, but which improves as the organism repeats it, is **learned behavior**.

A **habituation** is a *lack* of reaction to experiences that the animal has come to learn neither hurts nor helps the animal (*e.g.* the protozoan *Stentor coeruleus* that does not respond to repeated touches or sights while still responding to other touches or sights, a city pigeon not bothered by masses of people, a deer near an airport unbothered by planes taking off and landing).

reason—the memory, processing, and rearranging of information by an organism

instinctive behavior—behavior performed perfectly the very first time it is tried (vs. learned behavior)

learned behavior—behavior which is performed imperfectly the first time, but more perfectly as the organism repeats it (vs. instinctive behavior)

habituation—an organism's lack of response to experiences it has learned neither hurts nor helps the organism

imprinting—organismal behavior based upon early life experiences of that organism

conditioned response—behavior learned through repeated rewards or punishments

mental map—memory of the environment that an organism develops so as to move through it

imitative learning—behavior an organism copies from the behavior of another organism

Imprinting is behavior learned based on experiences at some early period of life (*e.g.* a baby songbird that develops its song upon hearing the song of other birds, a baby goose that follows along behind whatever organism leaned over it over when it first peeped, Coho salmon at the end of their life following the chemical signatures of waterways they swam down early in their life).

A **conditioned response** is behavior learned through repeated rewards or punishments (*e.g.* a dog automatically salivating when it hears a dinner bell[22], a slug[23] or other vertebrate[24] developing a distaste for something that created nausea when last eaten, a flatworm[25] or an octopus[26] or an ant[27] or a goldfish[28] or a mouse remembering routes through mazes that always yield food[29], a crow or a rat pressing levers that always yield food[30]). **Mental maps** of the environment are developed by animals so that they can make their way through the environment (*e.g.* newly launched honey bees circling in ever-winding arcs to determine and remember the position of their hive with respect to nearby landmarks).

In **imitative learning** an animal copies the behavior it observes in another animal (*e.g.* chicks learn to scratch and peck by watching adult chickens, raccoons learn to find eggs under sitting chickens by watching adult raccoons, chimps learn how to choose and use sticks to eat termites by watching other chimps, bottlenose dolphins learn stunts by observing other dolphins performing them[31]).

Some animals even copy behaviors that do not seem to give them any particular benefit. It is not uncommon, for example, for parrots to mimic human words. One parrot even mimicked the gestures that went with the words (*e.g.* voicing 'ciao' while waving with feet or wings, voicing 'forget it' while tossing something to the side, voicing 'whoops, dropped the peanut' while bending over as if to pick it up, voicing 'Remember Lloyd Morgan, do not forget' while using claw motions on the last two words)[32]. One bottle-nosed dolphin used a seagull feather to scrape its tank, copying the behavior of human divers who scraped the tank of algae. Another imitated every organism in its tank (a seal, turtles, skates, penguins)[33]. Orang-utans have also been known to mimic humans in the use of a toothbrush and insect repellant. One orang, 'Supinah', even ignited a fire with kerosene and tended the fire by waving a trash can lid over the burning embers[34].

6.4 | Biological Intelligence Continued

Memory, Counting. The most complicated behaviors require the organism to remember things. Learning, for example, requires memory, so each of the examples of learning given above provide evidence of an organism's memory. Some of them are quite impressive (*e.g.* a rat can learn to "…successfully climb spiral stair, push down and cross a drawbridge, climb a ladder, pull in and pedal a car through a tunnel, climb stairs, run through a tube, and take an elevator down to an area where finally, when a buzzer sounds, pressing a lever produces food."[35]).

Organisms that store food, and do so in separate locations (*e.g.* chickadees, tits, crows, ravens, nuthatches, squirrels, agoutis) also require considerable memory. Perhaps the most impressive of these is the Clark's nutcracker, which buries 22,000-33,000 seeds in up to 7500 different locations in order to be eaten throughout the winter, months after burial[36].

Some animals can even count. Pigeons, parrots, rats, porpoises, monkeys, and apes can put different numbers of things in the correct numerical sequence[37]. One African gray parrot 'Alex' can vocalize the numbers two through six and associate the vocalization with groups of objects with the proper number of objects[38]. A pigeon can learn and remember the exact number of pecks necessary to produce a food reward for any number between 1 and 50[39]. There is even evidence

that rhesus monkeys can add (staring an extra long time if four shapes disappear behind a wall, then four more shapes disappear behind the wall, and the wall moves aside to only reveal four shapes[40]).

Reason. The most sophisticated behavior involves reasoning, where information has to be remembered, processed, and rearranged in an unusual way for that organism. Fascinating examples of reasoning have been observed in the biological world. An octopus, for example, figured out how to take the lid off of a jar in order to eat the lobster inside[41]. In another example, alfalfa flowers dust a bumblebee with pollen by slamming its pollen-carrying anthers onto the bumblebee's back. This action harms smaller insects like honeybees, so honey bees learn to either visit only alfalfa flowers that have already 'sprung' or to chew a hole in the back of the flower to avoid the anthers[42].

In another example, hummingbirds usually vocalize a territorial warning immediately upon discovering an intruder in their territory. One male Anna's hummingbird, however, responded by silently circling around to the other side of the intruder from a mist net that he had discovered in his territory earlier that day. When he vocalized the territorial warning, the intruder flew right into the net[43]. Some house sparrows figured out how to fly in front of an electric eye in order to open the door of a cafeteria[44].

A famous example of intelligence involved British titmice figuring out how to pull the metal foil off the tops of milk bottles to access the milk inside[45]. A blue jay was observed to use available objects (piece of newspaper, feather, paper clip, piece of straw) to sweep up food pellets that had fallen outside his cage[46]. Since ravens avoid carcasses with dead ravens lying nearby, one raven figured out how to have a carcass all for himself by rolling onto its back and lying motionless when he saw another raven fly overhead[47]. Ravens that watch an experimenter place food in the middle of an opaque tube will look in both ends of the tube, and if they cannot reach the food at either end will pick up one end to slide the food out the other end[48]. When food was dangled from a perch on string in the presence of hand-raised ravens that had never seen string, ravens would eye the situation, then—without practicing—proceed to pull up the string, stand on it, pull up the string some more and stand on it until they obtained the food[49]. It is even more interesting that when ravens were shooed off the perch while this prize was in their mouth, ravens that had pulled up the string to get the food would drop the food before flying off, whereas ravens that had been given the food attached to the string would fly off with the food (and get the food wrenched from their beaks when the string became taught).

One crow figured out how to eat a palm nut by dropping the nut onto a busy street until it was broken open by a car[50]. A bottlenose dolphin who watched another dolphin trained to grab a ball and pull it through the water to raise a flag, created its own variation on the stunt by batting the ball in front of it[51]. Birdwatchers will acknowledge that squirrels seem infinitely clever at finding ways to access the food on bird feeders designed to prevent access by squirrels. Likewise, beavers have been found to be quite clever in outsmarting humans intent on stopping the beavers from constructing dams. One raccoon, after watching a wire twisted three turns to close its cage, and before the handler walked away, reached through the cage bars on either side of the wire and untwisted the wire, pushed open the cage door, then closed the cage door and returned to the back of the cage—as if to show the keeper that he could get out any time he wished.

A famous horse, 'Hans' would stomp out the answer to mathematical equations by following subtle visual cues in the people who were watching him[52]. Elephants figure out how to unscrew bolts holding their cages together[53], and some elephants use sticks to scratch inaccessible body regions[54]. Two different Anubis baboons, 'Asparagus' and 'Sage', independently figured out how to drink lake water when the lake was choppy by digging a hole near the high-water mark of a lake[55]. In 1952 a young female Japanese macaque 'Imo' began washing sweet potatoes in the water before eating them, a behavior that spread through her troop. In 1956 the same Imo separated wheat from sand by throwing handfuls into water and scooping the wheat off the water's surface[56].

6.5 | Organismal Will

It is characteristic of a 'person' to respond in unique ways. Rather than responding in a highly predictable, programmed manner, a person purposely chooses among options according to his or her preferences or will. A person also makes reasoned choices by considering the consequences of different options and choosing according to reason. God demonstrated His own will by providing organisms with different degrees of will.

Choice. Many animals make choices. They choose food, they choose mates, they even choose one behavior over another. The simpler of these choices may be fully programmed, such as birds choosing seeds to fit the size of their beak, or honey bees choosing more nectar-rich clover over apple blossoms, or squirrels choosing the path of escape to the closest tree.

Other choices look more like free choices, such as different house cats preferring different brands of cat food. Complex animal behaviors are more flexible and less uniform. This makes it easier for an animal capable of complex behavior to override its programming and make a true choice. Organisms capable of reasoning find it even easier to make choices. A dog, for example, has a strong instinctive behavior to protect itself from pain, but it is capable of choosing not to bite someone who is tending a wound. This suggests that at least some animals have a will.

Moral Sense. There is some evidence that some animals can make moral choices. Although the curse on the serpent in Eden included a prophecy about Satan (Gen. 3:15), the first part of the curse (Gen. 3:14) seems to be a curse on the specific animal that spoke to Eve, as if it was in some way morally responsible for participating in the temptation of Eve or for choosing Satan's control. Scripture also requires the killing of anything that kills a human, whether human or animal (*e.g.* Gen. 9:5; Exo. 21:28). More than once, God also condemned animals along with humans (*e.g.* in Noah's Flood according to Gen. 6:7 & 12-13; the

Amalekites according to I Sam. 15:3). Nearly all pet owners also insist that from time to time their pets consciously *choose* to obey or disobey them (and slink about as if they know punishment will result).

Although many examples exist of animals deceiving other organisms, it is often difficult to determine whether the animal understands the action to be morally wrong. The simpler examples may merely be programmed responses that permit an organism to survive in the midst of predators and competitors (*e.g.* a molting mantis shrimp being uncharacteristically aggressive so as to scare off invaders while it is most vulnerable to attack[57], a plover bird faking injury to draw predators away from her nest[58], a raven acting like it is caching food in one location in order to actually stash its food in a second location while other ravens are searching for the food they think was stored at the first site[59], a possum faking death; a hognose snake making a grand show of 'dying'—writhing, vomiting, bleeding at the mouth—to discourage a predator[60]).

Other deceptive behaviors seem to be less necessary, more selfish, and a bit more purposeful (*e.g.* a bird making false alarm calls to disperse other birds so as to monopolize a food source[61]; a captive orang-utan 'Unyuk', upon seeing some backpacks, gradually shifting position while being groomed by a keeper until she can finally reach the backpacks and run[62]; pigs, monkeys, and chimps faking a lack of interest in some food in order to get it for themselves[63]).

Other deceptions are only performed after the animal first makes sure there are no observers, suggesting the animal understands the consequences of being caught (*e.g.* a rooster, after checking that competing roosters are absent, offering a food call in order to lure an unsuspecting female to mate[64]; female monkeys and apes scanning for and hiding from higher-ranking males in order to mate with lower-ranking males[65]; a lower-ranking male vervet monkey is less aggressive with an infant if he can see the mother watching[66]). The animal world even has examples of 'tattle-taling' (*e.g.* the male chimp 'Dandy', upon witnessing a female friend 'Spin' mating with another chimp, brought the alpha male to the scene to deal with it[67]) and chastisement (*e.g.* When the elephant 'Hanako' refused to nurse her calf at its birth, two of the older female elephants in her herd "…picked Hanako up and bounced her against the wall… They did it two or three times—until Hanako allowed the calf to nurse…"[68]).

6.6 | Biological Emotions

Another unique characteristic of personhood is the variety of emotions that a person experiences and expresses. Happiness and sadness, pleasure and anger, grief and joy are emotions of the heart and soul that persons feel or express. God illustrated his own heart by placing a spectrum of emotions in the organisms He created.

Emotions are challenging to demonstrate but are suggested in a number of animals. An octopus, for example, turns bright red when it

encounters something with which it has had no previous experience. It does so as if it is angry or fearful. Honey bees seem to differ from one hive to another in meanness. Some hives are docile, allowing beekeepers to reach into the hive without the bees stinging them at all. The next hive, however, can be quite mean, for no apparent reason stinging anyone who comes within one hundred feet of the hive. Corvids (crows, ravens, jays) apparently mate for life and with the loss of a mate show evidence of days of bereavement (silence for a normally noisy bird or singing a unique song or repeatedly calling the identifying sound of the mate[69]). Rodents (mice, rats, guinea pigs, hamsters, gerbils) are routinely used, with some success, for laboratory research regarding anxiety, hostility, and depression[70]. Spotted dolphins released from nets often leap repeatedly in the air, as if in joyful celebration[71].

One elephant, 'Sadie', repeatedly unsuccessful at performing what was asked of her, sank to her knees, rolled onto her side and cried with tears and body-wrenching sobs[72]. Elephants seem to mourn as another elephant in the herd is dying (e.g. acting frantically, trying to lift the dying elephant, trying to stuff food into the mouth of the dying elephant[73]). In 1996, when a young boy fell into the gorilla enclosure and was knocked unconscious, the female 'Binty' went over to him, picked up the boy, cradled him in her lap, then walked over and gently laid the boy in front of the caretaker's door[74]. Following a reconciliation of chimpanzees in the Arnheim chimp colony, the chimps engaged in hooting, banging on metal drums, kissing, and embracing[75].

Finally, pet owners will insist that their pets display a wide spectrum of emotions, including fear when threatened, anger when abused, contentedness when petted, sadness when left, joy when played with, and compassion on the hurting.

A number of organisms play, or at least they seem to play. This is most common in the young of predator species, and often involves roughly the same behavior that the animal will use as an adult to kill its food or defend its territory, as if it is practicing (e.g. attacking from a hiding spot, chasing objects they set in motion, mocking fights).

Other behaviors, are not so obviously useful and seem to be done purely for the enjoyment of the animal. Dogs, for example, are known for chasing their tails and cats for chasing balls of string. Ravens are known to play 'catch the stick', slide down snow slopes, hang upside down on branches, hook talons with one another and spiral towards

earth, and drop objects on humans[76]. Otters slide down snowy slopes[77] and macaques roll snowballs[78]. One red fox was seen to catch a shrew, carry it to a road, play with it, then carry it back to its burrow and release it[79]. Raccoons seem to play practical jokes on other organisms (*e.g.* squirt faucet water at other animals, smear egg and dirt on a cat) and each individual raccoon plays a different kind of joke.

Some organisms show some interest in symmetry, or even art. To impress the female, the male bowerbird assembles quite an array of objects into an elaborate shelter (bower). Each bower is unique, many times the bowerbird shows a particular preference for objects of a particular color (*e.g.* blue), and often the bower is quite beautiful[80]. Crows tend to line their nests with shiny objects, such as jewelry, lugnuts, and crystals. Several orang-utans have shown a preference for symmetry (*e.g.* 'Paul' while breaking up a termite nest carefully arranged the fragments side by side along the length of his hand and arm, 'Siswi' took bark off a pile and restacked it with the inner bark all facing upwards, 'Princess' broke a half dozen sticks of like diameter to similar lengths and arranged them next to each other and parallel[81])

Self-Awareness

In the highest forms of relationship, persons purposely give of themselves to interact with others. Persons are aware of their own existence, person, and uniqueness. They are self-aware. God illustrates His own self-awareness by creating various levels of self-awareness among organisms of His creation. Of all the characteristics of personhood, however, self-awareness is the most difficult of all to measure.

Yet, it is very possible that animals that express emotion, play, and humor are actually self-aware. Their behavior is at least complex enough to allow some degree of personal relationship. One type of self-awareness is identified by using mirrors. While an animal is anesthetized, marks are placed on the face of the animal (*e.g.* on one eyebrow and one ear). If the animal looks in the mirror and proceeds to touch the marks on its own face, but not the mirror, the animal seems to be identifying the animal in the mirror with himself. Chimpanzees[82], organ-utans, and cotton-top tamarins[83] demonstrate this kind of self-awareness. Chimps even smell their fingers after touching the marks[84]. Of 25 gorillas tested only one responded in this way, and that was 'Koko', the gorilla that learned sign language. In fact, when the trainer asked Koko what she saw in the mirror Koko signed "Me, Koko". Elephants, marmosets, rhesus monkeys, and capuchin monkeys have not demonstrated this kind of self-awareness[85].

Biological Relationship

Of all the characteristics of personhood, the most telling are personal relationships. The three persons of the Godhead are in the greatest and most intimate of all relationships. God illustrates those

perfect relationships by degrees of relationship among the organisms of His creation.

A number of animals mate for life and at least *seem* to develop close emotional relationships. One example is the Canada goose, which seems to have an abbreviated lifespan following the death of a long-time mate, as if it dies of a broken heart. Other examples of bereavement include the corvids and elephants mentioned above. Other evidences that animals form relationships are found among domesticated animals—even members of different species—that spend time together as if they were bonded in friendship.

Still other examples are found between humans and their pets. Bonds develop between equestrians and their horses that are very difficult to distinguish from close friendships. The same can be said of many relationships between man and dog ('man's best friend'). Dogs are known to refuse food when a long-term dog-owner leaves, even when the dog is not left alone. It is as if the dog cannot eat because of depression caused by loss. Note that these examples are not due to any kind of instinct, because they are developed over time. Short-term relationships do not result in such interactions among animals.

6.7 | The Origin of Personality

In the midst of his great suffering, Job got to a point that he began to question God's wisdom. To convince Job that he had no right to even begin questioning God's wisdom, God asked Job a series of questions that he could not answer (Job 38-41). Included were a number of questions about the origin of animal behavior, from foraging (Job 38:39-41; 39:5-8) to reproductive (Job 39:1-4, 13-17) behaviors. The questions are quite interesting. From whence does animal behavior come? Many of the behaviors are so automatic (*e.g.* instincts) that it is as if they were programmed. In our experience, however, programs do not just arise spontaneously—at least programs that work. Even the simplest programs require intelligent programmers. So, who programmed behaviors into organisms? The way it was asked of Job, who taught the animals those behaviors? The only physical beings we know about, having any sort of intelligence, have behaviors programmed into them, so they could not have done the programming.

If the physical world is all there is, there are no candidates for the programmers. Naturalism, therefore, has no explanation for the origin of animal behaviors. Perhaps the behavior of animals is part of that non-physical component of biblical life that God creates in each animal (Psa. 104:30). Perhaps this is part of what the Bible means by the *nephesh* (or 'soul') of animals. Whether directly inspired of God or given to animals as part of their soul, the huge range of animal behaviors is best explained as coming from an intelligent God—the God of Scripture.

Even more challenging, to the naturalistic worldview, is the origin of emotion, will, humor, and self-awareness—personhood itself. For those who believe that the physical is all that exists each of these things are the consequence of chemical reactions and nothing more. Even if this were the case, no adequate explanation has been forthcoming about how physical processes (like naturalistic evolution) can generate such things. Although these characteristics of personhood have physical manifestations, each one of them is ultimately non-physical. The naturalistic worldview denies the existence of such things and certainly lacks explanation for their origin. In contrast, a personal God Who desires to illustrate His personhood through His creation would provide an explanation for the origin of emotion, will, humor, and self-awareness.

Animal Behavior: Our Responsibility

Our Responsibility to God

Organisms demonstrate a spectrum of perfection of uniqueness. Clones can show even less uniqueness than is found in non-living forms. Sexual organisms have substantial differences in their DNA but some species otherwise appear identical in form. In other species, each individual has a unique form, but behaves the same as every other individual in the species. In other species, each individual is unique in form and behavior. This spectrum of perfection of uniqueness in those things we can see points to the existence of something we cannot see that is even more unique, distinct, or set apart—the holy God Who created biological uniqueness to point us to His holiness.

Biological activity demonstrates several spectra of perfection. There is, for example, a spectrum of perfection in degree of motion. Some things appear to be completely inactive, such as rocks and dead bodies. Then there are non-living things like crystals that change but do so merely because of internal structure.

Organisms, in contrast, are much more active. Even fungi and plants that stay in one location metabolize and grow. Other organisms move about, some short distances and others significant distances across our planet. There is also a spectrum of perfection in degree of programming. There is highly predictable, programmed behavior such as metabolism, growth, development, and simple motions towards energy, nutrients, or mates and away from toxins and danger. A variety of more complex activities, such as trapping or stealthily approaching prey, scaring or fighting predators, attracting mates, developing and raising young, resting and hibernating, seem to be instinctive. Even though they are complex, they are also involuntary and result in very consistent and repetitive behaviors. They seem, then, to be due to more complex programs, but programs nonetheless.

Cooperative and learned behaviors are even more complex. The fact that there seem to be definite limits on how much organisms can vary from a basic pattern of behavior in every case suggests that these behaviors are also due to some sort of internal programming. However, the programs have to be complicated enough to allow for flexibility. When more than one individual is involved, any number of unforeseen events can occur, so cooperative behavior requires reactions to a wide

spectrum of possible experiences. The variety of possible experiences in the world also requires flexibility in learned behaviors. Reasoned behavior seems to involve the fewest programming constraints and results in the least predictable behavior. As the degree of programming decreases, an organism's behavior becomes more unpredictable and the organism is freer to behave as it might choose.

Finally, there is also a spectrum of perfection of intelligence, with none involved at all where there is no activity, to simple levels of intelligence needed for simple programs, higher levels for more complex programs, higher levels for flexible programs, and even higher for reasoning. These spectra of perfection of motion, freedom, and intelligence in those things we can see, point to the existence of something we cannot see that is even more dynamic, free, and intelligent—the all-powerful, completely free, all knowing God of creation.

Further spectra of perfection are found in will, emotions, and self-awareness. Many organisms lack any evidence at all of free will, emotions, or self-awareness, but various other organisms show different degrees of choice, emotion, and levels of self-awareness. The spectrum of perfection of will in biological organisms leads the human mind to extrapolate to a being with perfectly free will—having complete freedom to make all decisions.

The spectrum of perfection of emotions in biological organisms leads the human mind to extrapolate to a being with perfect emotions (*i.e.* feeling full joy, perfect peace, fully righteous anger, *etc.*). The spectrum of perfection of self-awareness leads the human mind to infer a being with perfect self-awareness (the 'I Am', who deservedly is jealous when anything or anyone is placed above or before him).

God also created a spectrum of perfection of relationship. There are items in God's creation that do not interact at all. There are other items that interact so briefly as to make relationships impossible. Within biological systems there is a spectrum of tightness of interaction, but most of the parts interact in such a consistent manner as to suggest that it is merely programmed interaction. This is more like the interaction among machines in a factory than beings in true relationships. Then there are animals that bond with other animals and increase their bonding with time. Humans also bond with animals.

Finally, the most profound of relationships among biological organisms are formed among and between humans. Human relationships themselves fall on a spectrum from no relationship at all, to casual acquaintances, to friendships, to close friendships, to family bonds, to the oneness potential between man and wife. This spectrum of perfection of relationship was created to lead us to infer the existence of a perfect relationship—namely, the perfect relationship that exists among God the Father, God the Son, and God the Holy Spirit.

The characteristics for which spectra exist—uniqueness, motion, freedom, intelligence, will, emotion, self-awareness, and relationship—are characteristics of personhood. The human mind extrapolates to

the infinite manifestation of person, God Himself. The human mind extrapolates to fully free, holy, all-powerful, all-knowing, willful, emotive, self-existent infinite being. The spectra of perfections in the behavior of millions of different organisms of the world help us understand what it means when Scripture claims God is a personal God.

The God of creation is not just a force or the source of all power or truth. He is a personal being. As a personal being he has the ability to relate to another being, and the freedom to choose whether or not to enter into such a relationship. He made us capable of relating to Him and He placed the characteristics of personhood in the creation so that we could better understand and relate to Him. The God of creation is not worshipped by the other monotheistic religions (*e.g.* Islam, modern Judaism[a]), for the god of each of these religions is distant and transcendent, but not immanent and intimate. The God described in Scripture desires a close relationship with humans. He created everything originally so that He could dwell among men and commune with them, and that is how it will ultimately be (Rev. 21:3).

Even when sin severed the original relationship, God created ways to dwell among men. This included descending between the cherubs in the tabernacle and the temple in the midst of humans for over a thousand years. This included taking on the form of a man and living among humans (John 1:14), and it includes today living within those who believe. Along the way there have been those who were able to walk with God (*e.g.* Enoch: Gen. 5:22-24; Noah: Gen. 6:9) and commune with God (*e.g.* Abraham: Gen. 18:33; Moses: Ex. 31:18). Moses was even the friend of God (Ex. 33:11; II Chr. 20:7; Isa. 41:8; James 2:23). Through the work of Jesus Christ, believers are adopted as children of God (John 1:12-13; Gal. 4:4-7; Eph. 1:5; I Jn. 3:1). Let us allow the contemplation of the spectra of perfection of biological personhood to lift our hearts in worship of the One we can know as Abba (father) (Rom. 8:15).

One of the reasons God created biological personality was to give us a visible and finite illustration of the invisible and infinite personhood of God. Because He did, study of biological personality can improve our personal understanding of God and draw us closer to Him. To help us fulfill our priestly role, God not only created us to respond to

---

a   This is a comment about *modern* Judaism, not the worship God encouraged of the Jews in the Old Testament, for many Old Testament Jews had an intimate relationship with God (*e.g.* Abraham and David).

illustrations of His nature, He also gave us the Holy Spirit to lead us into truth. Because of this, contemplation of biological personality can stimulate us to worship God, glorify God, and bring others into our worship of God.

Our Responsibility to the Creation

Preserving Personhood. God created the biological world with characteristics of personhood so that we could learn something about Him. He placed us in charge of that creation and commanded us to guard and keep it. This means that at the end of our lives we should 'return' His creation showing at least as much illustration of personhood as it had when we received it.

We should, for example, preserve biological uniqueness. We may appreciate certain characteristics so much that we want every member of a particular species to have those characteristics, but if we did this, uniqueness and diversity would be reduced. To give ourselves more room, or to protect ourselves, or to experience a thrill, or to hold organisms captive we may wish to control or restrict the activity of organisms. But to do this is to reduce the motion and freedom evident in the creation.

We may wish to modify the behavior of organisms to have them do what we want them to do, but many times this is an attempt to replace animal behavior with what are essentially human behaviors and restrict the organism's freedom of choice. We may wish to eliminate certain behaviors or emotional responses that we find undesirable, but to do so is to reduce the diversity of behavior and emotion illustrated in the creation. Finally, extinction reduces the message of personhood in the creation. We should be working to maintain the evidence of personhood we see throughout the creation.

Humane Treatment of Animals. If the biological creation were lacking any amount of personhood or characteristics of personhood, we would still have a substantial responsibility to preserve it. Personhood characteristics in the biological creation heighten that responsibility. Since animals can not only feel pain but (to various extents) experience emotion, animals can suffer. When we do not care for the biological creation properly, animal suffering can result. Not only do we have a responsibility to God in caring for the creation, but we have a responsibility to the animals themselves.

First, we should not cause suffering in animals when it is for human pleasure or because of human evil (*e.g.* sports fishing that returns fish to the water, rather than consuming them; using research animals in the development of cosmetics or weapons). Second, we should use as few animals as possible in medical research by using efficient research methods and using computer models or tissue cultures rather than animals whenever possible[a]. Third, when animals *must* be kept in captivity, such as when used in medical research to cure human diseases or when used to feed humans, they should be treated well. For example, since animals raised in captivity are less traumatized by captivity, we should avoid bringing wild animals into captivity. As much as possible, and according to the degree of personhood characteristics that an animal might have, animals should be given adequate room, cleanliness, lighting, companionship, and activities that prevent boredom. Fourth, when it is necessary to hurt animals, such as in medical research, they should be anesthetized or given pain killers so that they do not suffer. Finally, when it is necessary to kill them, animals should be killed mercifully, so that the animal does not suffer. For example, bow hunting of large animals is not appropriate—or any sort of hunting with weapons that may lead to a prolonged death when a more effective weapon could be used.

---

[a] Each year, an estimated 41 to 100 million animals are used in scientific experiments. In the early part of the twentieth century, an estimated 120,000 wild rhesus monkeys were killed just in the development of the Salk vaccine for polio.

Enhancing Personhood. Not only should we preserve the personhood evidenced in the creation, we should enhance it. Good stewards invest what they are given and then present the true owner with more than what they were given. God not only placed physical illustrations of personhood in the creation, He hid even more illustrations for us to discover and reveal—all to the glory of the One Who put it there.

Breeding has not only generated new body forms, it has also generated new behaviors. In the case of dogs, for example, even though each individual dog will have a unique personality, there is also a type of personality found in most or every individual of a particular dog variety. Hundreds of dog varieties have been generated by breeding in the last several centuries and with them just as many new personalities. Breeding has increased the physical evidence of personality in the creation. Most probably there is much more still to be revealed.

Summary of Chapter

A. God, as a personal God has all the attributes of personhood:

   a. uniqueness (He is holy)

   b. activity (He is all-powerful)

   c. intelligence (He is all-knowing)

   d. will (He is fully free)

   e. emotion (e.g. He is love)

   f. self-awareness (He is the 'I am'—the self-existent One)

g. relational (He adopts believers as His children)

B. To illustrate His invisible personhood, God placed a spectrum of perfection of each of the characteristics of personhood in the biological world:

a. uniqueness: Living things are quite distinct from non-living things; organisms show a spectrum of perfection of distinction (from clones to visibly distinct individuals); every living thing is distinct from every other living thing (especially sexual organisms); and organisms show a spectrum of perfection of recognition of individuals (some even recognizing individuals from other species).

b. activity: Metabolism is continually changing the molecular structure of every organism (storing or releasing energy, building molecules or breaking down); every organism grows (by increasing cell size in one-celled organisms or cell number in multicellular organisms) and develops (some merely by growing, some by developing new structures, some through multiple life cycle stages); and organisms show a spectrum of perfection of biological activity in locomotion, in acquisition of resources, in defense, in reproduction, in resting behavior, and in caring for young.

c. intelligence: Animals show a spectrum of perfection of instinctive to learned behavior (including habituation, imprinting, conditioned responses, mental mapping, imitative learning), of memory, of counting, and of reason

d. will: Animals show a spectrum of perfection of choice and morality.

e. emotion: Animals show a spectrum of perfection of emotions, of play, and of aesthetics.

f. self-awareness: Animals show a spectrum of perfection of self-awareness.

g. relational: Animals show a spectrum of perfection of relationships.

C. The naturalistic worldview has difficulty explaining the origin of biological behavior, especially the origin of instinctive behaviors (because in our experience programs have programmers), intelligence (because non-intelligence does not seem capable of generating intelligence), and non-physical characteristics like emotion, will, humor, and self-awareness (because physical causes do not seem capable of generating non-physical things).

D. The spectrum of perfection of the characteristics of personhood helps us to conceive of something possessing an infinite measure of each of these characteristics—in other words a fully free, holy, all-powerful, all-knowing, emotive, self-existent, relational, infinite being (the God of Scripture!).

E. Our responsibility as priests of God to the creation is to recognize the Creator in the personhood characteristics of organisms, worship Him, glorify Him, and bring others into our worship of Him.

F. Our responsibility as kings over the creation is to preserve the characteristics of personhood in the biological creation by maximizing the uniqueness, activity, and freedom of organisms. We should also minimize animal suffering by causing them pain only when necessary, minimizing their pain if they must be hurt, and treating them well when we hold them captive. We should also seek ways to increase the illustrations of personhood in the biological creation by discovering more of the organismal designs God hid in His creation.

Test & Essay Questions

1. Define metabolism / instinct / learned behavior / habituation / imprinting / direct development / indirect development / mental map / imitative learning / reasoning.

2. Compare and contrast growth of an individual cell and an organism made of multiple cells / direct and indirect development / imprinting /

conditioned response / habitation / mental maps / imitative learning / reasoning.

3. List three / four / five characteristics of personhood.

4. [essay] What suggests that God is person?

5. In what way is each biological organism unique?

6. In what ways are organisms active?

7. What resources must organisms seek, and why do they seek them?

8. List two / three / four / five different ways that organisms actively seek out resources.

9. Why must organisms defend themselves?

10. List two / three / four different ways that organisms defend themselves.

11. List three / four different kinds of reproductive activity in organisms.

12. List two / three / four different types of rest behaviors in organisms.

13. List two / three / four different ways in which organisms determine orientation.

14. How does one distinguish instinctive from learned behaviors?

15. What evidence is there for organisms having will / emotion / relationship?

16. Describe the spectrum of perfection of uniqueness / freedom / intelligence / will / emotion / self-awareness / relationship and what does it indicate?

17. [essay] How does the Christian worldview provide a better explanation for animal behavior / personality than the naturalistic worldview?

18. In what way might animal training / cloning / captivity be wrong in the light of the personhood that God placed into the creation.

19. How might we enhance the evidence of personhood in the creation?

References

1. Caldwell, M. C., D. K. Caldwell, and P. Tyack, 1990, Review of the signature whistle hypothesis for the bottlenose dolphin, in Leatherwood, S., and R. R. Reeves, *The Bottlenose Dolphin*, Academic, New York, NY; Marler, P., and S. Peters, 1989, Species differences in auditory responsiveness in early vocal learning, pp. 243-73 in Dooling, R. J., and S. H. Hulse, editors, *The Comparative Psychology of Audition: Perceiving Complex Sounds*, Lawrence Erlbaum, Hillsdale.

2. Alexander, Shana, 2000, *The Astonishing Elephant*, Random House, New York, NY, 300 p. (p.57); Hauser, Marc D., 2000, *Wild Minds: What Animals Really Think*, Henry Holt, New York, NY, 315 p. (p.181).

3. Cousteau, Jacques-Yves, and Philippe Diolé, 1973, *Octopus and Squid: The Soft Intelligence*, Doubleday, Garden City, NY [transl. from the French by J. F. Bernard], 304 p. (pp.125-31); Heinrich, Bernd, 1999, *Mind of the Raven: Investigations and Adventures with Wolf-Birds*, Cliff Street Books, New York, NY, 380 p. (p. 273).

4. Lloyd, J. E., 1966, Studies on the flash communication system in Photinus fireflies, Miscellaneous Publications of the Museum of Zoology, University of Michigan 130:1-95.

5. Cousteau, Jacques-Yves, and Philippe Diolé, 1973, *Octopus and Squid: The Soft Intelligence*, Doubleday, Garden City, NY [transl. from the French by J. F. Bernard], 304 p. (pp.36, 66, 97, 147-165).

6. Lucas, J. R., 1989, The structure and function of antlion pits: Slope asymmetry and predator-prey interactions, *Animal Behavior* 38:318-30.

7. Lloyd, J. E., 1975, Aggressive mimicry in *Photuris* fireflies: Signal répertoires by *Femmes fatales*, *Science* 187:452-453.

8. Fellers, J., and G. Fellers, 1976, Tool use in a social insect and its implications for competitive interactions, *Science* 192:70-72.

9. Chisholm, A. H., 1972, Tool-using by birds: A commentary, *Bird Watcher* 4:156-9.

10. Isack, H. A., and H.-U. Reyer, 1989, Honeyguides and honey gatherers: Inter-specific communication in a symbiotic relationship, *Science* 243:1343-6.

11. Meinertzhagen, R. 1959, *Pirates and Predators: The Piratical and Predatory Habits of Birds*, Oliver & Boyd, Edinburgh, Scotland (p.47).

12. Griffin, Donald R., 1992, *Animal Minds*, University of Chicago, Chicago, IL, 310 p. (pp.63-66).

13. Heinrich, Bernd, 1999, *Mind of the Raven: Investigations and Adventures with Wolf-Birds*, Cliff Street Books, New York, NY, 380 p. (pp. 257-65).

14. Blest, A. D., 1957, The function of eyespot patterns in the Lepidoptera, *Behavior* 11:209-56.

References

15. Cousteau, Jacques-Yves, and Philippe Diolé, 1973, *Octopus and Squid: The Soft Intelligence*, Doubleday, Garden City, NY [transl. from the French by J. F. Bernard], 304 p. (pp.29-31, 41).

16. Duerden, J. E., 1905, on the habits and reactions of crabs bearing actinians in their claws, *Proceedings of the Royal Zoological Society of London* 2:494-511.

17. Boinske, S., 1988, Use of a club by a wild white-faced capuchin (*Cebus capucinus*) to attack a venomous snake (*Bothrops asper*), *American Journal of Primatology* 14:177-9.

18. Emlen, S. T., 1975, The stellar-orientation system of a migratory bird, *Scientific American* 233(2):102-111.

19. Wehner, R., B. Michel, and P. Antonsen, 1996, Visual navigation in insects: Coupling of egocentric and geocentric information, *Journal of Experimental Biology* 199:129-40; Keeton, W. T., 1974, The orientational and navigational basis of homing in birds, *Advances in the Study of Behavior* 5:47-132.

20. Waterman, T. H., 1981, Polarization sensitivity, pp. 281-469 in Autrum, H., editor, *Handbook of Sensory Physiology, Vol. VII/6B*, Springer-Verlag, Berlin, Germany.

21. Wiltschko, W. and R. Wiltschko, 1988, Magnetic orientation in birds, pp. 67-121 in Johnson, R. F., editor, *Current Ornithology, Volume 5*, Plenum, NY.

22. Pavlov, I. P., 1927, Conditioned Reflexes: An Investigation of the Physiological Activity of the Cerebral Cortex, Oxford University, New York, NY.

23. Gelperin, A., 1975, Rapid food-aversion learning by a terrestrial mollusk, *Science* 189:567-70.

24. Warren, J. M., 1965, Comparative psychology of learning, *Annual Review of Psychology* 16:95-118.

25. Best, J. B., 1965, Behavior of planaria in instrumental learning paradigms, *Animal Behavior 1965 Suppl.* 1:69-75.

26. Andrew Packard acc. to Cousteau, Jacques-Yves, and Philippe Diolé, 1973, *Octopus and Squid: The Soft Intelligence*, Doubleday, Garden City, NY [transl. from the French by J. F. Bernard], 304 p. (p.135).

27. Schneirla, T.C., 1946, Ant learning as a problem in comparative physiology, pp. 276-305 in Harriman, P. L., editor, *Twentieth Century Psychology*, Philosophical Library, New York, NY.

28. Flood, N. B., and J. B. Overmier, 1971, Effects of telencephalic lesions and olfactory lesions on appetitive learning in goldfish, *Physiology and Behavior* 6:35-40.

29. Warren, J. M., 1965, Comparative psychology of learning, *Annual Review of Psychology* 16:95-118.

30. Powell, R. W., 1973, Operant responding in the common crow (*Corvus brachyrhynchos*), *Bulletin of the Psychonomic Society* 1:401-3; Skinner, B. F., 1938, *The Behavior of Organisms: An Experimental Analysis*, Appleton Century Crofts, New York, NY.

31. Herman, L. M., 1980, *Cetacean Behavior: Mechanisms and Functions*, Wiley, New York, NY (p.406).

32. Moore, B. R., 1992, Avian movement imitation and a new form of mimicry: Tracing the evolution of a complex form of learning, *Behaviour* 122:231-63; Moore, B. R., 1996, The evolution of imitative learning, pp. 245-66 in Heyes, C. M., and B. G. Galef, Jr., editors, *Social Learning in Animals: The Roots of Culture*, Academic, San Diego, CA.

33. Tayler, C. K., and G. S. Saayman, 1973, Imitative behavior of Indian Ocean bottlenose dolphins (*Tursiops aduncus*) in captivity, *Behaviour* 44:286-98.

34. Russon, A. E., and B. M. F. Galdikas, 1993, Imitation in free-ranging rehabilitant orangutans (*Pongo pygmaeus*), *Journal of Comparative Psychology* 107:147-61; Russon, A. E., and B. M. F. Galdikas, 1995, Constraints on Great Apes' imitation: Model and action selectivity in rehabilitant orangutan (*Pongo pygmaeus*) imitation, *Journal of Comparative Psychology* 109:5-17

35. Pierrel R., and J. G. Sherman, 1963, Barnabus, the rat with college training, *The Brown Alumni Monthly*, Brown University, acc. to Boden, Margaret A., 1995, Creative creatures, pp. 45-65 in Roitblat, Herbert L., and Jean-Arcady Meyer, editors, *Comparative Approaches to Cognitive Science*, MIT, Cambridge, MA, 533 p.

36. Balda, R. P., 1980, Recovery of cached seeds by a captive *Nucifraga caryocatactes*, *Z. Teirpsychol.* 53:331-46.

37. Dehaene, S., 1997, *The Number Sense*, Oxford University, New York, NY; Boysen, S. T., and E. J. Capaldi, 1993, *The Development of Numerical Competence: Animal and Human Models*, Lawrence Erlbaum, Hillsdale.

38. Pepperberg, I. M., 1994, Numerical competence in an African gray parrot (*Psittacus erithacus*), *Journal of Comparative Psychology* 108:36-44.

39. Hauser, Marc D., 2000, *Wild Minds: What Animals Really Think*, Henry Holt, New York, NY, 315 p. (p.55).

40. Carey, Susan, 2009, *The Origin of Concepts*, Oxford University, New York, NY, 598 p. (pp.121-3).

41. Cousteau, Jacques-Yves, and Philippe Diolé, 1973, *Octopus and Squid: The Soft Intelligence*, Doubleday, Garden City, NY [transl. from the French by J. F. Bernard], 304 p. (pp.109-115).

42. Inouye, D. W., 1983, The Ecology of nectar robbing, in Bentley, B., and T. Elias, editors, *The Biology of Nectaries*, Columbia University, New York, NY.

43. Kamil, A. C., 1988, A synthetic approach to the study of animal intelligence, in Leger, D. W., editor, *Comparative Perspectives in Modern Psychology*, University of Nebraska, Lincoln, NB (pp.257-8).

44. Breitwisch, R. and B. E. Breitwisch, 1991, House sparrows open an automatic door, *Wilson Bulletin* 103:725.

45. Fisher, J., and R. A. Hinde, 1949, The opening of milk bottles by birds, *British Birds* 42:347-57; Hinde, R. A., and J. Fisher, 1951, Further observations on the opening of milk bottles by birds, *British Birds* 44:393-6.

46. Jones, T., and A. Kamil, 1973, Tool-making and tool-using in the northern bluejay, *Science* 180:1076-8.

47. Manchester, England "Guardian", 25 June 1995 acc. to Heinrich, Bernd, 1999, *Mind of the Raven: Investigations and Adventures with Wolf-Birds*, Cliff Street Books, New York, NY, 380 p. (pp.305-6).

48. Heinrich, Bernd, 1999, *Mind of the Raven: Investigations and Adventures with Wolf-Birds*, Cliff Street Books, New York, NY, 380 p. (pp.307-9).

49. Heinrich, Bernd, 1999, *Mind of the Raven: Investigations and Adventures with Wolf-Birds*, Cliff Street Books, New York, NY, 380 p. (pp.312-321).

50. Savage, Candace, 1995, *Bird Brains: The Intelligence of Crows, Ravens, Magpies, and Jays*, Sierra Club, San Francisco, CA, 134 p. (p.110).

51. Herman, L. M., 1980, *Cetacean Behavior: Mechanisms and Functions*, Wiley, New York, NY (p.406).

52. Pfungst, O., 1911, *Clever Hans*, Henry Holt, New York, NY.

53. Alexander, Shana, 2000, *The Astonishing Elephant*, Random House, New York, NY, 300 p. (p.57).

54. Griffin, Donald R., 1992, *Animal Minds*, University of Chicago, Chicago, IL, 310 p. (p.106).

55. Goodall *in* Kummer, Hans, and Jane Goodall, 2003, Conditions of innovative behavior in primates, pp. 223-235 *in* Reader, Simon M., and Kevin N. Laland, editors, *Animal Innovation*, Oxford University, New York, NY, 344 p.

56. Kawai, M., 1965, Newly-acquired pre-cultural behavior of the natural troop of Japanese monkeys on Koshima Islet, *Primates* 6:1-30.

57. Caldwell, R. L., 1986, The deceptive use of reputation by stomatopods, *in* Mitchell, R. W., and N. S. Thompson, editors, *Deception: Perspectives on Human and Nonhuman Deceit*, SUNY, Albany, NY.

58. Gochfeld, M., 1984, Antipredatory behavior: Aggressive and distraction displays of shorebirds, *in* Burger, J., and B. L. Olla, editors, *Shorebirds: Breeding Behavior and Population*, Plenum, New York, NY.

59. Heinrich, Bernd, 1999, *Mind of the Raven: Investigations and Adventures with Wolf-Birds*, Cliff Street Books, New York, NY, 380 p. (pp. 300-301).

60. Burghardt, G. M., 1991, The cognitive ethology of reptiles: A snake with two heads and hog-nosed snakes that play dead, *in* Ristau, C. A., editor, *Cognitive Ethology: The Minds of Other Animals*, Erlbaum, Hillsdale, NJ.

61. Moller, A. P., 1988, False alarm calls as a means of resource usurpation in the great tit *Parus major*, *Ethology* 79:25-30; Munn, C. A., 1986, The deceptive use of alarm calls by sentinel species in mixed-species flocks of neotropical birds, p.174 *in* Mitchell, R. W., and N. S. Thompson, editors, *Deception: Perspectives on Human and Nonhuman Deceit*, SUNY, Albany, NY.

62. Russon, A. E., 2002, Pretending in free-ranging rehabilitant orang-utans, pp. 229-40 *in* Mitchell, R. W., editor, *Pretending in Animals, Children, and Adult Humans*, Cambridge University, New York, NY.

63. Menzel, E. W., 1974, A group of chimpanzees in a 1-acre field: Leadership and communication, pp. 83-153 *in* Schrier, A. M., and F. Stollnitz, editors, *Behavior of Nonhuman Primates*, Academic, NY; Coussi-Korbel, S., 1994, Learning to outwit a competitor in mangabeys (*Cercocebus t. torquatus*), *Journal of Comparative Psychology* 108:164-171; Held, S., M. Mendl, C. Devereux, and R. W. Byrne, 2002, Foraging pigs alter their behavior in response to exploitation, *Animal Behaviour* 64:157-66.

64. Marler, P., A. Dufty, and R. Pickert, 1986, Vocal communication in the domestic chicken, II: Is a sender sensitive to the presence and nature of a receiver? *Animal Behaviour* 34:194-8.

65. Byrne, Richard W., 2003, Novelty in deceit, pp. 237-259 *in* Reader, Simon M., and Kevin N. Laland, editors, *Animal Innovation*, Oxford University, New York, NY, 344 p.

66. Keddy-Hector, A. C., R. M. Seyfarth, and M. J. Raleigh, 1989, Male parental care, female choice and the effect of an audience in vervet monkeys, *Animal Behavior* 38:262-271.

67. Byrne, Richard W., 2003, Novelty in deceit, pp. 237-259 *in* Reader, Simon M., and Kevin N. Laland, editors, *Animal Innovation*, Oxford University, New York, NY, 344 p.

68. Alexander, Shana, 2000, *The Astonishing Elephant*, Random House, New York, NY, 300 p. (pp.62-63).

69. Savage, Candace, 1995, *Bird Brains: The Intelligence of Crows, Ravens, Magpies, and Jays*, Sierra Club, San Francisco, CA, 134 p. (p.75).

70. Brain, Paul F., and Lynne Marrow, 1999, Rodent models of Human Neuroses and Psychoses, pp. 59-73 *in* Haug, Marc, and Richard E. Whalen, editors, *Animal Models of Human Emotion and Cognition*, American Psychological Association, Washington, DC, 341 p.

71. Griffin, Donald R., 1992, *Animal Minds*, University of Chicago, Chicago, IL, 310 p. (pp.214).

72. Sanderson, G. P., 1882, *Thirteen Years Among the Wild Beasts of India... With an Account of the Modes of Capturing and Taming Elephants*, W. H. Allen, London, England.

73. Moss, C., 1987, *Elephant Memories*, William Morrow, New York, NY, p.73.

74. Hauser, Marc D., 2000, *Wild Minds: What Animals Really Think*, Henry Holt, New York, NY, 315 p. (p.218).

75. de Waal, F. M. B., 2000, The first kiss, pp. 15-33, *in* Aureli, F., and F. M. B. de Waal, editors, *Natural Conflict Resolution*, University of California, Berkeley, CA, p.16.

76. Heinrich, Bernd, 1999, *Mind of the Raven: Investigations and Adventures with Wolf-Birds*, Cliff Street Books, New York, NY, 380 p. (pp. 280-94).

77. Bourliere, F., 1964, *The Natural History of Mammals*, 3rd edition, Knopf, NY.

78. Eaton 1976, G. G., 1976, The social order of Japanese macaques, *Scientific American* 234:96-106.

79. Henry, J. D., 1986, *Red Fox: The Catlike Canine*, Smithsonian Institution, Washington, DC (pp. 136-140).

References

80 Boden, Margaret A., 1995, Creative creatures, pp. 45-65 *in* Roitblat, Herbert L., and Jean-Arcady Meyer, editors, *Comparative Approaches to Cognitive Science*, MIT, Cambridge, MA, 533 p. (p.53).

81 Russon, Anne E., 2003, Innovation and creativity in forest-living rehabilitant orang-utans, pp. 279-306 *in* Reader, Simon M., and Kevin N. Laland, editors, *Animal Innovation*, Oxford University, New York, NY, 344 p.

82 Gallup, G. G., Jr., 1970, Chimpanzees: Self-recognition, *Science 167*:86-7.

83 Hauser, M. D., J. Kralik, C. Botto, M. Garrett, and J. Oser, 1995, Self-recognition in primates, *Quarterly Review of Biology 63*:377-412.

84 Povinelli, D. J., A. B. Rulf, K. R. Landau, and D. T. Bierschwale, 1993, Self-recognition in chimpanzees (*Pan troglodytes*): Distribution, ontogeny, and patterns of emergence, *Journal of Comparative Psychology 107*:347-72.

85 Povinelli, D. J., 1987, Monkeys, apes, mirrors, and minds: The evolution of self-awareness in primates, *Human Evolution 2*:493-507; Povinelli, D. J., 1989, Failure to find self-recognition in Asian elephants (*Elephas maximus*) in contrast to their use of mirror cues to discover hidden food, *Journal of Comparative Psychology 103*:122-32.

CHAPTER 7

PROVIDER GOD

Bioprovision and the Anthropic Principle

"O Lord my God, You are very great! ...set the earth on its foundations, ...springs in the valleys ...give drink to every beast of the field,... cause the grass to grow for the livestock and plants for man to cultivate,... made the moon to mark the seasons,... O Lord how manifold are your works! In wisdom you have made them all..."

Psa. 104:1-24, ESV

7.1 | God is Love

One of God's attributes is love. David wrote "I will praise Your name for Your lovingkindness..." (Psa. 138:2). His 'everlasting love' (Jer. 31:3) is referred to as 'lovingkindness' in the Old Testament[a] and as 'love' in the New Testament[b]. In fact, God *is* love (II Cor. 13:11; I John 4:7, 16, 19). He loves His people, both the people of Israel ("...the Lord loved Israel for ever...": I Ki. 10:9[c]) and the people of His church ("Behold, what manner of love the Father has bestowed upon us...": I John 3:1[d]), in fact *nothing* can separate anyone from the love He has for them (Rom. 8:35-39).

Love can be demonstrated for another by providing for that person's needs. This can be done by meeting the needs as they arise, such as God providing forgiveness for individual sins as we ask for it (I John 1:9). Or, needs can be anticipated and provided for ahead of time, such as God showing His love for us by sending His Son to die for our sins nearly two thousand years ago (Rom. 5:8).

To assure us that God takes care of our physical needs, Jesus pointed out that God provides for the physical needs of plants and animals (Matt. 6:25-33). As a physical illustration of His love, an attribute we cannot see, God provides for His physical creation. This chapter will focus on the ways in which God designed the *non*-biological world thousands of years ago to provide for the needs of organisms for the rest of earth history.

Provision and the Anthropic Principle

In the course of the last century, scientists have discovered a rather long list of characteristics of the universe that indicate it is a very special place, and that we occupy a special place in it. The list of these characteristics is so long and diverse that it draws most people—even unbelievers—towards a concept called the cosmological anthropic principle, or **anthropic principle** (which we will abbreviate in this text

> **anthropic principle**—the universe has characteristics that suggest it was fashioned with man in mind

---

a For example: Psa. 17:7; 25:6; 26:3; 36:7, 10; 40:10-11; 42:8; 48:9; 51:1; 63:3; 69:16; 92:2; 103:4; 107:43; 119:88, 149, 159; 138:2; 143:8; Isa. 63:7; Jer. 9:24; 32:18; and Hos. 2:19.

b For example: II Thess. 3:5; I John 2:5; and Jude 21.

c Other references indicating God's love for the people of Israel include Deu. 7:7-8, 13; 10:15; 23:5; 33:3; II Chr. 2:11; 9:8; Psa. 47:4; Isa. 43:4; 48:14; Jer. 31:3; Hos. 3:1; 11:1; Mal. 1:2; and Rom. 9:13.

d References indicating God's love for the church include Psa. 108:6; 146:8; Prov. 3:12; 15:9; Rom. 1:7; 5:5; 8:35-39; Gal. 2:20; Eph. 2:4; 3:19; 5:2, 25; II Thess. 2:13, 16; Titus 3:4; Heb. 12:6-7; I John 3:1, 16; 4:10-12, 16; and Rev. 1:5.

as AP). According to the anthropic principle, the universe appears as if it was fashioned with man in mind. Let us now consider some of the characteristics of the universe that suggest—even to unbelievers—that the universe was designed with humans in mind.

The list of anthropic principle (AP) characteristics is so long that we will not consider all of them here. In fact, we will consider AP characteristics of two general types—those *necessary* for biological life, and those handy for understanding the universe[a]. In the case of the AP characteristics necessary for life, we will group them into ten categories and examine at least one AP characteristic in each category. These are designs that God planned and placed into the universe to make it livable by organisms (Isa. 40:22; 45:18).

We will review these ten categories before looking at the AP characteristics that are *not* necessary for life, but handy for humans to understand the universe. This category is reserved for last because in some ways it is even more amazing than the first ten. Of the ten categories of necessary AP characteristics, seven apply to very large objects in the universe, or involve the structure of the entire universe. These seven categories will be dealt with first. The three remaining

---

[a] There is another type of AP characteristic that must be true for naturalistic evolution to occur. Since the creationist does not believe that the universe was generated by naturalistic evolution, the creationist does not think that these are AP characteristics at all.

necessary AP characteristics apply to objects too small to be seen by humans. These three categories will be dealt with after the first seven and before the AP characteristics *not* necessary for life.

Necessary Large-Scale AP Characteristics

We begin our brief survey of anthropic principle (AP) characteristics with the seven categories of AP characteristics that are necessary for biological life and which apply to very large objects. Of those, we begin with those characteristics most familiar to you and move to categories dealing with successively larger and larger objects.

Earth Structure

God designed the earth in a very special way to allow organisms to flourish. The conditions found in most of the universe would not permit life to exist at all. First of all, the earth has just the right atmosphere to absorb heat from the sun and keep the surface of the earth warm in the midst of very cold space at hundreds of degrees below zero (somewhat like the glass of a greenhouse keeps the plants inside warm, even in the midst of the winter). This is because the earth is just the right size to have enough gravity to hold onto what are called **greenhouse gases** such as water vapor (H_2O), carbon dioxide (CO_2), methane (CH_4), and a form of oxygen known as ozone (O_3). These are gases that absorb sunlight and re-emit the energy from that sunlight

greenhouse gas—atmospheric gas that absorbs sunlight, thus heating the earth's surface

as infrared radiation. Not only is the infrared radiation trapped inside the earth's atmosphere, but it heats up objects on the earth's surface, thereby heating the earth's surface by scores of degrees.

At the same time, the earth is not so big that its gravity holds onto lighter gases[1], such as the poisonous gases of helium (He) and hydrogen (H_2) which make up most of the atmospheres of Jupiter, Saturn, Uranus and Neptune. Besides having the right gravity to hold greenhouse gases, the particular mixture of greenhouse gases is just what is needed for biological life[2]. There is enough carbon dioxide (CO_2) for plants, but not so much as to overheat the earth (such as the CO_2-caused 400°C surface temperature of Venus). There is enough oxygen (O_2) for animals, but not so much that forest and grass fires burn uncontrollably. Finally, there is enough non-reactive nitrogen (N_2) to provide the air pressure necessary for organisms to take in the gases they need. Other special designs of the earth (such as the tilt of the earth's axis[3] and the distribution of oceans and continents[4]) make sure that the heat that is absorbed by the atmosphere is spread around evenly enough to make life possible across most of the earth.

water cycle—continuous natural process supplying organisms with water, by evaporation from the oceans, precipitation to the earth's surface, and flow back to the oceans

God also designed a **water cycle** on the earth so that water is continually delivered to all the organisms of the planet that need it. Energy from sunlight evaporates water from the earth's surface—especially the ocean. Water is then returned to the earth's surface by precipitation. Water that falls on the continents then flows downhill towards the oceans in ground water and streams. "All the rivers run

into the sea; yet the sea is not full; unto the place from whence the rivers come, thither they return again" (Ecclesiastes 1:7). The shape of the earth's continents, along with the earth's gravity and temperature, are designed to continuously provide water to all the organisms of the planet.

The earth is also specially designed to block harmful radiation from the sun. Ozone in the atmosphere absorbs harmful ultraviolet radiation[5]. The earth's magnetic field takes harmful charged particles from the sun and deflects them to the poles[6]. And, so that the earth can have a magnetic field, the earth has a core made of iron and nickel—rocks very unlike the rocks making up the surface of the earth.

7.2 | Solar System Structure

The earth is part of a larger structure of the universe known as the Solar System. The Solar System includes the sun and everything revolving about it—namely the earth, the rest of the planets, asteroids, comets, *etc*. God placed several special designs into the Solar System to permit biological life on earth.

First of all, life can only thrive in a somewhat narrow range of temperatures—*much* warmer than the average temperature of the universe (hundreds of degrees below zero) and *much* cooler than the average temperature of the heat producers of the universe (the coolest stars have temperatures in the thousands of degrees). This means that life can only thrive in what is called the 'goldilocks zone'—a particular

range of distances around a star where the temperature is neither too hot nor too cold for life. The earth is located in the 'goldilocks zone' around our sun[7]—Venus being closer to the sun and too hot, and Mars being farther from the sun and too cold. The earth is held more consistently within that zone by its huge moon (compared to the size of the earth[a])[8].

Another special design of the Solar System is the fact that our sun is a 'calm' star[9]. Many stars emit their radiation rather violently. Other stars have radiation that varies on a regular cycle—some over the course of minutes, others over the course of hours, days, or years. Life would not be possible around stars whose radiation varies too much. A majority of the stars in the universe are too variable to support life. Another important feature of our Sun is its color. Of all the different colors that stars are in the universe, our Sun produces the best kind of radiation for plants to pick up energy through photosynthesis[10].

Another important feature of our Solar System is that there is only one sun[11]. If another star orbited our sun—such as is the case with most of the stars of the universe—the two stars would 'compete' with each other for the planets, regularly or permanently drawing the earth out of its 'goldilocks zone'. Although some science fiction accounts would like you to believe otherwise, multiple-sun systems are not friendly to life.

Galactic Structure

At a larger scale, the earth and its solar system is included in the Milky Way Galaxy—a huge spiral arrangement of at least one hundred billion stars. The galaxy also has AP characteristics that allow organisms to live on earth. This includes having enough space between stars to allow the earth to orbit the sun without interference and thus stay in its 'goldilocks zone'. It also includes there being no stars in the vicinity that are ready to explode.

Basic Universe Structure

The remaining four categories of necessary, large-scale, AP characteristics are characteristics of the entire universe. For example, the universe needs to 'behave' consistently enough for organisms to persist through time. There must be consistent patterns or regularities

---

a The earth's moon is by no means the largest moon in the solar system, but *compared to the size of the planet around which it is traveling*, our moon is the largest.

in the universe[12]. Our universe has such regularities and each regularity, as it is discovered, is described by means of a **natural law**.

Without some sort of natural laws, life in the universe could not exist. Another necessary AP characteristic is the three-dimensionality of the universe[13]. If the universe were only one-dimensional (a line) or two-dimensional (a plane), objects in the universe could not be complex enough to build organisms. If space had more than three dimensions, a planet would be unable to maintain a stable orbit around a sun[a]. In such a universe, a planet could not remain in the 'goldilocks zone', so life would not survive very long. A similar problem would occur if we could move both forward and backward in time the way we can move forward and backward in all three dimensions of space[14].

natural law—a regularity of the universe

Natural Laws

Not only does the universe have to have some sort of natural laws for life to persist, it seems that there are some *particular* natural laws that have to be in place. One such law is the **Second Law of**

second law of thermodynamics— the tendency of everything in the universe to spontaneously move towards areas where that something is in lower concentration

---

a Thinking multi-dimensionally is challenging. A substantial portion of the human population actually cannot even visualize in three dimensions! Very, very few indeed can—or even want to—think about four or more dimensions. But of those who are able, some actually enjoy determining what the universe would be like with four, or five, or more dimensions. Those people assure the rest of us that stable orbits are not possible. Also, to be technically accurate here, these stable orbits are possible as long as space has no more than three large and more or less equal dimensions. More dimensions are possible as long as there are only three major dimensions and all other dimensions are miniscule compared to them. In fact, it is common for physicists to postulate that our *present* universe actually has higher dimensions, but that these dimensions are infinitesimally small compared to the three dimensions we are used to.

Thermodynamics[15]. The Second Law seems to drive just about every event in the physical universe, determining the direction of change for most of those events. Explained one way, it is the tendency of natural processes to lose energy and not get it back.

The tendency was originally observed in thermodynamics, a field of physics dealing with the motion of heat. It has long been known that heat energy tends to move from warmer places to cooler places—from areas where heat energy is found at higher concentrations to areas where heat energy is found at lower concentrations. When an ice cube (containing a low concentration of heat energy) is placed in warm water (with a high concentration of heat energy), heat energy tends to move from the water—where it is in high concentration—to the ice—where it is in low concentration. Hot water added to one end of a bathtub of lukewarm water tends to cool as the heat energy at greater concentration in the hot water moves into the cooler water, where the heat energy is in low concentration.

It turns out that the Second Law applies to a lot more than just the movement of heat. This tendency of things to move from areas where it is in high concentration to areas where it is in lower concentration seems to work for just about everything in the physical universe, from energy to sub-atomic particles and beyond. The Second Law causes many things to change and move and determines the direction of those movements and changes. The Second Law, for example, is absolutely

essential for organisms that need to get oxygen from the atmosphere and nutrients from food, and to get rid of wastes[a]. Water moves into an organism by **osmosis**, a process driven by the Second Law (where water in areas with a high concentration of water move across membranes to areas with a low concentration of water). Molecules move from one place to another by **diffusion**, yet another process driven by the Second Law (where molecules in areas with a high concentration of that molecule move to areas where the concentration of that molecule is low).

> **osmosis**—the spontaneous motion of water across a semi-permeable membrane towards areas of lower concentration of water
>
> **diffusion**—spontaneous motion of a molecule towards areas of lower concentration of that molecule

Two other laws of the universe that seem to be necessary for biological life describe how particles of the universe are attracted to each other or repelled from one another. The law of gravity, for example, describes how particles are attracted to one another based on how much matter they contain. The law of gravity describes how organisms stay affixed to the surface of the earth and how the moon stays in orbit about the earth and how the earth stays within the 'goldilocks zone' of the sun. Another law, Coulomb's law, describes how particles with opposite electrical charges are attracted to one another—and how particles with the same electrical charges are repelled from one another. Coulomb's law describes how the earth stays together, how the bodies of organisms stay together, and how molecules of organisms stay together. Both the law of gravity and Coulomb's law are necessary for organisms to exist.

Mathematical Form of Laws

Not only must certain natural laws exist, but it seems that life can only exist when those particular laws follow special mathematical relationships. The force of gravity between two objects ($F_{gravity}$), for example, equals a particular number (G, called the gravitational force constant) times the mass of the first object (M_1) times the mass of the second object (M_2), all divided by the square of the distance between the

---

[a] Since oxygen is in a higher concentration in breathed-in air than the oxygen in the blood, oxygen breathes into the lungs tends to move into the blood, oxygenating the blood in the lungs. When that oxygen-rich blood gets to cells that have used oxygen to generate energy, it tends to move from the blood into the cells where the oxygen concentration is low, thus supplying cells with oxygen. At the same time, carbon dioxide, which is a waste product left over after the cell generates its energy, builds up to high concentrations in cells. By the Second Law it will tend to pass out of the cell into the blood where carbon dioxide levels are low. By the time the blood gets to the lungs, it contains a high concentration of carbon dioxide. In the lungs, then, the carbon dioxide passes from the blood into the air of the lungs, which has a lower carbon dioxide level, which, in turn, is exhaled into the atmosphere.

two objects (r^2). In other words, the law of gravity follows a particular mathematical equation, represented by

$$F_{gravity} = \frac{GM_1M_2}{r^2}$$

Because of the r^2 in the denominator of this equation, gravity decreases as distance increases—in fact it decreases with the *square* of distance. For this reason, gravity is an *inverse-square* relationship. It turns out that this inverse-square relationship allows planets to follow stable orbits around the sun, something necessary for biological life. If the denominator of the equation was r or r^3, rather than r^2, those stable orbits would not be possible[16]. Coulomb's law, concerning the attraction of electrical charges, follows a similar inverse-square relationship

$$F_{Coulomb} = \frac{\kappa Q_1 Q_2}{r^2}$$

where κ is Coulomb's constant and Q is the electrical charge. This law permits stable electron orbits around atoms. Again, such orbits would not be stable if the denominator of the equation had r or r^3 rather than r^2. Because Coulomb's law is necessary for biological life, the particular mathematical form of both the law of gravity and Coulomb's law has to be the way it is for life to exist in this universe.

Physical Constants

As if the particular laws and the particular mathematical form of each of those laws was not sufficient evidence of special design, even the constants of those laws seem to be specially chosen. In the law of gravity, for example, there is a very specific gravitational force constant (G) that determines the strength of gravity across the universe[17]. Similarly, Coulomb's law has a very specific constant—Coulomb's constant (κ)—that determines the strength of attraction between charged particles.

In the case of Coulomb's constant, if it was merely 1% *stronger*, atoms would hold onto their electrons so strongly that they would never share electrons. And, if atoms did not share electrons, molecules would not be possible, and without molecules, organisms would not be possible[18]. At the same time, if the electromagnetic constant was just 1% *weaker*, atoms could not hold onto their electrons, so there would be no electrons to share. Once again, molecules and organisms would not be possible. Similarly, each natural law necessary for life has both a particular mathematic form and a particular narrow range of constants that make life possible in this universe.

7.3 | Necessary Small-Scale AP Characteristics

In the previous section we touched upon the AP characteristics of the large-scale design of the universe. Let us now consider some of the small-scale features of the universe that make life possible.

Subatomic Particles

Even the smallest particles of the universe have to be the way they are to permit life to exist. These particles are called 'subatomic' because they are smaller than an atom. There are scores of different kinds of subatomic particles, each with a unique combination of characteristics such as mass and charge. Of these particles, the most well-known are **neutrons**, **protons**, and **electrons**. Neutrons and protons are among the most massive of all sub-atomic particles, partly because they are made up of even smaller particles. A neutron has no electric charge (so named because having no charge is also known as having a neutral charge). A proton has about the same mass as the neutron but has a positive electric charge. We define the charge of a proton as +1. An electron has a negative charge equal in strength to the positive charge of a proton, so the charge of an electron is -1. However, electrons are *much* less massive than protons. It takes about 1836 electrons to equal the mass of one proton!

neutron—large, chargeless subatomic particle usually found in the nucleus of an atom

proton—large positively-charged subatomic particle usually found in the nucleus of an atom

electron—small negatively-charged subatomic particle often found orbiting the nucleus of an atom

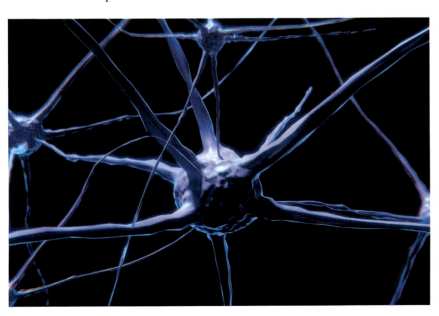

Since they have opposite charges of the same strength, protons and electrons are drawn towards each other according to Coulomb's law. This allows protons and electrons to construct larger objects, which in turn makes biological life possible[19]. At the same time, the ratio of masses between a proton and an electron (1836 to 1) results in electrons orbiting rapidly (near the speed of light) around protons, rather like the planets of our Solar System orbit around the much larger sun. This arrangement ultimately allows the existence of atoms and molecules that are necessary for biological life[20]. Special nuclear forces hold protons and neutrons together to form a **nucleus** (pl. *nuclei*). Composed of only neutrally-charged neutrons and positively-charged protons, all nuclei are positively charged.

Because the proton and electron have opposite charges of the same strength, an equal number of electrons and protons creates an electrically neutral object—an atom. If the number of electrons is *not* equal to the number of protons, the result is an electrically-charged ion. If there are fewer electrons than protons the ion is positively-charged. If there are more electrons than protons the ion is negatively-charged. All biological life is made up of, and dependent upon, the existence of atoms, positive ions, and negative ions. Consequently, even the characteristics of protons and electrons are essential for biological life.

Elements

A very large number of different atoms and ions are possible, and each type has a unique set of properties. Many of the properties are determined by the 'face' that the atom or ion offers to the outside world. Because neutrons are neutral, neutrons determine very few of these properties[a]. Most of the properties seem to be determined by the number of protons in the nucleus (what is called the **atomic number**). The charges of the protons determine the kind of electron 'surface' that is facing the outside world.

As a consequence, all atoms and ions having the same number of protons have similar enough properties to be considered the same **element**[b]. Although a very large number of different combinations of

---

nucleus—double-membrane-surrounded organelle that houses the cell's DNA

atomic number—number of protons in an atom's nucleus (determining the properties of that atom)

element—atoms and ions having similar properties because they have the same number of protons

ion—an electrically charged arrangement of proton(s) with a different number of orbiting electron(s)

---

a   Neutrons do affect mass (because they are such large particles) and stability (for if the number of neutrons differs too much from the number of protons the nucleus is unstable, or radioactive).

b   **Ions** result when the number of electrons and protons is different. Different isotopes result when there are different numbers of neutrons. Because the element is determined by the number of protons, multiple isotopes and ions are possible for each element.

neutrons, protons, and electrons are possible, the number of naturally occurring elements is less than one hundred[a].

In 1869, the chemist Dmitry Mendeleyev discovered that elements could be arranged by chemical properties into a table where elements of similar properties are arranged in each column and elements increase in mass from left to right and top to bottom. So, if one begins at any element in the table and considers elements of progressively greater mass (moving across a row from left to right, then stepping down a row and moving across the next row from left to right), one will encounter an element lower in the same column having similar properties as the beginning element. In this way, as one considers elements of greater and greater mass, properties repeat regularly (or periodically). The modern form of Mendeleyev's table arranges elements according to their atomic number[b]. And since the properties of the elements repeat periodically with atomic number, it is known as the **periodic table**.

periodic table—tabular arrangement of elements by atomic number, which places elements with similar characteristics in the same column

Periodic Table

Group I	Group II											Group III	Group IV	Group V	Group VI	Group VII	Group VIII
1 H Hydrogen																	2 He Helium
3 Li Lithium	4 Be Beryllium											5 B Boron	6 C Carbon	7 N Nitrogen	8 O Oxygen	9 F Fluorine	10 Ne Neon
11 Na Sodium	12 Mg Magnesium											13 Al Aluminum	14 Si Silicon	15 P Phosphorus	16 S Sulfur	17 Cl Chlorine	18 Ar Argon
19 K Potassium	20 Ca Calcium	21 Sc Scandium	22 Ti Titanium	23 V Vanadium	24 Cr Chromium	25 Mn Manganese	26 Fe Iron	27 Co Cobalt	28 Ni Nickel	29 Cu Copper	30 Zn Zinc	31 Ga Gallium	32 Ge Germanium	33 As Arsenic	34 Se Selenium	35 Br Bromine	36 Kr Krypton
37 Rb Rubidium	38 Sr Strontium	39 Y Yttrium	40 Zr Zirconium	41 Nb Niobium	42 Mo Molybdenum	43 Tc Technetium	44 Ru Ruthenium	45 Rh Rhodium	46 Pd Palladium	47 Ag Silver	48 Cd Cadmium	49 In Indium	50 Sn Tin	51 Sb Antimony	52 Te Tellurium	53 I Iodine	54 Xe Xenon
55 Cs Cesium	56 Ba Barium	57 La Lanthanum	72 Hf Hafnium	73 Ta Tantalum	74 W Tungsten	75 Re Rhenium	76 Os Osmium	77 Ir Iridium	78 Pt Platinum	79 Au Gold	80 Hg Mercury	81 Tl Thallium	82 Pb Lead	83 Bi Bismuth	84 Po Polonium	85 At Astatine	86 Rn Radon
87 Fr Francium	88 Ra Radium	89 Ac Actinium	104 Rf Rutherfordium	105 Db Dubnium	106 Sg Seaborgium	107 Bh Bohrium	108 Hs Hassium	109 Mt Meitnerium	110 Ds Darmstadtium	111 Rg Roentgenium	112 Uub Ununbium	113 Uut Ununtrium	114 Uuq Ununquadium	115 Uup Ununpentium	116 Uuh Ununhexium	117 Uus Ununseptium	118 Uuo Ununoctium

		58 Ce Cerium	59 Pr Praseodymium	60 Nd Neodymium	61 Pm Promethium	62 Sm Samarium	63 Eu Europium	64 Gd Gadolinium	65 Tb Terbium	66 Dy Dysprosium	67 Ho Holmium	68 Er Erbium	69 Tm Thulium	70 Yb Ytterbium	71 Lu Lutetium
90 Th Thorium	91 Pa Protactinium	92 U Uranium	93 Np Neptunium	94 Pu Plutonium	95 Am Americium	96 Cm Curium	97 Bk Berkelium	98 Cf Californium	99 Es Einsteinium	100 Fm Fermium	101 Md Mendelevium	102 No Nobelium	103 Lr Lawrencium		

---

a More than 100 elements are possible, but as far as we know, only about 90 or so are found naturally in the universe.

b Since electrons are so small, the mass of an atom or ion is almost entirely determined by the number of protons and neutrons. Since the number of protons and neutrons tends to be roughly the same, the mass of an element tends to increase with atomic number. Thus, when Mendeleyev arranged the elements by mass, he was arranging them by atomic number, which is currently the way in which we arrange Mendeleyev's table today.

The repeating pattern of chemical properties in the periodic table seems to be due to the kind of orbits that electrons follow as they revolve about the protons in the nucleus. Those orbits are determined by rules of the universe that seem to operate only for objects of the size of atoms and ions. The rules about what orbits particular electrons are to follow are somewhat analogous to the way in which people might be seated in a special kind of theater.

Imagine a theater where there were different seating sections, the first section being closest to the stage, the second being a bit farther out, and so on. Imagine further, that people are seated in the order they enter the theater, and that, at least for the first few sections, the sections are filled one section at a time beginning with the first. Finally, imagine that the theater prefers each of its occupied seating sections to be full, encouraging more people to enter the theater if a seating section is nearly full and discouraging people from entering or staying in the theater if they would be alone or nearly alone in a seating section. The rules for electron 'seating' around a nucleus are similar. Each 'seating section' around a nucleus is called an **electron shell**. The first electron shell has room for 2 electrons, and the second, third, and fourth shells each have room for eight electrons. The rules become more complicated for further electron shells, but that is more detail than we need to consider here.

electron shell—a region of space around an atomic nucleus where electrons might orbit the nucleus

The element helium has two protons, so the helium atom has two protons and two electrons. Since a helium atom has exactly the right number of electrons (two) to fill the first shell, it does not readily give up electrons to other atoms and it does not readily accept electrons from other atoms. Consequently, helium atoms tend not to enter into chemical reactions with other atoms. They are thus non-reactive. They exist as isolated atoms, so helium is a gas. Since helium does not react with other elements, it is called a noble gas[a].

Generalizing from the specific example of helium, *any* element with only full electron shells will be a noble gas like helium. Neon, for example, is a noble gas because it has ten protons, giving its atom exactly the right number of electrons (ten) to fill the first electron shell (with 2 electrons) and the second electron shell (with 8 electrons). The third noble gas is argon, for its eighteen protons result in exactly the

---

a This is a reference to the fact that in European society, it was not appropriate that nobility associate with non-nobility.

right number of electrons in the argon atom (eighteen) to fill the first (two), second (eight), and third (eight) electron shells. Noble gases are found in the far-right column of the periodic table and are considered elements in Group VIII.

Elements in Group I (the far-left column) of the periodic table have one electron in their outermost electron shell. The first three of these elements are hydrogen with one proton (so the hydrogen atom has one electron), lithium with three protons (so the lithium atom has two electrons in the first shell and a single electron in the second shell), and sodium with eleven protons (so the sodium atom has two electrons in the first shell, eight in the second, and one in third). Atoms in Group I can readily donate an electron to another atom, either by completely giving up the electron and becoming an ion with a +1 charge, or by sharing the electron with another atom.

Elements in Group VII (the next to the last column on the right) on the periodic table have one *less* electron than is necessary to fill their shells. These are metals like fluorine with nine protons (so the fluorine atom has 2 electrons in the first shell and 7 in the second shell) and chlorine with seventeen protons (so the chlorine atom has 2 electrons in the first shell, 8 in the second, and 7 in the third). Group VII atoms can readily accept an electron from another atom, either by completely accepting the electron and becoming an ion with a -1 charge, or by sharing an electron donated by another atom.

In a similar manner, elements in Group II (the second column) can readily donate two electrons, elements in Group VI can readily accept two electrons, elements in Group III can readily donate three electrons, and elements in Group V can readily accept three electrons. The number of electrons a particular atom can accept or donate determines the chemical properties of that element.

Of the scores of elements in the universe, a number of them have properties essential for biological life. Carbon, for example, has 6 protons, four electrons less *and* 4 electrons more than is needed to fill its shells. Each carbon atom can accept or donate four electrons—usually by sharing—and thus can react with up to four other atoms to produce quite a variety of molecules. And, because of the shape of carbon's outermost electron shell, it can connect in up to four different directions. As a result, a carbon atom can be connected with other carbon atoms to produce almost any shape. Carbons can be connected in chains, branched chains, rings, sheets, and three-dimensional networks. Finally, because it is the smallest of the Group IV elements, carbon forms fairly strongly bonds with other atoms. This makes the molecules made from carbon rather stable. For these reasons carbon forms the 'backbone' of virtually every molecule of which organisms are made. This makes carbon essential for organisms[21]. But carbon is not the only element essential for biological life. Other essential elements include hydrogen[22], oxygen[23], nitrogen[24], phosphorus[25], and sulfur[26].

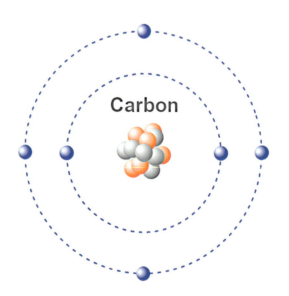

7.4 | Compounds

When two or more different elements are combined to produce a substance with its own unique properties, that substance is called a **compound**. The atoms or ions that make up these compounds can be joined either by electrical charges, or they can share electrons. The strongest bond is formed when atoms share electrons. Such a bond is called a **covalent bond** and any combination of atoms (even two atoms of the same element) joined by a covalent bond is called a molecule.

The attraction between ions with opposite electrical charges is an **ionic bond**. Ionic bonds are weaker than covalent bonds. The compound that results from atoms joined by ionic bonds is called an **ionic compound**. An example of an ionic compound is table salt, NaCl, formed between ions of sodium (Na^+) and ions of chlorine (Cl^-). The number of electrons an atom can accept or donate determines what atoms combine and in what ratios. Each atom in Group I, for example, readily combines with one atom in Group VII. Each atom in Group II can combine with two atoms in Group VII or one from Group VI, and so on. A whole host of different molecules can be formed by means of covalent bonds. A number of these different molecules are necessary for biological life. This includes such molecules as water (H_2O), carbon dioxide (CO_2)[27], carbonic acid (H_2CO_3)[28], ammonia (NH_3)[29], and phosphoric acid (H_3PO_4)[30].

VIDEO 7.4

compound—a combination of elements that exhibits properties different from the component elements

covalent bond—(strong) attraction between atoms caused by sharing electrons

ionic bond—attraction between oppositely-charged ions

ionic compound—a compound composed of ions bound together by ionic bonds

The most celebrated of the molecules necessary for biological life is water. Water has a suite of characteristics that are not only unique to water, but also essential for biological life[31]. Each water molecule is made up of an oxygen atom with two hydrogen atoms stuck onto one side of the oxygen atom (see below).

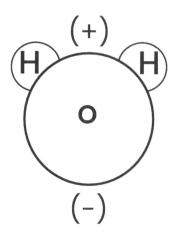

Because of the nature of the oxygen atom, the electrons shared between the oxygen atom and the two hydrogen atoms spend more time around the oxygen atom[a]. This gives the hydrogen end of the molecule a slight positive charge and the oxygen end of the molecule a slight negative charge. Since opposite charges attract, the hydrogen end of a water molecule is attracted to negative charges, including the negatively-charged ends of *other* water molecules. These charges are not full charges so the bond formed is not even as strong as an ionic bond, so this kind of bond is the weakest of chemical bonds. Because it is often associated with a hydrogen atom, it is called a **hydrogen bond**. The hydrogen bond in water, especially between one water molecule and another water molecule, provides water with a whole suite of interesting characteristics essential for all or some organisms:

*Water is a very good **solvent**, meaning it dissolves things well.* More specifically, water is a very good polar solvent, being very good at dissolving substances with electrical charges (**polar substances**). A substance is dissolved in a liquid when its components are suspended in that liquid in such a way that the components of substance cannot

hydrogen bond—(weak) attraction between slight electrical charges on molecules

solvent—a substance into which other substances dissolve. Types of solvents: polar solvent dissolves polar substances; non-polar solvent dissolves non-polar substances

polar substance—a substance with electrical charges (vs. non-polar substance)

a This is a result of what is called the electronegativity of an atom, which is a measure of how strongly a nucleus holds onto an electron. The electronegativity of an oxygen nucleus (3.44) is so much greater than the electronegativity of a hydrogen nucleus (2.20) that an electron shared between them spends more time around the oxygen nucleus than around the hydrogen nucleus.

be seen in the liquid. A polar substance dissolves in water when its charged molecules or ions are separated from one another and water molecules gather around the charges.

As the other end of each of these water molecules is attracted to other water molecules, the molecules or ions of the polar substance are kept suspended in the water (in other words, the substance dissolves in water). The more charges a substance has, the heavier the molecule or ion that water can keep suspended. Because of its rather strong hydrogen bonds, water can dissolve molecules quite a bit more massive than water molecules.

Since it is such a great polar solvent, organisms use water (*e.g.* sap in plants and blood in animals) to transport all sorts of things that can dissolve in water. Even water found outside of animals—such as in streams, ponds, and oceans—carries a host of ions and molecules essential for organisms that live in that water. Furthermore, because water is such a great polar solvent, cells are designed to hold a lot of water (more than half the weight of any given cell). Water's polar solvent properties allow the molecules needed by the cell to be stored and moved around as required. The same properties allow chemical reactions to occur that are needed by the cell to survive.

Water has a very high specific heat. The **specific heat** of a substance is the amount of heat that must be absorbed by that substance to raise the temperature of one gram of that substance one degree Celsius (which is equal, in turn, to the amount of heat that must be released by that

specific heat—heat required to raise the temperature of 1 gram of a substance 1° C

substance to drop the temperature of one gram of that substance one degree Celsius). To increase the temperature of a substance you must make the particles of that substance vibrate faster.

Since water molecules are attracted to one another, it is as if they are connected to each other by springs. So, to heat up one water molecule, you have to get it to vibrate faster. But each water molecule is connected to others, so you have to move more than one of them at a time. That is why it takes much more heat to speed up water molecules than is does to speed up the molecules of most other substances. Thus, water has a much higher specific heat than most other substances. For example, water has a higher specific heat than air, rocks and soil, so bodies of water heat up, and cool off, much more slowly than the air or the land.

You have probably seen the effect as you observe swimming pools remaining uncomfortably chilly for a while into the summer and remaining rather warm into the cool days of autumn. This characteristic of water results in cities on the shores of very large bodies of water to have a more moderate climate than inland cities[a]. The greatest value of this characteristic of water to life on earth is how it evens out the temperatures of the entire earth. Whereas the land cools off rapidly at night and heats up rapidly during the day, 70% of the earth's surface is covered by oceans. Ocean water holds heat through the night hours

---

[a] In locations with large temperature differences between summer and winter, coastal cities are typically 10° Fahrenheit cooler in the summer and 10° Fahrenheit warmer in the winter.

and even resists seasonal temperature changes for weeks or more. The resulting evening out of earth temperatures makes biological life possible on this planet.

Water has a very high heat of vaporization. The **heat of vaporization** of a substance is the amount of heat that a liquid needs to absorb to change one gram of that substance into a gas when it is already at its boiling temperature (which is equal, in turn, to the amount of heat that a substance releases to change one gram of that substance into a liquid when it is already at its temperature of condensation).

For water to go from a liquid to a gas, the water molecules must pull away from surrounding water molecules and be released into the atmosphere as isolated molecules. While water is a liquid, it is attracted to other water molecules. It takes quite a bit of energy to release any of those molecules. Thus, water has a much higher heat of vaporization than most other substances. As we pointed out in the previous paragraph, a lot of energy is required to get water molecules vibrating fast enough to be *ready* to evaporate[a]. The high heat of vaporization requires a large amount of *additional* energy to evaporate that water.

Since water needs so much heat to evaporate, and that heat can come from things located next to the water, the evaporation of water can be used to cool other things down. It is this feature that organisms

heat of vaporization— heat required to evaporate 1 gram of a liquid that is already at its boiling point

---

a Since temperature is related to how fast molecules are vibrating, to be *ready* to evaporate, water molecules must be vibrating at the speed that corresponds to the boiling point of water. If water is starting out at a lower temperature, heat has to be added to the water to get them to their boiling point speed. The high specific heat of water means that a lot of heat is required to get those molecules vibrating that fast.

use for cooling. Honey bees, for example, use this characteristic to air condition their hives. Developing honey bees are so sensitive to heat and cold that the temperature of the hive must be kept within just a few degrees of one temperature in order for the young to develop into healthy adult bees. If the temperature of the hive gets too low, bees burn honey to release body heat and warm the hive. If the temperature of the hive gets too high, bees collect water, spread it on the combs of the hive, and fan air into and out of the hive in order to evaporate the water and cool the hive.

Many large mammals also cool themselves off by using this amazing feature of water. They release water onto the outside of their skin, so that the water evaporates and cools their skin.

Water has high surface tension. Water molecules are so attracted to each other that the surface of the water can be rather resilient. This characteristic causes water to 'bead' on surfaces and form raindrops as water falls through the air. The bird's nest fungus is designed to take advantage of this raindrop-forming feature of water to disperse their spores[a]. A number of other small organisms (like water striders) are designed to take advantage of water's surface tension to walk on water.

Capillary action draws water up the walls of tubes. The smaller the diameter of the tube, the higher the water rises. Plants are designed

---

a   Four two-millimeter-diameter spores sit inside a centimeter-diameter cup (looking so much like a miniature bird's nest to give the fungus its name) until a raindrop falls just inside the cup and kicks out spores. A well-placed raindrop can propel seeds many cup diameters away.

with microscopic tubes (xylem) that run from the roots (where the water is drawn into the plant) to the leaves (where the water is needed for photosynthesis). Xylem tubes have a small enough diameter for water to rise up a couple feet by capillary action alone. By this process short plants can supply their leaves with water without the plant having to use energy to pump the water up from the roots.

In taller plants, like trees, water is thought to be pulled up the rest of the way by what is called the 'transpiration-cohesion' or 'cohesion-tension' process. To picture how this process works, imagine each water molecule in the xylem connected to the next water molecule by means of a tiny spring. At the upper end of the xylem tubes, photosynthesis or evaporation pulls a water molecule out of the top of the xylem tubes. As the water molecule is pulled out, the 'spring' attaching it to the next molecule is stretched until it has enough tension on it to suddenly pull up the next molecule. That molecule does the same thing to the next water molecule, and so it continues, until each water molecule in the xylem has been pulled up one water molecule higher than it was before.

Amazingly enough, in a large tree on a sunny summer day, this process operating in thousands of xylem tubes can pull up gallons of water scores or even hundreds of feet from the roots to the top of the tree—all without using any of the tree's energy.

Unlike every other liquid known to us, water expands upon freezing[a]. For most substances, molecules are closer together when the substance is frozen than when it is melted.

For water, however, it is the other way around. At the freezing temperature of water, water molecules are farther apart in ice than they are in liquid water. Therefore, there are fewer molecules in a given volume (space) of ice than there are in the same volume of water. This means ice has a lower density (mass divided by volume) than water. Since ice is less dense, it floats on water. Consequently, as a body of water freezes, a layer of ice forms over the top and remains on top. Continued freezing causes the ice to gradually thicken. The insulation

---

a Actually, for the entire temperature range of most substances, the substance expands when it is heated and contracts when it is cooled. For a small number of substances in the universe there is a certain temperature range where this is reversed—where heating the substance causes it to contract and cooling the substance causes it to expand. Water is one of those rare substances. Water, however, is the only one of those substances where that odd temperature range is close to a freezing point or boiling point for the substance. Thus, water is the only substance that we know of in the universe which happens to expand as it freezes.

from already formed ice slows the freezing of water, allowing water organisms to survive in cold environments.

In fact, if water was like other substances and ice was denser than water, then ice would sink to the bottom as it formed and water would quickly freeze from the bottom up. Under these conditions it is likely that most of the bodies of water on earth would freeze and life as we know it would not be possible. Again, this is a unique property of water. Every other liquid we know becomes *denser* upon freezing.

In a number of different ways, then, water is crucial for life to exist and persist on this planet.

God designed scores of features of the universe in such a way that organisms could thrive. From features of the universe's basic structure, to the nature of the smallest and largest objects of the universe, to the arrangement of entities in the universe, everything has been crafted to permit biological life.

7.5 | Unnecessary AP Characteristics

The AP characteristics that are *necessary* for biological life, such as were discussed above, *must* be in place or we would not exist at all. In some sense, the AP characteristics that are even more impressive are those that are *not* necessary for physical life, but which do permit humans to observe and understand the universe.

For example, though it is not necessary for our existence, it is sure convenient that the universe can be described by the human language of mathematics. It is even more useful that the mathematical regularities of the universe are simple enough for us to understand[32]. There is even more value in the tendencies and regularities of the universe being crafted in such a way as to permit humans to describe them in the space and time of a single human lifetime. It is even handy that our atmosphere is designed in such a way as to block those wavelengths of radiation that are harmful to us, but not block the wavelengths that we use for communication (such as microwaves and radio waves). Even our position in the galaxy is useful, as it allows us to see both into the structure of our own galaxy and to see out of our galaxy to the universe beyond. These universe designs seem to have been placed there by God so that we can see the attributes of God illustrated by those things that He made.

The Origin of Bioprovision

Those who accept the naturalistic worldview cannot accept what the Anthropic Principle *seems* to be suggesting. Since naturalism does not accept the existence of anything non-physical, nothing with the ability to design could have been capable of fashioning the entire universe. Therefore, the naturalist is forced to assume that the design implied by the Anthropic Principle is an illusion—it just *seems* as if the universe was designed with humans in mind. Typically, those with the naturalistic worldview will discount the AP characteristics *necessary* for life, by saying it could not be any other way. After all, if any of these necessary characteristics *were not* true, we would not be here to observe them! Therefore, for us to even notice them, the AP characteristics necessary for life *have* to be in place.

However, without anything guiding the origin of the universe in some specific way, the universe could have been different in a host of ways that would *not* have allowed biological life. In fact, there may be only one way to build a life-friendly universe, but billions upon billions of ways to build a universe where biological life was impossible. If the structure of our universe was not designed, but came about instead by some sort of random, undirected process, our universe would not be expected at all, even if it were tried over and over again millions of times. Our universe is a very, very improbable one and that is not at all the expectation of the naturalistic worldview. It seems much more

reasonable to explain our universe as a result of the intelligent design of a wise God.

The AP characteristics which are *not* necessary for biological life may be even more informative. We certainly do not *have* to be located in such a perfect spot where we can effectively observe the universe, but we do happen to be located in such a spot. And, if the non-physical world does not exist, then humans are being arrogant indeed to say that a language that humans created (mathematics) happens to be able to characterize the entire universe. How does this happen to be true? And what right has a human being to claim that something discovered in the course of his or her very limited experience works across the enormous extent of the entire universe?

It seems to be an extremely large stretch to suggest that the universe just *happens* to have a structure humans can understand. It is much more reasonable to claim that the universe and humans were designed *so that* humans could understand the universe. And this is exactly what the God of Scripture did. He created the physical world so that humans could understand the invisible God through it. The universe has multiple evidences of being designed by the God described in Scripture.

Bioprovision: Our Responsibility

Our Responsibility to God

When we observe a living thing, we should recognize the extraordinary designs necessary to make that life possible. The very structure of the universe itself had to be specially designed with biological life in mind. Our galaxy and sun and earth and moon had to be fashioned in a very particular manner, and the earth's atmosphere, oceans, and land had to be crafted properly. Even the objects of the world too small for us to see had to be created in a very special manner. Particular subatomic particles and rules by which they behave, particular elements, and particular molecules, all had been crafted in very special ways. There are so many things we take for granted that have to be the way they are for us to exist all—such as the water we interact with in so many ways throughout a given day. The very existence of biological life should be a graphic reminder to us of the extraordinary wisdom and power of God as Creator.

If God can design the entire universe thousands of years ago in such a way as to meet the needs that you have today, how much more can God be trusted to orchestrate your circumstances and experiences to provide what is best for you. Furthermore, considering the extent to which God provided for the creation which He cares for, but does not love, imagine how much more God cares for you, whom He actually loves.

Finally, the fact that the AP characteristics of the universe stand as physical illustrations of His invisible love reminds us that that incredible God desires a relationship with us. Contemplation of the AP characteristics of creation should lead us inexorably into worship. As we learn more about these characteristics, our lives should be increasingly filled with worship. And, at some point in this process we cannot help but draw others into that same worship, and so fulfill our role as priests of the creation.

Our Responsibility to the Creation

Preserving Provisionary Designs

When humans were given dominion over the animals (Gen. 1:26-28), they were given specific authority over whatever was created to serve the animals. This would include plants, the earth, and the universe itself. The scriptural commentary on the creation account found in Psalm 8 clarifies that humans have authority over *all* the creation. We were called upon to 'guard and keep' the entire creation, including its AP characteristics.

Most of these AP characteristics have been created in such a way that humans cannot change them (*e.g.* the rules, regularities and basic structure of the universe; the position of the earth in the Milky Way and Solar System; the subatomic particles, elements, and molecules of which we are constructed; the makeup of the solid earth and moon). These seem to be features of the creation that God reserved the sole right to control. However, there are a few of the AP characteristics that man *can* impact—namely those relating to the air we breathe and water we drink. As rulers of the creation, we have a responsibility to preserve life by preserving the life-supporting nature of these AP characteristics.

For example, it would seem that on Day Two of the Creation Week, God created the earth's atmosphere. That atmosphere was carefully designed to keep the earth's surface at temperatures friendly to biological life, protect biological life from harmful radiation, and constantly provide biological life with clean water. It also seems as if God designed the earth in such a way as to re-establish any of these life-giving functions if and when any of them are damaged. Humans can alter several of these life-giving functions, and perhaps we can alter them faster than they can be repaired by the restoration processes that God created.

Humans can alter the water cycle—that process God designed to provide organisms all across the earth with water that they need to survive. To supply themselves with water, for example, humans have built canals and dams to redirect the path of water as it flows across the earth's surface to return to the sea. By means of wells and pipelines, humans have also drawn out ground water. Many of these alterations are actually good, for they allow more humans, animals and plants to live in places that otherwise have too little water to support that many organisms. Other such alterations may not be, as water can be directed away from areas that in the past supported more organisms, but no longer can[a].

We need to be careful as we use water, that we promote life in the process. Humans can not only affect the *quantity* of water in an area; humans can also affect the *quality* of that water. God designed the water cycle to provide clean and pure water to the earth's organisms. As it is evaporated, for example, water is evaporated in nearly pure form. Water evaporated from polluted water is evaporated as clean water and water evaporated from salty water is evaporated as fresh water. God even created organisms to clean water that gets polluted (something we'll learn more about in Chapter 8). Humans, however, can pollute streams, rivers, and the ocean, and sometimes may be able to do so faster than organisms can clean it. To preserve the creation over which we rule, we should clean up polluted waters and control the pollutants we add to water so that fresh, clean water is available to organisms everywhere.

Humans can also alter the atmosphere's ability to protect life from harmful radiation. For example, chlorofluorocarbons (CFCs) used to be widely used as aerosol propellants, coolant fluids (*e.g.* Freon), solvents, and the construction of plastic foam. In the 1970s, however, laboratory experiments showed that CFCs can destroy ozone. Ozone (O_3), as we saw above, is a greenhouse gas in the atmosphere that protects organisms on the earth from destructive ultraviolet light. The fact that the amount of ozone in the atmosphere varies up and down suggests that God created some sort of processes (that we still do not fully understand) to keep ozone levels at proper levels.

---

[a] So much water is drawn from the Colorado River, for example, that the states of California and Arizona fight over the water and only a trickle of water crosses the border into Mexico.

However, in the 1970's many people thought that CFCs produced by humans were destroying ozone faster than ozone could be produced by these natural processes. As a result of these fears, beginning in 1987, laws were introduced to stop the production of CFCs and other ozone-destroying chemicals. Now it may not be that CFCs are as capable of destroying ozone as was once thought. Although CFCs do destroy ozone, it was never demonstrated that the human production of CFCs at the earth's surface was actually responsible for substantial destruction of ozone in the earth's upper atmosphere. In fact, the amount of ozone in the atmosphere has changed rather independently of both CFC concentrations on the earth's surface and CFC production rates.

It still might be a good thing that we halted CFC production, but it could also be true that we did so before we understood enough about the earth's atmosphere to have done so for the best reasons. Either way, it is a good idea that we carefully monitor anything that might damage the atmosphere's ability to protect biological life from harmful radiation.

Humans can also alter the atmosphere's ability to control the temperature of the earth's surface. For example, slash and burn agriculture is destroying tropical rain forest much faster than the rain forest is able to regrow. This is releasing carbon dioxide (CO_2) into the atmosphere faster than the CO_2 is being put back into plants. Burning **fossil fuels** (fuels like coal, oil, and natural gas which are made of fossils of dead organisms) is also releasing large amounts of CO_2 into the atmosphere that is not being returned to the earth or plants. Consequently, slash and burn agriculture and burning of fossil fuels is increasing the amount of CO_2 in the atmosphere.

Since CO_2 is a greenhouse gas trapping heat from the sun on the earth's surface, slash and burn agriculture and burning of fossil fuels should raise the surface temperature of the earth. And, since earth temperatures *are* rising (**global warming**) and the CO_2 concentration in the atmosphere is rising, it is commonly believed that humans are responsible for the warming. If it continues, this global warming will melt the polar ice caps and melt all the mountain glaciers. The loss of ice threatens organisms that currently live on that ice (*e.g.* polar bears). The water from the melted ice would raise sea level nearly 300 feet, destroying many of the world's cities, inundating something like 10% of the earth's land, and probably displacing more than a billion people.

---

fossil fuels—fuels like coal, oil, and natural gas which are made of fossils of dead organisms

global warming—rise of Earth's surface temperature due to greenhouse gas increase (especially CO2)

A warmer earth would lessen the temperature differences between the equator and the poles, which would substantially affect the earth's winds. This, in turn, would change rainfall and ocean circulation worldwide. Global climate change would have many profound and lasting effects on many aspects of biological life on earth. As a result of global warming concerns, various efforts are under way to reduce slash-and-burn agriculture and to reduce the burning of fossil fuels worldwide.

Now it may not be that humans are most of the cause of global warming. After all, humans are not the only producers of CO_2 gas, and CO_2 is not the only greenhouse gas in the atmosphere. Furthermore, we actually do not know what the best CO_2 levels are for life on earth. It may be that global warming has more positive long-term effects on biological life than it has negative effects. For example, there is evidence that biological life may have been both more abundant and more diverse during periods of earth history when the earth was too warm to sustain ice caps. The higher sea level that would result if the earth's ice were to melt would probably increase our harvest of seafood, thus allowing us to feed more people in the world. The warming of the planet that would result would widen the earth's tropical zone and probably stretch the temperate zones of the earth to the poles.

This, in turn, would allow farming in areas now too cold for agriculture, further increasing our ability to feed more people. Finally, most of the predictions about global warming assume that the earth is billions of years old. If different assumptions are made (*e.g.* if the earth is assumed to be only thousands of years old, as indicated in Scripture), different predictions result[a]. There is no doubt that global warming will substantially affect biological life on earth, but we currently do not know enough about the earth to predict exactly what those effects will be. We have a lot of work to do in order to learn enough about our atmosphere to properly fulfill our role as rulers.

Enhancing Bioprovision

Besides preserving the earth's special design for physical life, it should also be our responsibility to enhance it, to 'return it' better able to support organisms than when it was given to humans in the first place. It is likely that God created the earth with even more potential than is actually currently realized in supporting biological life on earth. We ought to be aware of such possibilities and even search for them in God's amazing creation.

Summary of Chapter

A. As a physical illustration of His love, God designed the entire universe in the Creation Week in such a way that organisms could exist for thousands of years to follow.

B. Anthropic Principle (AP) characteristics are those features of the universe that make it look like it was created with man in mind—even to those having a naturalistic worldview.

  a. Several AP characteristics are not necessary for our existence but permit humans to observe and understand the universe. These include

   i. the close correspondence between the structure of the universe and the human language of mathematics,

   ii. the structure of the universe being simple enough for us to understand,

   iii. the regularities of the universe being understandable in the course of a single human lifetime, and

   iv. our position in the galaxy allowing us to see most of the rest of the universe.

  b. Scores of AP characteristics are necessary AP characteristics—those that must be in place in the universe for biological life to exist at all:

---

a   For example, most predictions about the rate of climate change are based upon how fast climate has changed in the past—especially how fast ice was thought to melt after the Ice Age. Since most scientists believe the earth is very old, they also believe that this happened much more slowly than young-age creationists believe it happened. Consequently, young-age creationists believe global warming and the associated sea level rise is likely to happen many times faster than estimated by other scientists.

i. The second law of thermodynamics (something in an area of high concentration tends to move to areas of lower concentration) is essential for all the chemical processes of organisms (including osmosis, the movement of water across membranes, and diffusion, the movement of molecules in water);

ii. For biological life to exist on the earth, the earth must have a stable orbit about the sun. For that to happen, God created three-dimensional space, set in place a law of gravity according to an inverse-square relationship, and established a gravitational constant of the proper strength.

iii. For biological life to exist, electrons must have stable orbits around atomic nuclei. For that to happen, God created three-dimensional space, set in place Coulomb's law according to an inverse-square relationship, and established the Coulomb's constant of the proper strength.

iv. God placed our earth far enough from other stars in the Milky Way Galaxy so we are not close to exploding stars and so that the earth can orbit the sun without interference from other stars.

v. God designed our Solar System with a number of features necessary for life, including
    - a stable sun to warm the earth consistently,
    - a sun without a companion star that would pull the earth out of orbit,
    - a sun producing the right kind of radiation for photosynthesis,
    - the earth at a 'goldilocks' distance from the sun to be the right temperature, and
    - a moon large enough to stabilize the earth's orbit.

vi. God designed our Earth with many features necessary for life, including
    - the correct mass to hold onto useful gases and release poisonous gases,
    - the correct combination of atmospheric greenhouse gases to maintain the proper surface temperature,
    - an axial tilt that evens out surface temperatures,
    - a distribution of oceans and continents that would even out surface temperatures and supply water to organisms through the water cycle,
    - atmospheric ozone that blocks harmful ultraviolet radiation, and
    - a core of iron and nickel that maintains a magnetic field to deflect harmful charged particles.

vii. God designed subatomic particles so that stable atoms and ions would be possible. This includes creating three-dimensional space, the electromagnetic force in an inverse-square relationship, protons and electrons with equal but opposite charges, and protons 1836 times as massive as electrons.

viii. God established rules about how electrons orbit nuclei that allow the generation of special elements (*e.g.* carbon, oxygen, nitrogen) and define how atoms combine to produce molecules (no molecules for Group VIII elements; equal pairings of Group I and VII elements and Group II and VI elements; twice as many Group I elements with Group VI, Group VII elements with Group II, and Group II elements with Group IV elements), and the generation of special molecules (*e.g.* water and carbon dioxide) essential for life.

ix. Because of partial charges on each molecule, water molecules bond together with hydrogen bonds. This feature results in a long list of characteristics of water that are essential for biological life. For example,
    - water is such a good polar solvent (of polar substances—those with electrical

- charges) that all cells are full of water and nutrients are transported in water,
  - water has such a high specific heat (heat needed to heat up a substance) that its abundance on the earth evens out the temperature of the earth,
  - water has such a high heat of vaporization (heat needed to evaporate a substance) that it is used to cool organisms or their surroundings,
  - water has a high enough surface tension to allow organisms to live on its surface,
  - water exhibits enough capillary action to get water from roots to the leaves of short plants,
  - water expands upon freezing, allowing water organisms to live in water when the air temperature is low enough to freeze the water.
C. A universe with as many necessary AP characteristics as our universe has, is too improbable to have arisen by naturalistic process, but expected in a universe designed by God with man in mind.
D. The AP characteristics that are unnecessary for life but convenient for our observation and understanding of the universe are not expected at all in a naturalistic worldview, but are expected of a God Who desired us to recognize His invisible attributes through it.
E. Consideration of the AP characteristics of the universe give us insight into God's love, ought to stimulate us to worship God, glorify God by acknowledging His role in providing for us, and bring others into the worship of God.
F. As rulers of the creation we have a responsibility to preserve any designs that permit life to exist on earth. This includes controlling anything that might lessen the life-sustaining properties of the atmosphere (such as drawing off too much water or polluting water in the water cycle, releasing CFCs or other substances that might destroy ozone, and producing so much CO_2 as to overheat the earth).

Test & Essay Questions

1. Define anthropic principle.
2. Compare and contrast neutrons, protons, and electrons / subatomic particle/ atom / ion / molecules / ionic, covalent, and hydrogen bonds.
3. How does God physically illustrate His invisible love?
4. List three / four / five anthropic principle characteristics necessary for biological life.
5. What do anthropic principle characteristics that are necessary for biological life tell us about God?
6. What is a simple statement of the second law of thermodynamics?
7. Why is the second law of thermodynamics essential for biological life?
8. Label neutrons/protons/electrons/nucleus on a drawing of an atom.
9. What are the characteristics of the noble gases and why do they have those characteristics?
10. From a particular element on the Periodic Table, indicate the number of electrons / number of protons / the number of electrons usually accepted or donated.
11. In the case of two different elements on the Periodic Table, what is the most likely chemical formula of the molecule formed by the two atoms?
12. What about the structure of water explains many of the unique characteristics of water.
13. List three / four / five / six unique characteristics of water.

14. Explain how the electric charges on a water molecule explain why water is a good solvent / has a high specific heat / has a high heat of vaporization / has a high surface tension.

15. Explain what it means for water to be a good solvent / to have a high specific heat / to have a high heat of vaporization / to have high surface tension / to exhibit capillary action / to expand upon freezing and explain why that feature is important for biological life.

16. Arrange ionic bonds, covalent bonds, and hydrogen bonds from the strongest to the weakest

17. List three / four anthropic principle characteristics that are *not* necessary for biological life.

18. What do anthropic principle characteristics that are *not* necessary for biological life tell us about God?

19. Short Essay: Compare and contrast the explanation of the anthropic principle in the naturalistic and Christian worldviews.

20. Short Essay: How is the anthropic principle better explained in the Christian worldview than in the naturalistic worldview?

21. Short Essay: Why are the anthropic principle characteristics that are *un*essential for biological life more difficult for a naturalist to explain than the anthropic principle characteristics that are essential for biological life?

22. Why were CFCs banned in the United States?

23. What is global warming?

24. What is a fossil fuel?

25. How do humans contribute to an increase in carbon dioxide in the atmosphere?

26. How does carbon dioxide warm up the earth's surface?

27. Short Essay: Why is there uncertainty about humans causing global warming / ozone depletion?

28. List three / four / five different impacts global warming would have on the earth.

References

(used repeatedly in this chapter and useful for further student research)

Barrow, John D., and Frank J. Tipler, 1986, *The Anthropic Cosmological Principle*, Oxford University Press, New York, NY, 706 p.

Breuer, Reinhard, 1991, *The Anthropic Principle: Man as the Focal Point of Nature*, Birkhäuser, Boston, MA, 261 p. [transl. by Harry Newman and Mark Lowery of Breuer, Reinhard, 1981, *Das Anthropische Prinzip: Der Mensch im Fadenkreuz der Naturgesetze*, Meyster Verlag Gmbh, Wien - München.]

Corey, M. A., 1993, *God and the New Cosmology: The Anthropic Design Argument*, Rowman & Littlefield, Lanham, MD, 332 p.

Darling, David, 1993, *Equations of Eternity: Speculations on Consciousness, Meaning, and the Mathematical Rules That Orchestrate the Cosmos*, Hyperion, New York, NY, 190 p.

Gribbin, John, and Martin Rees, 1989, *Cosmic Coincidences: Dark Matter, Mankind, and Anthropic Cosmology*, Bantam, New York, 302 p.

Henderson, Lawrence J., 1913, *The Fitness of the Environment: An Inquiry Into the Biological Significance of the Properties of Matter*, Peter Smith, Gloucester, MA, 317 p. [Reprinted, 1970 with same page numbers].

Leslie, John, 1989, *Universes*, Routledge, New York, NY, 228 p.

Ross, Hugh, 1994, Astronomical evidences for a personal, transcendent God, pp. 141-172 *In* Moreland, J. P. (ed.), *The Creation Hypothesis: Scientific Evidence for an Intelligent Designer*, InterVarsity, Downers Grove, IL, 335 p.

Schroeder, Gerald L., 1990, *Genesis and the Big Bang: The Discovery of Harmony Between Modern Science and the Bible*, Bantam, New York, NY, 212 p.

Shklovskii, I. S., and Carl Sagan, 1966, *Intelligent Life in the Universe*, Holden-Day, San-Francisco, CA, 509 p.

Wald, George, 1958, Introduction, pp. xvii-xxiv *in* 1970 reprint of Henderson 1913.

Wilkinson, Denys Haigh, 1991, *Our Universes*, Columbia Univ. Press, New York, NY, 213 p.

Wise, Kurt P., and Matthew S. Cooper, 1998, A compelling creation: A suggestion for a new apologetic, pp. 633-644 *In* Walsh, Robert E. (ed.), *Proceedings of the Fourth International Conference on Creationism...*, Creation Science Fellowship, Pittsburgh, PA, 658 p.

References Continued

1. Henderson 1913:52-3, 58; Shklovskii and Sagan 1966:347-8; Barrow and Tipler 1986:309; Ross 1994:166; Wilkinson 1991:172-3; Corey 1993:78-9, 80, 112; Wise and Cooper 1998.

2. Barrow and Tipler 1986:548, 567; Ross 1994:168; Corey 1993:103, 113; Wise and Cooper 1998.

3. Newman, Michael J., and Robert T. Rood, 1977, Implications of the solar evolution for the earth's early atmosphere, *Science* 198(4321):1035-7; Ross 1994:167; Corey 1993:112; Wise and Cooper 1998.

4. Henderson 1913:52-3; Ross 1994:168, 169; Wise and Cooper 1998.

5. Ross 1994:168; Schroeder 1990:124; Corey 1993:113; Wise and Cooper 1998.

6. Breuer 1981(1991:192-3); Ross 1994:167; Schroeder 1990:125; Corey 1993:110-1; Wise and Cooper 1998.

7. Henderson 1913:52-3; Shklovskii and Sagan 1966:247, 344; Newman, Michael J., and Robert T. Rood, 1977, Implications of the solar evolution for the earth's early atmosphere, *Science* 198(4321):1035-7; Ross 1994:166-7, 169; Barrow and Tipler 1986:337-8; Schroeder 1990:123-4; Corey 1993:57, 110-1; Wise and Cooper 1998.

8. Ross 1994:168-9; Murray, Carl D., 1993, Seasoned travelers, *Nature* 361(6413):586-7; Laskar, J., and P. Robutel, 1993, The chaotic obliquity of the planets, *Nature* 361(6413):608-612; Corey 1993:112; Kerr, Richard A., 1994, The Solar System's new diversity, *Science* 265:1360-2; Wise and Cooper 1998.

9. Shklovskii and Sagan 1966:343, 345; Breuer 1981(1991:192-3); Wilkinson 1991:173, 186; Corey 1993:75-6; Wise and Cooper 1998.

10. Breuer 1981(1991:228); Ross 1994:166; Wilkinson 1991:185-6; Corey 1993:75-6; Darling 1993:122-3; Wise and Cooper 1998.

11. Henderson 1913:58-9; Shklovskii and Sagan 1966:345-7; Ross 1994:166; Corey 1993:111-2; Wise and Cooper 1998.

12. Leslie 1989:58-9; Wilkinson 1991:173-4; Wise and Cooper 1998.

13. Breuer 1981(1991:53-4); Barrow and Tipler 1986:12, 15-16, 247, 259-275; Hawking, Stephen W., 1988, *A Brief History of Time: From the Big Bang to Black Holes*, Bantam, New York, NY, 198 p. (pp. 163-5); Gribbin and Rees 1989:259-63; Leslie 1989:46-7; Wilkinson 1991:174-6; Corey 1993:19, 58, 94-100; Darling 1993:118-9; Ross, Hugh, 1996, *Beyond the Cosmos: The Extra-Dimensionality of God: What Recent Discoveries in Astronomy and Physics Reveal About the Nature of God*, NavPress, Colorado Springs, CO, 231 p. (pp. 32-3); Wise and Cooper 1998.

14. Linde, A. D., 1984, The inflationary universe, *Reports on Progress in Physics*, 47(8):925-86; Leslie 1989:47; Wise and Cooper 1998.

15. Gribbin and Rees 1989:256-7; Hawking, Stephen W., 1988, *A Brief History of Time: From the Big Bang to Black Holes*, Bantam, New York, NY, 198 p. (pp. 151-2); Leslie 1989:63; Corey 1993:100-1; Wise and Cooper 1998.

16. Corey 1993:58.

17. Leslie 1989:36-7, 44.

18. Barrow, John D., and Joseph Silk, 1980, The structure of the early universe, *Scientific American* 242(4):118-128; Rozental, I. L., 1980, Physical laws and the numerical values of fundamental constants, *Soviet Physics: Uspekhi*, 23:293-305 (p. 298); Gale, George, 1981, The anthropic principle, *Scientific American* 245(6):154-171; Templeton, John M., 1984, God reveals Himself in the astronomical and the infinitesimal, *Journal of the American Scientific Affiliation*, 36(4):194-200 (p. 196); Neidhardt, W. Jim, 1984, The anthropic principle: A religious response, *Journal of the American Scientific Affiliation* 36(4):201-7 (p. 202); DeYoung, Donald B., 1985, Design in nature: The anthropic principle, *Impact* 149:i-iv (p.iii, iv); Barrow and Tipler 1986:298, 326-7; Leslie 1989:45-6; Templeton, John M., and Robert L. Herrmann, 1989, *The God Who Would Be Known: Revelations of the Divine in Contemporary Science*, Harper & Row, San Francisco, CA, 214 p. (p. 68); Ross 1994:161, 162; Wise and Cooper 1998.

19. Rozental, I. L., 1980, Physical laws and the numerical values of fundamental constants, *Soviet Physics: Uspekhi*, 23:293-305 (p. 298); Gribbin and Rees 1989:6-7; Leslie 1989:44-5; Wilkinson 1991:180-1; Corey 1993:62-3; Wise and Cooper 1998.

20. Rozental, I. L., 1980, Physical laws and the numerical values of fundamental constants, *Soviet Physics: Uspekhi*, 23:293-305 (p. 303); Templeton, John M., 1984, God reveals Himself in the astronomical and the infinitesimal, *Journal of the American Scientific Affiliation*, 36(4):194-200 (p. 196); Neidhardt, W. Jim, 1984, The anthropic principle: A religious response, *Journal of the American Scientific Affiliation* 36(4):201-7 (p. 202); Barrow and Tipler 1986:297; Leslie 1989:5, 44; Ross 1994:161; Wilkinson 1991:185.

21. Henderson 1913:191-248, 251-3, 254-5, 256, 264-267, 267-273; Wald 1958:xx-xxiv; Shklovskii and Sagan 1966:229; Breuer 1981(1991:5); Barrow and Tipler 1986:143-4, 545-8; Barrow, John D., 1988, *The World Within the World*, Clarendon Press, Oxford, England, 398 p. (p. 354); Schroeder 1990:120-2; Wilkinson 1991:170-1; Corey 1993:103-4; Wise and Cooper 1998.

22. Henderson 1913:191-248, 251-3, 254-5, 256, 267-273; Wald 1958:xx-xxiv; Barrow and Tipler 1986:143-4, 541-5; Wise and Cooper 1998.

23. Henderson 1913:191-248, 251-3, 254-5, 256, 267-273; Wald 1958:xx-xxiv; Barrow and Tipler 1986:143-4, 541-5; Wise and Cooper 1998.

24. Wald 1958:xx-xxiv; Barrow and Tipler 1986:549-556; Corey 1993:104-5; Wise and Cooper 1998.

25. Barrow and Tipler 1986:553-5; Corey 1993:104-5; Wise and Cooper 1998.

26. Barrow and Tipler 1986:553-5; Corey 1993:104-5; Wise and Cooper 1998.

27. Henderson 1913:133-163, 251-3; Barrow and Tipler 1986:548; Corey 1993:103; Wise and Cooper 1998.

28. Henderson 1913:vi-vii, 133-163, 251-3, 257, 259-260, 266-7, 267-273; Barrow and Tipler 1986:143-4, 548; Wise and Cooper 1998.

29. Henderson 1913:263-5; Barrow and Tipler 1986:550-2; Wise and Cooper 1998.

30. Henderson 1913:vi-vii; Wald 1958; Barrow and Tipler 1986:143-4; Wise and Cooper 1998.

31. Henderson 1913:vii-viii, 72-132, 164-190, 251-3, 255-260, 262-4, 266-7, 267-273; Shklovskii and Sagan 1966:228; Breuer 1981(1991:5, 209-218); Barrow and Tipler 1986:143-4, 524-541; DeYoung, Donald B., 1987, [book review of Barrow & Tipler, 1986], *Creation Research Society Quarterly 23*(4):182; Gribbin and Rees 1989:270; Schroeder 1990:120-2; Wilkinson 1991:171-2; Corey 1993:105-8; Darling 1993:122; Ross 1994:163; Wise and Cooper 1998.

32. Breuer 1981(1991:233-7); Templeton, John M., and Robert L. Herrmann, 1989, *The God Who Would Be Known: Revelations of the Divine in Contemporary Science*, Harper & Row, San Francisco, CA, 214 p. (p. 39); Gribbin and Rees 1989:21, 238, 284-5; Leslie 1989:50-51, 59; Wise and Cooper 1998.

CHAPTER 8

THE SUSTAINING GOD

Understanding the Biomatrix

"Consider the lilies of the field... even Solomon in all his glory was not arrayed like one of these... if God so clothes the grass of the field... will He not much more clothe you ...?"
Matt. 6:28-30, NKJV

8.1 | God is Sustainer

After God provided a ram so that Abraham would not have to sacrifice his son, Abraham called that place *Jehovah-jireh*, which is Hebrew for 'God is provider' (Gen. 22:14). A God of love is a God Who provides, not only the requirements needed for life in the first place, but also those things needed for life every day. To assure us that God takes care of our physical needs, Jesus pointed out that God provides for the on-going physical needs of plants and animals (Matt. 6:25-33).

God upholds all things (Heb. 1:3), preserves all things (Neh. 9:6; Psa. 36:6), and by Him all things consist and hold together (Col. 1:17). God is responsible for weather[a] and the rising and setting of the sun (Psa. 104:19-20; Matt. 5:45). He creates the life of each individual sea animal (Psa. 104:29-30), He is responsible for animal instincts (Job 39:26-30), He makes plants to grow (Psa. 104:14-17; 147:7-8; Matt. 6:28-30), and He provides food for animals of the sea (Psa. 104:25-28), the air (Job 38:41; Psa. 147:7-9; Matt. 6:26), and the land (Job 38:39-40; Psa. 104:21; 147:7-9).

Provision and the Biomatrix

In the last chapter we examined features of the *non*-biological world that make it possible for organisms to exist. In this chapter we will examine ways in which God designed the *biological* world to provide for the on-going, daily needs of plants and animals. Animals require food and other nutrients that are not found in the right form in the non-biological world. In the case of the land animals, most of that food and many of the nutrients are provided by plants (Gen. 1:29-30).

However, even the plants need nutrients that are not supplied in usable form by the non-biological world. So, even with all the AP characteristics of the universe in place, plants and animals still could not exist on the earth by themselves. To make life possible for plants and animals, and to further illustrate His attribute of love, God created a complex network of unnoticed organisms (called a **biomatrix** or **organo-substrate**[b]) that provide what animals and plants need to exist.

biomatrix—system of bacteria, algae, protozoa, and fungi required for plants and animals to live on earth (aka organo-substrate)

organo-substrate—system of bacteria, algae, protozoa, and fungi required for plants and animals to live on earth (aka biomatrix)

---

a God is responsible for springs (Psa. 104:10) as well as the evaporation of water (Psa. 135:5-7), the formation of clouds (Job 37:10-11; Psa. 147:7-8), thunder & lightning (Job 37:5; Psa. 135:5-7), wind (Psa. 135:5-7), and the dropping of frost (Job 37:10), snow (Job 37:5-6) and rain (Job 5:10; 37:5-6; 38:36; Psa. 147:7-8; Matt. 5:45).

b The word 'biomatrix', suggested by Joseph Francis, is derived from the Greek word *bios* meaning life (both because it is necessary for animal (and plant) life and because it is made up of biologically living organisms)

bacterium (pl. bacteria)—single-celled organisms small enough not to need organelles (aka prokaryote)

protozoan (pl. protozoa)—eukaryotic organism that is generally a one-celled consumer

Many of the biomatrix organisms are unnoticed because they can only be seen with a microscope: the **bacteria** (sing. *bacterium*), most of the algae (sing. *alga*), and most of the **protozoa** (sing. *protozoan*). Other biomatrix organisms (the fungi and the remainder of the protozoa and algae) are large enough to be seen (including the largest organisms on this planet[a]), but are rarely seen because they live in places we do not frequent, or appear at times we normally do not observe, or stay *inside* other things for most or all of their life.

At this point, we know relatively little about the biomatrix. We have only begun to study this long unnoticed part of biology. Let us spend a little bit of time considering this critically important part of the biological world.

Food Providers

All biological organisms need energy to do what they do. Many organisms—including all animals, fungi, protozoa, and some bacteria—get that energy from food (energy-containing molecules formed by other organisms). The energy is first put into organic molecules by **producers**, organisms specially designed for that purpose. These

producer—an organism that gets its energy from the physical environment (vs. consumer)

and 'matrix' (both because it is a rather complex system of organisms and because it is an unnoticed basis of the reality—as the 'matrix' was, as in the 1999 American-Australian science fiction action film 'the Matrix'). 'Organo-substrate' is derived from 'organic' (both because it is part of carbon-based or organic life and because it is necessary for the existence of carbon-based plant and animal organisms) and 'substrate' (because it is the substrate or ground or foundation for animal and plant life).

a   Some fungi in the soil and some algae in the sea are spread over square miles of earth surface, and are in the running for the largest individual organisms on the planet.

molecules then become food for all other organisms (the **consumers**). Organisms of the biomatrix play a very important role in providing food to many of the animals of the world, especially those in water environments.

consumer—an organism that gets its energy by consuming organic molecules (vs. producer)

Producers in Water Environments

The main producers on the land are plants[a]. All other producers (the algae and some of the bacteria) are part of the earth's biomatrix. Some algae[b] and photosynthetic bacteria produce on the land what the plants do not. The remainder of the algae and photosynthetic bacteria are the main producers for the streams, lakes, and oceans of the world. There are thousands of species of algae and thousands of species of photosynthetic bacteria in an amazing variety of forms—a vast percentage of which are microscopic[c].

As in the case of plants, algae and photosynthetic bacteria construct organic molecules by the process of **photosynthesis** (Greek words *phōs*, 'light', + *synthesis*, 'putting together'). One group of algae, the green algae, are green because they collect energy from the same colors of sunlight that plants do. Other algae and photosynthetic bacteria use different **pigments** to absorb different colors of sunlight. Consequently, they display different colors and are often named accordingly (*e.g.* 'red algae', 'brown algae', 'golden algae', 'yellow-green algae', 'cyanobacteria', 'purple bacteria').

photosynthesis—cellular process that fixes carbon and stores sunlight energy in organic molecules

pigment—that which absorbs some color and reflects the rest; photosynthesis molecule that collects sunlight energy

Producers in Extreme Environments

Plants, algae, and photosynthetic bacteria produce almost all the food needed by the consumers of the world, but not quite all. There are some environments where plants, algae and photosynthetic bacteria cannot live, and where their energy-containing organic molecules cannot reach.

Algae and photosynthetic bacteria, for example, cannot perform photosynthesis without light, so cannot live their entire life in the soil or in rocks, inside organisms, or in deep water where light cannot penetrate. They also cannot live in 'extreme' environments—namely wherever there is too little water or where there are too few nutrients, or wherever temperatures are too hot or too cold, or wherever the pH

---

a   Although nearly all plants are producers, there are a very few plants that are not (such as the Indian pipe, *Monotropa uniflora*).

b   As in the case of plants, although nearly all algae are producers, a few species are not.

c   All bacteria, and all but a few species of algae (*e.g.* the kelp and seaweeds) are microscopic.

is too high or too low, or wherever the salinity is too high, or wherever the pressure is too high.

All the producers of these extreme environments are members of the biomatrix. Lichens, for example, are the main producers in land environments where nutrients are hard to find or water is in short supply. The fungi that make up lichens are capable of absorbing nutrients when there are not enough of them to support plants, and capable of pulling water out of the air—even the dry air of deserts. The algae or photosynthetic bacteria that cooperate with the fungi to make up the lichens produce the energy-containing organic molecules needed by both organisms. The producers in most other extreme environments are bacteria—each specially designed to live in a particular kind of extreme environment.

In extreme environments where light is absent, producers construct their molecules by the process of **chemosynthesis**. Rather than using light energy, chemosynthesis uses 'chemo', or chemical energy, to synthesize energy-containing organic molecules. Around hot springs on the bottom of the ocean, for example, chemosynthesizing bacteria are producers for a fairly diverse group of animals, including tube worms, shrimp, crabs, fish, and octopuses. Chemosynthesizing bacteria often inhabit environments where no other organism can survive.

In Yellowstone National Park, for example, the waters flowing out of hot springs support a fairly diverse community of biomatrix organisms, creating, in turn, the beautiful colors seen in Yellowstone

chemosynthesis—cellular process that extracts energy from chemicals in the physical environment

hot springs. In the extremely hot water at the source of the springs, the producers are chemosynthesizing bacteria that thrive in hot, near-boiling water. In the progressively cooler waters found at progressively greater distances from the springs, the producers are photosynthesizing bacteria and algae.

Other chemosynthesizing bacteria are designed to thrive in very acidic (low pH) environments. Others still live in very alkaline (high pH) environments. Others live in very salty environments—some planted directly on crystals of salt! Still others live miles beneath the surface under the extreme pressures that are caused by miles of overlying rock.

First consumers

Plants come in such a wide variety of sizes that nearly every size of land animal has a suite of plants that they can consume directly. Most of the biomatrix producers, on the other hand, are too small to directly provide food for animals. For example, most of the animals of the water environments of the world are fed actually fed by protozoa. Protozoa are biomatrix consumers that eat the tiny food particles produced by smaller bacterial producers, and in turn, become the food for progressively larger animals.

Break Down Molecules

Producers use energy from their environment to build energy-containing organic molecules. The energy stored in these molecules can be extracted by breaking them back down again. In fact, the breaking down of energy-containing organic molecules is how consumers get their energy, including many of the organisms of the biomatrix.

Digestion

Animals must consume food (energy-containing molecules) to survive. Animals digest the food by breaking down molecules into simpler molecules. However, not all the food that animals consume can actually be digested by those animals. To help animals with digestion, an entire community of bacteria and fungi inhabit the digestive systems of animals—perhaps all animals. It is this gut flora[a] that is largely responsible for the digestion of grass in ruminants like cattle and for

---

a   When first discovered, microscopic organisms were classified among the 'plants'. The plants living in a particular area were known as the 'flora' of that area, so the microorganisms of the stomach and intestines of animals were referred to as the 'flora' of the gut, or 'gut flora'.

the digestion of wood in termites. The gut flora in humans may involve hundreds of species and include 100 trillion individuals—more than ten times the total number of human cells in the human body. Without biomatrix organisms that help in digestion, most animals would not be able to survive.

Decomposers

When an organism dies, it is important that the body of that organism be broken down (decomposed) into the simple molecules from which it was constructed. There are a couple reasons for this. First, if bodies were not decomposed, we would quickly find ourselves walking on a very deep pile of dead bodies. Second, if organic molecules were not returned to the molecules from which they were made, the earth would run out of these molecules.

Decomposition cleans the earth of dead bodies and dangerous wastes and replenishes the supply of nutrients needed by the earth's organisms. Almost all decomposition is done by bacteria and fungi of the earth's biomatrix. Without biomatrix decomposers the earth would be polluted with dead bodies and waste, and organisms would run out of nutrients.

Convert Molecules

Fixation

A number of elements are needed by organisms (*e.g.* oxygen, carbon, hydrogen, nitrogen, calcium, phosphorus, potassium, sulfur, sodium, chlorine, magnesium, boron, cobalt, copper, fluorine, iodine, iron, manganese, molybdenum, selenium, silicon, zinc). Outside of organisms, each of these elements is available in one or more inorganic forms.

The most common elements of life are found in the atmosphere: oxygen in the form of oxygen gas (O_2), water in the form of water vapor (H_2O), carbon in the form of carbon dioxide (CO_2), methane (CH_4), hydrogen in the form of hydrogen gas (H_2), and nitrogen in the form of nitrogen gas (N_2). Most of the remaining elements are found in minerals in rocks. In the case of most of these elements, the natural, inorganic form is not in a form that can be easily used by organisms. Most of the molecules have to be broken apart to release the needed elements, and some of the bonds are extremely difficult to break. In the case of several of the elements, even the atoms have to be altered by taking away or adding electrons.

If the inorganic form of an element has to be substantially changed for it to be usable by organisms, it is said that the element must be fixed. The process that accomplishes the 'fixing' of that element is called **fixation**. The element carbon, for example, must be fixed to take it out of carbon dioxide and construct organic molecules from it. Almost every organic molecule is storing more energy in its bonds than carbon dioxide, so building organic molecules from carbon dioxide is a rather difficult process that requires considerable energy. Certain organisms had to be provided with rather complex designs in order to fix carbon. In plants, algae, and photosynthetic bacteria, for example, carbon fixation has been designed as part of the process of photosynthesis. In fact, carbon fixation is a process God designed into *all* producers, whether they build their organic compounds by photosynthesis or chemosynthesis. Consequently, much of the carbon fixation that occurs on the earth is performed by algae and bacteria of the biomatrix.

Another life-essential element that requires fixation is nitrogen. With 81% of the earth's atmosphere made up of nitrogen gas, nearly every organism on earth has easy access to inorganic nitrogen. Very few organisms, however, can use this form of nitrogen. Each particle (molecule) of nitrogen gas is made up of two atoms of nitrogen bound together by one of the strongest chemical bonds known. Because it is

fixation—extraction of an element from its physical world reservoir and 'fixing' it in an organic molecule

very difficult to break that bond, nitrogen gas does not enter into very many chemical reactions.

On the one hand, this makes nitrogen gas a great choice for making up most of the atmosphere. On the other hand, it is difficult to separate the nitrogen atoms so that they can be used by organisms. Extremely hot fires and billion-volt lightning bolts are some of the very few things outside organisms that can break these bonds. To make nitrogen available to organisms, nitrogen must be fixed. The only organisms capable of nitrogen fixation are some bacteria in the biomatrix known as nitrogen-fixing bacteria.

The inorganic form of a majority of the life-essential elements is available only in rocks. In several cases (such as for molybdenum) the element is found in very low concentrations, even in rocks where it is most abundant. In rocks, the element is often available in parts per thousand or parts per million. For any of these elements to be used by organisms, it must somehow be mined from the rocks.

In the case of a couple of the elements, bacteria have been discovered that mine that element from the rock. In each case a special type of bacterium living in water releases a molecule into the water that 'lands' on the rock surface and bores into the rock until it encounters an atom of that element. It then jettisons a portion of the molecule attached to the element into the water to be picked up by the mining bacterium. It is likely that for each of the elements necessary for life that is found in minerals, there is at least one bacterium species of the biomatrix that is designed to mine that element from rock and fix it for use by organisms.

Remediators

Bacteria of the earth's biomatrix also work at keeping the earth safe for life. There are times when organisms release waste that is unhealthy for other organisms. There are other times when non-biological processes like volcanoes, hot springs, and storms expose organisms to dangerous substances that have been buried in the rocks of the earth. Our experience is that if we wait long enough these dangerous substances disappear. Closer examination usually reveals that there is a particular bacterium that transforms that dangerous substance into something that is not hazardous to other organisms.

It would appear that for each dangerous substance that might appear in the world, God designed a bacterium to remediate—to take the toxicity out of dangerous molecules. Thus, some of the biomatrix bacteria have been created for the purpose of **bioremediation**. As in the case with other aspects of the biomatrix, we are only just beginning to learn about bioremediation. New bacteria are being discovered all the time that make toxins non-toxic, place radioactive elements in safer molecules, *etc*.

bioremediation—transformation of a harmful substance into a less toxic form

Provide Nutrients

Develop Soil

"Every spoonful of garden soil contains some 10^{10} bacteria..."[1]—that is 10 billion bacteria! What many of those bacteria are doing—in a non-stop fashion, 24 hours a day, 7 days a week, 365 days a year—is generating nutrients that plants need to survive. Fixers are putting needed elements into forms the plants can use, decomposers are breaking down dead plant and animal material for re-use by the plants, and bioremediators are making dangerous molecules safe.

Most plants require soil to thrive and biomatrix bacteria are mostly responsible for making soil 'fertile'. That is why when a sterile landscape is formed—such as after a very hot fire or after lava flows across the earth's surface—bacteria are the very first organisms to inhabit that landscape. They immediately begin the process of building soil so that plants can eventually thrive in that area.

Feed and Water Plants

The role of the biomatrix organisms does not end with creating the nutrients. Some organisms of the biomatrix also deliver the nutrients. In the case of trees, for example, studies indicate that tree roots alone are not capable of absorbing water fast enough to supply the needs of the tree. Closer examination reveals that the roots of the tree are *not* alone. Fungi have been discovered to be associated with the roots to help in absorbing the needed water and nutrients. These fungi are called **mycorrhizal fungi** (Gk. *mykós*, 'fungus' + *riza*, 'roots'). Fungi have been specially designed by God to be more effective at pulling substances into their bodies than any other kind of organism.

Combined with the absorbing ability of the plant roots, mycorrhizal fungi can, in fact, absorb enough water and nutrients to supply the needs of even the very largest trees. Research suggests that these fungi may be used by all woody species of plants, and perhaps by a vast majority of all plants.

Vitamin Source

Biomatrix bacteria also produce vitamins that are necessary for the health of plants and animals. Vitamin B_{12}, for example, cannot be produced by any animal or plant—it can *only* be produced by biomatrix bacteria. In humans, bacteria in the large intestine produce Vitamin K and Vitamin B_7 (also known as biotin or vitamin H).

> **mycorrhizal fungi**—fungi in mutualism with plant roots that provide water and nutrients to the plants

Protection

To keep them healthy after the Fall, God designed organisms with rather complex systems of protection from pathogens. Part of this involves keeping the pathogens from getting into an organism's body. This includes surrounding organisms with tough coverings that are hard for pathogens to get through. But it also includes planting friendly microorganisms on the outside of those tough coverings so that pathogens are prevented from living on those coverings. Being microscopic, these friendly microorganisms are part of the biomatrix, and created to protect larger organisms. Humans, for example, have a skin flora designed to provide our first line of defense against disease.

8.2 | Biogeochemical Cycles

As mentioned already, a whole suite of elements is necessary for biological life to survive on earth. 98% of the bodies of organisms are made up of oxygen, carbon, hydrogen, and nitrogen. Despite the rarity of the remaining elements, biological life cannot exist without a number of them, including (in alphabetical order) boron, calcium, chlorine, cobalt, copper, fluorine, iodine, iron, magnesium, manganese, molybdenum, phosphorus, potassium, selenium, silicon, sodium, sulfur, and zinc. Many of these elements have to be provided to organisms on a continuous basis, to build growing bodies, to repair damage, and to replace worn out parts.

Furthermore, God originally intended His creation to persist indefinitely. It would seem that God created processes to supply His organisms with a continual supply of these life-necessary elements. Similar to the water cycle that was created by God to make water continually available, God also created a **biogeochemical cycle** for each element needed by organisms. Each biogeochemical cycle is a cyclical process so that the element is always 'in stock'. Most of the biogeochemical cycles include a **reservoir** or sink for storing the element so that there is enough to provide all the organisms with the element.

As noted above, the common form of many of the elements needed by organisms are not usable by organisms. So, for many of the biogeochemical cycles God has designed biomatrix organisms to fix the desired element so that organisms can use it. Once the element is fixed, it is usually picked up first by producers and then passed on

biogeochemical cycle—organism-driven process continuously supplying organisms with a required element by fixing the element from an inorganic reservoir, passing it through all organisms and then returning the element to the reservoir

reservoir (elemental)—location where an element essential to organisms is stored in sufficient quantity to provide all organisms with the element

through consumers to all the remaining organisms. And, for many of the biogeochemical cycles, there is another set of biomatrix organisms designed to return the element to the reservoir.

Consequently, biomatrix organisms play critical roles in most of the biogeochemical cycles. Without those tiny organisms, life would run out of the elements they need to survive—either because the reservoirs would empty or because we would be unable to get the element out of the reservoir... or both. And, once again, as with other aspects of the biomatrix, our understanding of the biogeochemical cycles is incomplete. We are learning more about them every day. A few of the better-known cycles are summarized below.

The Carbon Cycle

Carbon is the backbone of every organic molecule, so organisms cannot exist without it. Of all the places where inorganic carbon is found, only a very small percentage is in the form of carbon dioxide (CO_2) gas in the earth's atmosphere.

In fact, even there, carbon dioxide only makes up about 0.04% of the atmosphere. Yet, it is atmospheric carbon dioxide that is the inorganic source of almost all the carbon found in organisms. So, CO_2 in the atmosphere is the reservoir for the carbon cycle. And, as noted above, carbon cannot be used by many organisms in the form of CO_2. Carbon is one of the elements that has to be fixed so that it can be used.

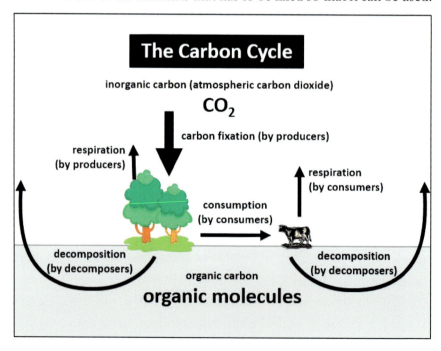

God created producers to use carbon fixation to develop organic molecules—either as part of photosynthesis in plants, algae, and photosynthesizing bacteria, or as part of chemosynthesis in chemosynthesizing bacteria. Consumers then distribute the carbon to the remaining organisms of the planet. Some of the organic carbon is returned to CO_2 in the atmosphere as organisms respire (break down organic molecules to get energy from them). Most of the remaining carbon is returned as decomposers break down organic molecules.

The Nitrogen Cycle

Nitrogen is used in DNA (the genetic molecule of nearly all organisms) and most proteins (used by all organisms to make chemical reactions go fast enough to benefit organisms). The reservoir for inorganic nitrogen is nitrogen gas in the earth's atmosphere. About 81% of the earth's atmosphere is nitrogen gas (N_2). So that the atmosphere is safe, the nitrogen atoms of nitrogen gas are so strongly bonded that they are extremely difficult to separate. God created special nitrogen-fixing bacteria to break those bonds and produce ammonia (NH_3). He also created ammonia-oxidizing bacteria to transform ammonia into nitrites and nitrite-oxidizing bacteria to transform nitrites into nitrates (the ion NO_3^-).

In places where oxygen is abundant—such as on land and in shallow water—producers can readily take up nitrates to get their nitrogen. In

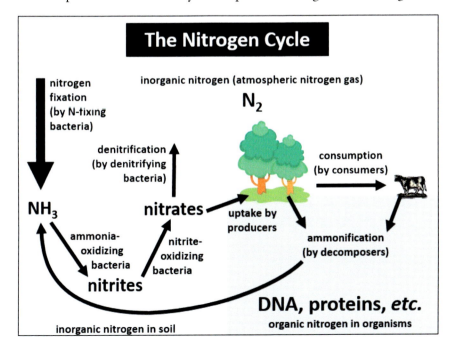

places where oxygen is lacking, producers can get their nitrogen from ammonia. Consumers then distribute the nitrogen to the remaining organisms of the planet. Decomposers convert most of the nitrogen in dead bodies and wastes back to ammonia, and denitrifying bacteria convert nitrates back to nitrogen gas.

The Sulfur Cycle

All organisms use proteins, and a couple of the common **amino acids** from which proteins are built utilize sulfur. Consequently, sulfur is an essential element for biological life. Minerals in the rocks of the earth function as the reservoir for inorganic sulfur. That sulfur is in the form of sulfides and elemental sulfur (S). God created sulfur-oxidizing bacteria to transform mineral sulfur to sulfate (ions of $SO_4^=$). Producers can take in sulfate to get the sulfur they need. Consumers then distribute the sulfur to the remaining organisms of the planet.

If plenty of oxygen is available to decomposers, then the sulfur in dead bodies is returned to sulfate for used again by organisms. If oxygen is not so readily available, sulfur-reducing bacteria transform the sulfur in organic molecules into sulfur and sulfides.

We do not have as complete an understanding of how the other life-essential elements are provided to organisms. However, it is most probable that for each element, biomatrix organisms have been designed to form a biogeochemical cycle that continually provides that

> **amino acid**—monomer for proteins

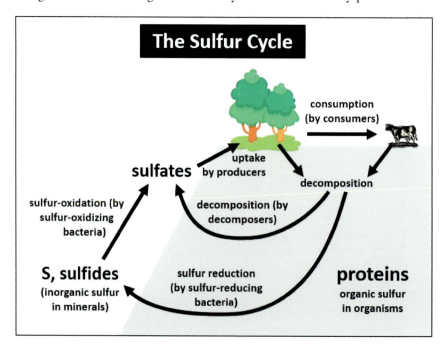

element to organisms. When all the cycles are considered together, the biomatrix is always active in altering the earth in major ways. Biomatrix organisms seem to control what gases are found in the earth's atmosphere and in what concentrations. The biomatrix also affects—and perhaps determines—the chemistry of the world's oceans.

Finally, the biomatrix may be responsible for many of the earth's rocks, concentrating elements into minable ore deposits, creating huge accumulations of oil and natural gas, and probably forming most of the carbonates (limestones and dolomites) of the world.

It is apparent that God created a complex network of organisms to make life possible on this planet. None of these organisms are specifically mentioned in Scripture, but they were clearly created by God to make it possible for plants and animals to thrive. All of these organisms—the producers and consumers, the digesters and decomposers, the fixers and remediators, the developers of soil, the mycorrhizal fungi, and the protectors—have been created to make biological life possible. The biomatrix bridges the gap between the living and the non-living. When the biomatrix is healthy, it is designed to provide everything needed for life and to eliminate everything harmful to life. If levels of any substance are too high the biomatrix works to bring them down. If levels are too low, the biomatrix works to bring them up.

Because plants and animals are dependent upon the biomatrix for their survival, the biomatrix organisms in soil and water were probably created before the plants on Day Three of Creation Week. The biomatrix organisms more closely associated with organisms (gut florae, mycorrhizal fungi, *etc.*) were probably created at the same time as the organisms with which they are associated.

The Origin of Biological Sustenance

In 1972, the radical environmental organization 'The Club of Rome' published a book entitled *Limits to Growth*. They warned that we were running out of non-renewable resources. Some resources were considered **renewable**, either because they were continually available (like sunlight) or because they could be replaced (such as trees). Other resources, however, such as iron, copper, silver, gold, oil, and natural gas, were considered **non-renewable**, because they were being used but not replaced. *Limits to Growth* predicted that the costs of non-

renewable resource—a substance needed by organisms that is continually made available or continually generated (vs. non-renewable resource)

non-renewable resource—a substance needed by organisms that is continually made available or continually generated (vs. renewable resource)

renewable resources would skyrocket as these things became scarce and we would run out of them by the early 2000's.

In the 40 years since the book was published, their predictions were shown to be quite inaccurate. Not only have most of these resources *increased*, despite increasing use and huge increases in human population, most of them are *cheaper* today than they were in 1972. In part this is because the earth contained more of these resources than was believed at the time, and in part this is because we have become better at finding these resources. But it is also due to the fact that very few things are actually truly non-renewable, because God created the earth capable of sustaining life.

Biogeochemical cycles seem to be cleverly designed. The combined characteristics of a reservoir large enough to supply the needs of organisms, specially designed organisms that fix the element, and still other organisms specially designed to return the element to the reservoir, makes each cycle simple and yet effective at supplying the everyday needs of billions of organisms.

It is hard to imagine a simpler system, and it is hard to comprehend how a system could do more. In human experience, this type of **elegance**—simultaneous simplicity and efficiency—is not generated without intelligence. Elegance is due to careful design. Naturalism would expect something like what The Club of Rome expected—that we should be running out of our resources.

elegance—beautiful simplicity of a design that accomplishes a complex result

Even the existence of a single biogeochemical cycle would suggest careful design. But there is not just one such cycle, there are many—probably as many as there are different elements in organisms. Then there are the thousands of different kinds of molecules that the biomatrix is able to bioremediate, and hundreds of thousands of different kinds of plants to which mycorrhizal fungi are able to provide water and nutrients. How such a complex and elegant biomatrix came to be is difficult to explain with the purposeless process of naturalistic evolution. It makes much more sense that the God of love specially designed the biogeochemical cycles and the other features of the biomatrix to provide for the ongoing needs of His biological creation.

After Adam sinned, he and Eve were kicked out of the Garden of Eden and denied reentry so that they would not eat of the tree of life and live forever. This suggests that before the Fall—and if the Fall had never occurred—Adam and Eve would have lived forever.

For this to happen, the original creation must have been fashioned with cycles of provision for everything in the universe. The cyclical nature of these processes would permit existence without end. Part of God's curse on the creation in response to man's sin probably involved the ending and/or inefficiency of one or more of these cycles[a].

---

a As we shall see in Chapter 10, not all of the changes that came with the curse are bad in an absolute sense. They would all be bad in a perfect creation, but some of the changes that came with the curse were designed to minimize the evil in a cursed creation (*e.g.* so that the creation would not last forever, always degenerating, but never able to be restored).

Biological Sustenance: Our Responsibility

Our Responsibility to God

When we observe a living thing, we should remember the extraordinary designs necessary to make that life possible. Hundreds of organisms had to be created in special ways to supply the elements that that organism needs every day to survive. The design necessary to supply all the needs of billions of organisms across this planet should remind us of the extraordinary wisdom of God.

The elegance of the cycles of provision should impress us even more with the wisdom and efficiency of God. The fact that the cycles of provision are too elegant to have come about by naturalistic evolution should direct our thoughts to God as Creator. But the cycles of provision were not just designed in the mind of God. They were actually created, and created in such a way as to successfully provide for every organism on this planet. This argues for the remarkable power of God to make His designs happen.

Furthermore, the fact that so many organisms are provided for ought to remind us that our God is *Jehavah-jireh*, the providing God. And, if God can provide for the biological world, should He not also be capable of providing for our needs? "If God so clothe the grass of the field, which today is, and tomorrow is cast into the oven, shall He not much more clothe you, O you of little faith?" (Mat. 6:30). Even, more, God's provision should provide insight into what it means for God to have the attribute of love.

Finally, the fact that cycles of provision were created to be physical illustrations of that love should impress us that such an awesome God desires that we know Him. It follows then, that contemplation of the biomatrix cannot help but prompt us into worship of the One Who created it. On-going contemplation of the biomatrix should prompt a life of worship. And, in the midst of such worship, how can we help but bring others into that worship? In so doing, we fulfill our calling as priests of the creation.

8.3 | Our Responsibility to the Creation

Preserve Our Body's Biomatrix

When God created the human body, He created it covered with and containing hundreds of species of biomatrix organisms. Some live on our skin and protect our bodies from disease. Others live in our intestines helping us digest our food and providing us with vitamins we cannot get from any other source. Others probably help us in a host of other ways we do not yet understand. As kings of the creation, we have a responsibility to preserve our body and the biomatrix He created as part of it.

Towards that end, although it is good idea to keep our bodies clean, completely sterilizing our skin or cleaning ourselves excessively, will destroy the organisms of the biomatrix created on our skin to protect us from pathogens. And, although taking an antibiotic may help our immune system destroy a pathogen that is infecting our bodies, taking an antibiotic when we are not sick is *not* a good idea. Antibiotics kill more than just pathogens; antibiotics also kill organisms of the biomatrix that are beneficial to us.

Taking antibiotics when we are not sick kills organisms that are of benefit to us—organisms that are protecting us, helping us digest food, providing vitamins, and otherwise helping in a host of other ways. Rather than making us healthier, taking excessive antibiotics weakens us. Furthermore, it destroys, rather than preserves, the biomatrix over which God made us rulers.

Restoring Provisionary Cycles

Humans have not only failed to preserve the biomatrix, time and time again they have upset its balance. Humans have introduced

pollutant—harmful chemical introduced into the environment

bioaccumulation—accumulation and storage of pollutants by an organism

magnification, biological—more rapid accumulation and storage of pollutants by higher consumers

pollution—harmful chemicals accumulated more rapidly than remediated by the biomatrix

harmful chemicals (**pollutants**) into the environment at a more rapid rate than the biomatrix has been designed to remediate. Because the biomatrix is designed to make life possible, pollution can kill organisms or even extinguish entire species.

Lacking a way to rid themselves of pollutants, some organisms accumulate the pollutants from the environment, a process called **bioaccumulation**. Because longer-lived species collect the pollutants for longer, bioaccumulation is a greater problem for longer-lived species. It can even be worse for species that consume other species. If a bioaccumulating consumer eats bioaccumulating species, the levels of that pollutant can increase faster in the consumer than in the species it eats.

This process, where organisms higher in the food chain accumulate toxins more rapidly, is called **biological magnification**. This process puts the highest predators in a given community at the greatest risk of being negatively affected by pollutants. A few examples of different types of **pollution** include the following:

Acid Rain

Because of carbon dioxide picked up from the atmosphere, precipitation (*e.g.* rain, snow, dew) is naturally slightly acidic from the carbonic acid formed when carbon dioxide dissolves in water ($CO_2 + H_2O \rightarrow H_2CO_3$).

However, precipitation can be made much more acidic by various types of pollution. Coal-burning power plants, for example, release sulfur dioxide particles into the atmosphere. Gas-burning vehicles, gas-burning and oil-burning power plants, and nitrogen-rich fertilizers release nitrogen oxide particles into the atmosphere. In the atmosphere, sulfur and nitrogen oxide particles affix onto dust grains and form dust grains themselves. When water condenses onto the dust grains, the oxides produce sulfuric and nitric acids, creating precipitation much more acidic than normal—something called '**acid rain**'.

Sometimes the acid rain is too acidic to be controlled by the earth's biomatrix in a given area. When lakes and rivers become too acidic, many creatures are harmed or die. For example, young fish may not develop correctly and adult fish can be killed. In fact, acid rain is thought to be responsible for over 200 lakes in the Adirondack Mountains having no fish at all. On land, acid rain burns tree leaves and acidic ground water depletes the soil of needed positive ions (like Ca^{++}) and releases positive ions that are toxic (like Al^{+++}).

These things weaken trees and make them susceptible to stress and disease. As a consequence, acid rain is thought to be responsible for widespread tree death in the Great Smoky Mountains. In an attempt to control acid rain, low-sulfur fuels have been sought and filters have been required for power plant stacks and automobile exhausts. There is even reason to believe that since pollution controls have been put in place, some of the damaged ecosystems in the Adirondacks and the Great Smokies are recovering.

Synthetic Materials

Humans have created a variety of substances never before known in the natural world, such as nylon, teflon, and plastic. At the time when many of these substances were created, there were no decomposers available to break the substances down and allow the components to be **recycled**. Substances like this are **non-biodegradable** and are 'disposed of' as trash. Being undigestible, some of this trash can accumulate in an animal's stomach until it kills the animal.

In landfills some of this trash can release non-biodegradable chemicals that pollute ground water. Such problems can be minimized by reusing (recycling) non-biodegradable substances. Or, biodegradable substances can be invented to replace the non-biodegradable substances.

acid rain—rain made more acidic than normal due to acid-causing pollutants in the atmosphere

recycle—clean and reform trash (especially non-biodegradable trash) so as to reuse it

non-biodegradable—not degradable into components that organisms can use (vs. biodegradable)

For example, biodegradable plastics have been created from trees to replace some non-biodegradable plastics made from petroleum. Alternatively, organisms can be sought that decompose the new substances, such as a fungus discovered in South America that can break down polyurethane plastic.

Environmental Chemicals

Whereas many non-biodegradable substances were not produced for the express purpose of adding them to the environment, humans have created some substances with the express purpose of applying them to the environment about them. But some of these substances, added to the environment for one purpose, have unintended side effects.

For example, humans have created many herbicides, insecticides, and fungicides to control diseases and promote agriculture. Many of these substances were unknown in the natural world until humans created them, so many of them were not decomposed or otherwise controlled by organisms of the biomatrix. For example, beginning in the 1940s, dichlorodiphenyltrichloroethane (DDT) was widely used as an insecticide, and saved millions of people from malaria, especially during World War II.

Over the following decades, DDT, and later a DDT break down product called dichlorodiphenyldichloroethylene (DDE), were found to biomagnify in the tissues of ospreys, eagles, and other birds of prey at the top of the food chain, causing them to get sick and their egg shells

to thin. These thin egg shells caused chick death which threatened raptor populations[a].

Several kinds of mining activities release toxins in the environment at a faster rate than the biomatrix cleans up the toxins. Heavy metal mining in the region of Ducktown, Tennessee, for example, destroyed most of the organisms in the area. Storage of nuclear wastes presents another challenge, as nuclear wastes can remain radioactive (and thus dangerous) for thousands of years, or even longer.

It is our responsibility to carefully study the earth and do everything we can to understand it, make sure its balance is not upset, and promote technology that will benefit both man and the creation. This should be done because the biomatrix is necessary for life on this planet. But it should also be done because God gave us the responsibility to preserve the creation, and because a healthy creation better illustrates the character of God.

Enhancing Bioprovision

Besides preserving the earth's biomatrix, it should also be our responsibility to enhance it, to 'return it' better able to support life than when it was given to humans in the first place. In 1975 a variety of *Flavobacterium* bacteria was discovered that could consume some previously non-biodegradable molecules associated with the production of nylon. This suggests that in the original creation God may have created organisms in such a way that they were ready to decompose nylon when humans finally created nylon.

This, in turn, suggests that God may have done this for every molecule that humans would ever create. If so, then God may have also given us the ability to express that information and thus enhance the earth's biomatrix. And, since doing this would only reveal what was already there, we would have no right to brag about our ability. Instead, our proper response would be to glorify God because He was not only

---

a   Although DDT is capable of harming some organisms, it got an undeserved reputation in the 1960's when its harm was exaggerated by activists. For example, in 1962 Rachel Carson published *Silent Spring*, a wildly popular propaganda piece. The premise of the book was a fictional community waking up to a silent spring, where the sounds of birds are no longer heard because they had been wiped out by pesticides. Carson actually distorted the scientific research and claimed that DDT caused human cancer, endangered birds with extinction, and threatened life in the oceans. Each of these statements was refuted in years to come, but the damage was done. In 1972 the United States banned DDT, and countries around the world followed suit. Since malaria spread as DDT was outlawed, it is likely that millions of people have suffered and died from malaria as a result of DDT's premature ban.

the One Who hid the information, but He was also responsible for our ability to reveal it.

A similar example of these hidden abilities has been shown in bacteria that can remediate DDT and DDE. At this point, however, we know so little about the biomatrix, that it will take much more research to understand its hidden qualities and use them to clean polluted environments. Yet, such research would fulfill our responsibility to enhance the biomatrix and glorify God.

Summary of Chapter

A. As a physical illustration of His love, God created the biomatrix and associated biogeochemical cycles to provide the daily, ongoing needs of the earth's plants, animals, and humans.

B. The biomatrix is made up of the bacteria, algae, protozoa, and fungi of the earth. The biomatrix makes it possible for plants, animals, and humans to exist on earth by

  a. algae and photosynthesizing bacteria being the main producers (the organisms constructing the energy-containing molecules that all organisms must have to survive) in water environments;

  b. being the main producers in extreme environments (*e.g.* lichens in dry and nutrient-poor land environments, chemosynthesizing bacteria in deep oceans, soils and rocks, and very hot, very acid, very alkaline, and very salty environments);

c. protozoa being the first consumers in water environments so animals can eat the energy-containing molecules that are otherwise in too-small bacteria

d. gut flora (various bacteria and fungi) helping animals digest food;

e. fungi and bacteria decomposing dead plant and animal material to clean the earth and return nutrients to the environment for its organisms;

f. bacteria fixing elements that organisms need (*e.g.* carbon, nitrogen, molybdenum, nickel) into a form that organisms can use;

g. bacteria returning elements to the reservoirs where those elements are stored;

h. bacteria remediating (making dangerous molecules safe);

i. soil flora building, maintaining, and fertilizing the soils of the world;

j. mycorrhizal fungi absorbing water and nutrients to provide for the needs of plants;

k. some bacteria producing vitamins (*e.g.* Vitamin B12) needed by plants and animals; and

l. skin flora protecting us from pathogens that might land on our bodies.

C. Each of the elements necessary for biological life (especially oxygen, carbon, hydrogen, and nitrogen, but also boron, calcium, chlorine, cobalt, copper, fluorine, iodine, iron, magnesium, manganese, molybdenum, phosphorus, potassium, selenium, silicon, sodium, sulfur, zinc) is part of a biogeochemical cycle that continually (and because it is a cycle, without end) provides that element to organisms. Many biogeochemical cycles include 1) a reservoir where that element is stored, 2) biomatrix organisms that take the element out of the reservoir and provide it to producers, 3) consumers that spread the element to the remaining organisms, and 4) biomatrix organisms that return the element to the reservoir. Examples include:

a. The carbon cycle, where producers (plants, algae, photosynthesizing and chemosynthesizing bacteria) fix CO_2 from the atmosphere, consumers spread the carbon to remaining organisms, and biomatrix decomposers (bacteria, fungi) return the carbon to the atmosphere;

b. The nitrogen cycle, where nitrogen-fixing bacteria fix $N2$ from the atmosphere into ammonia, ammonia- and nitrite-oxidizing bacteria create nitrates, producers take up the nitrates and consumers spread the nitrogen to remaining organisms, decomposers return the nitrogen to ammonia, and denitrifying bacteria return nitrate to N_2.

c. The sulfur cycle, where sulfur-oxidizing bacteria transform sulfur and sulfides in rocks into sulfates that producers can take up, consumers spread the sulfur to remaining organisms, decomposers return the sulfur to sulfates, and sulfur-reducing bacteria return sulfates to the sulfur and sulfides of rocks.

  i. The elegance (complex tasks performed with simplicity and efficiency) of biogeochemical cycles suggests that they were designed by a wise creator and are inadequately explained in the naturalistic worldview.

  ii. So as to preserve the biomatrix God created for our bodies we should not sterilize or overwash our skin, and we should take antibiotics only to help our bodies kill a pathogen that is making us sick.

  iii. Humans pollute when they introduce harmful chemicals into the environment faster than the biomatrix can remediate them. Pollutants are more harmful to those organisms that bioaccumulate (accumulate a given pollutant), and, by biological magnification, are most harmful to the bioaccumulators at the top of the food chain. We have a responsibility as rulers of God's creation to control our behavior to minimize or eliminate pollution. Examples of pollution:

- Acid rain is rain water that is made acidic enough to harm water animals and trees. Humans can cause acid rain by burning coal, oil, and gas. Such burning produces nitrogen and sulfur oxides that become nitric and sulfuric acid when dissolved in rain water. The problem can be minimized by reducing energy consumption, finding alternative energy sources, and filtering acid-causing particles out of exhaust smoke.
- When humans create artificial substances that cannot be broken down by biomatrix organisms, those non-biodegradable substances accumulate in landfills, and can potentially pollute the soil and harm organisms near those landfills. The problem can be minimized by recycling, replacing non-biodegradable substances with biodegradable substances, or finding, creating, or promoting organisms that can break down those substances.
- Mining can bring substances from beneath the earth's surface that are dangerous to organisms living on the surface. The problem can be minimized by preventing mining refuse from entering the environment around the mine.

D. Humans can enhance the biomatrix by revealing the organism designs that God placed in the creation for the purpose of cleaning up man's messes.

Test & Essay Questions

1. How does God demonstrate His love?
2. What is the purpose of a biogeochemical cycle?
3. What are the components of a biogeochemical cycle?
4. Why is a biogeochemical cycle cyclical?
5. What does it mean for an element to be 'fixed'?
6. Explain what a reservoir is in a biogeochemical cycle.
7. Describe the major components of the carbon / phosphorus / nitrogen / water cycle and explain what the cycle is designed to do.
8. Define bioremediation.
9. Define the biomatrix.
10. List three / four / five different components of the biomatrix.
11. What is the function of the biomatrix?
12. The observation of biological sustenance should cause what sort of worship?
13. Short Essay: Why should we restore provisionary cycles?
14. Define pollutant.
15. Define bioaccumulation / biological magnification. / Why does pollution affect some species more than others?
16. List three / four different types of pollution.
17. Explain what acid rain is / what causes acid rain / how acid rain negatively affects life / how to reduce acid rain.
18. Explain why artificial substances disrupt the biomatrix / how artificial substances can negatively affect life / how to reduce the negative effects of artificial substances.
19. Explain what it means for something to be biodegradable / why biodegradable substances should be preferred.
20. Short Essay: How can humans enhance bioprovision? / How might hidden information help us enhance bioprovision?

Reference

1. Margulis, Lynn, and Karlene V. Shwartz, 1998, *Five Kingdoms: An Illustrated Guide to the Phyla of Life on Earth*, Third Edition, W. H. Freeman, New York, NY, 520 p. (p. 44).

CHAPTER 9

GOD IS ONE

Biochemistry and Systems Biology

"I pray ...for those who believe in Me through their message. May they all be one, as You, Father, are in Me and I am in You... May they be one as We are one."
Jesus in John 17:20-22, HCSB

9.1 | One God

"The Lord our God is one Lord" (Deu. 6:4; Mark 12:29). "There is one God…" (I Tim. 2:5[a]) and there is no other God[b]. He is called by the names One (*e.g.* Isa. 30:29[c]) and Holy One (*e.g.* Job 6:10[d]). It is because of the unity of the Godhead—even though He exists in three persons (God the Father, God the Son, God the Holy Spirit)—that He desires His people to be one, even though many people are involved (John 17:20-26). While the Church reaches out to all people and all nations (Mat. 28:19; Mark 16:15), the church is to be unified as God is unified (Rom. 12:5; I Cor. 1:10; 12:12; Gal. 3:8).

The Unity of Life

Even before God created the first human, He created evidence of His oneness. He cannot be seen, so God's unity cannot be seen. Yet, since God wanted us to see His invisible attributes in those things He made (Rom. 1:20), He created illustration of His oneness in the visible creation. He did so by creating things out of the same building blocks, by creating similarities in all organisms, and by creating biological systems. Let us look more closely at each of these.

Common Monomers

God chose to construct the objects of the universe out of the same building materials—the same subatomic particles and the same atoms. This is both efficient—to allow transformation of one thing to another—but also an illustration of the fact that there is one God. At the same time, to illustrate His uniqueness and personhood, when He designed biological life, He created each individual distinct from every other. Each organism was given a unique set of organic[e] molecules. Most of these organic molecules are many times larger than the typical molecules found outside organisms, so they are called **macromolecules**. As unique as these billions of different

macromolecules—organic molecules many times larger than typical molecules found outside organisms

---

a Other references that indicate there is only one God include Psa. 86:10; Rom. 3:30; Gal. 3:20; I Tim. 2:5; and James 2:19.

b Reference that indicate that there is no God besides God include Deu. 32:39; II Sa. 7:22; I Chr. 17:20; Isa. 43:10; 44:6; 45:18; and Mark 12:32.

c Additional references where the name of God is One: Isa. 49:26; 57:15; and Zech. 14:9.

d Additional references where the name of God is Holy One: Psa. 71:22; Isa. 1:4; Hos. 11:9; Hab. 1:12; and 3:3.

e These molecules are 'organic' because they are constructed around a backbone of carbon atoms.

monomer—molecular building block of (biological) macromolecule

carbohydrate—organic molecule with carbon, hydrogen, and oxygen atoms in a ratio of 1:2:1

cellulose—complex carbohydrate making plant structure (fiber); earth's most abundant organic molecule

chitin—complex carbohydrate making up cell walls of fungi and exoskeletons of arthropods

starch—carbohydrate macromolecule for storing energy in plants

glycogen—carbohydrate macromolecule for storying energy in animals

monosaccharide—5- or 6-carbon sugar molecule that is the monomer for biological carbohydrates

protein—(biological) macromolecule constructed of amino acids

enzyme—protein molecule that speeds ups (catalyzes) biological chemical reactions

macromolecules are, however, most of them are constructed from just a handful of different **monomers**.

Most organic molecules can be grouped into four different categories:

- **Carbohydrates** are organic molecules where the carbon, hydrogen, and oxygen atoms are in a ratio of 1:2:1. They come in a host of different forms in different organisms. The most abundant organic molecule on our planet is **cellulose**, a complex carbohydrate macromolecule that makes up most of the structure (fiber) of plants. Other common carbohydrate macromolecules include **chitin** (which makes up the cell walls of fungi and the hard, external skeletons of creatures like insects, lobsters, and crabs), **starch** (energy storage molecule in plants), **glycogen** (energy storage molecule in animals), and the various sugars (for shorter term storage of energy). As different as carbohydrates are, they are all constructed from a handful of different simple sugar monomers known as **monosaccharides** (*mono* for 'one' + *sacchar* for 'sweet'). The most common monosaccharide is *glucose* (blood sugar). Others include galactose and fructose which are common sugars in fruit. Lactose (the common sugar found in milk) and sucrose (table sugar) are disaccharides, each being built of two monosaccharides. Lactose is made of one glucose and one galactose. Sucrose is made of one glucose and one fructose. Cellulose, chitin, glycogen, and starch are polysaccharides, each being constructed of many glucose molecules.

- **Proteins** are an extremely diverse group of organic macromolecules constructed of monomers of amino acids. Although organisms only use 20 different amino acids, millions of different proteins can be made from them. Most proteins do what they do because of the shapes they have. Proteins that are long and strand-like are used as cables and columns to hold cells together or as winches to drag chromosomes around inside cells, or as whips or oars to move cells. Tube-shaped proteins can function as doors through cell membranes. Clasping proteins that dissolve in water are used to carry molecules that do not dissolve in water. An extremely important group of proteins are **enzymes**, proteins that speed up the breakdown or buildup of other molecules. Many of them doing

so by having two 'arms' that grab two items and pull them together or stretch them apart. Without enzymes chemical reactions would take too long for physical life to be possible.

- **Nucleic Acids** are macromolecules made of extremely long chains of nucleotide monomers. Each nucleotide is made of a monosaccharide, a phosphate group, and a nitrogen-containing group. When the monosaccharide is ribose, the molecule is called Ribo-Nucleic Acid (RNA). When the monosaccharide is deoxyribose, the molecule is called **Deoxyribo-Nucleic Acid (DNA)**. The long sequence of nucleotides allows DNA and RNA to carry genetic information.

- **Lipids** are organic molecules that are **hydrophobic** ('water-fearing'—*hydros* for 'water' + *phobos* for 'fear'). These are the fats, oils, waxes, phospholipids, and steroids. These molecules are non-polar, meaning they either do not have electrical charges at all or the charges are extremely weak. Water molecules are so much more attracted to each other than they are to lipid molecules that they tend to associate with other water molecules and push lipid molecules away. Therefore, not only do not these molecules dissolve in water, they tend to separate from it. Bees' wax, for example, repels water, and cooking oil floats on water. Most lipids are constructed from **fatty acids**, which are long chains, or tails, of carbons with hydrogen atoms attached to them. In a long string of

nucleic acid—(biological) macromolecule constructed of nucleotides, designed for carrying information

DNA—large nucleic acid that stores genetic information

lipid—organic molecules that, because they are non-polar, do not dissolve in water

hydrophobic—water-repelling (because molecule is non-polar)

fatty acid—long carbon chains with hydrogen atoms; monomer for non-steroid biological lipids

carbons, each is attached to two other carbons, so it can form two more bonds. To stay in a chain form, any given carbon can either bond with two hydrogens or it can bond with one hydrogen and double up one of its bonds with a carbon next to it. When all the bonds between carbons are single bonds, the molecule is 'saturated' with the most hydrogens possible and it is called a saturated fatty acid. An unsaturated fatty acid has one or more double bonds along its length.

- Fats and oils are composed of one, two, or three fatty acid tails attached to a 3-carbon backbone[a]. Most animal fats are solid at room temperature because they are mainly constructed of saturated fatty acids that can be packed closer together. Most vegetable oils are mainly constructed of unsaturated fatty acids, so they tend to be liquid at room temperature. Because organisms can get energy from carbon-to-hydrogen bonds, fats and oils are important energy-storing molecules in organisms. In animals, fats are used around vital organs for cushioning and under the skin for insulation (*e.g.* 'blubber' in whales).

- Phospholipids make up another group of lipids important to organisms Phospholipids are similar to fats and oils by having the same 3-carbon backbone, to which is attached two fatty acid tails *and* a phosphate group. Phospholipids look like a globular head or body with two tails or legs. The phosphate group has electrical charges, so the head end of the molecule is **hydrophilic** (*hydros* or 'water' + *phileo* or 'loving') and the tail end is hydrophobic. In the midst of water, the tails of phospholipids naturally gather together to exclude water, producing double layers (bilayers) of phospholipids. These bilayers tend to create water-filled spheres and spontaneously create cell membranes.

- Waxes are made from fatty acid tails attached to other organic molecules.

- Some lipid molecules are not built from fatty acid monomers. The most important of non-fatty-acid lipids are the **steroids**, such as cholesterol, testosterone, progesterone, and estrogen.

hydrophilic—water-'loving' (because molecule is polar)

---

a This backbone is made from glycerol ($C_3H_8O_3$), which has three carbons connected in a chain with single bonds, and an OH group attached to each carbon at the oxygen (leaving room for 2 hydrogens attached to each end carbon and a single hydrogen attached to the middle carbon.

Steroids are built from four connected rings of carbons. Because steroids pass readily through cell membranes, steroids are common chemical messengers in our bodies.

Every organism has a unique set of organic molecules in each of these four categories. Yet, the diversity of organism chemistry is built from the same few dozen simple molecules. This similarity is the result of all biological organisms being created by the same Creator. It also permits organisms to interact more efficiently. Not only can any one of the macromolecules be built from monomers, but any macromolecule—no matter how unique—can be broken down into monomers.

This makes it possible for an organism to break down its own damaged macromolecules and use the pieces to make replacement molecules. It also makes it possible for a consumer to break down (digest) the macromolecules of another organism and use the pieces to build the molecules unique to the consumer. Finally, it makes it possible for decomposers to break down (decompose) organisms and the macromolecules of organisms so that the earth can be kept clean of dead bodies. And, in the process, decomposers are generating monomers. These monomers can be used by the decomposer to build its own macromolecules or released into the environment to be nutrients for other organisms to build their molecules.

9.2 | Similar Structures

Not only did God create organisms of the same building blocks (monomers), He also built similar structures in different organisms. All organisms contain a few similar structures (DNA to store genetic information, mRNA to carry genetic information, ribosomes to build proteins, ATP to carry energy in cells, *etc.*). Successively smaller groups of organisms can have more and more similarities. For example[a]:

- All non-bacterial organisms (eukaryotes[b]) have the common features of all organisms (*e.g.* DNA, ribosomes, plasma membrane) *plus* DNA in the form of linear chromosomes, cells containing nuclei, *etc.*;

- All animal eukaryotes (and humans) have the common features of all eukaryotes plus they develop from embryos through a blastula stage of development;

- All chordate animals (and humans) have the common features of all animals plus a notochord, a hollow dorsal nerve cord, and pharyngeal slits—at least at some point during their life;

---

a Many of the structures listed in the example may be unfamiliar to the student. It is not important that the student understand what the structures actually are. Rather, it is important that the student understand that members of each successively smaller group of organisms have a longer list of similarities. Thus, organisms within each group are more similar to one another than they are to organisms outside that group. Consequently, the similarity among organisms in successively smaller groups *increases*.

b 'Eukaryotes' (*eu* meaning 'true' + *karyon* meaning 'body'—referring to at least one nucleus 'body' in each cell).

- All vertebrate chordates (and humans) have the common features of all chordates plus vertebrae;
- All mammal vertebrates (and humans) have the common features of all vertebrates plus hair, milk-producing glands in females, and three inner ear bones;
- All primate mammals (and humans) have the common features of all mammals plus forward-facing eyes and three-dimensional vision;
- All hominoid primates (and humans) have the common features of all mammals plus five cusps on their lower molars;
- All humans have the common features of all hominoids plus large brains, *etc.*

This pattern of similarities, which occurs across all of life (not just in the example given above) creates a spectrum of perfection of similar structures. This spectrum suggests an overall unity to life, even in all its variety. From this spectrum, humans also naturally extrapolate to something infinitely one—to one God and one Creator of all things.

Systems Biology

A third way in which God demonstrates His oneness is the way He created things to work together to accomplish common tasks. In spite of the huge variety of things found in the biological world, they work together in integrated wholes. For example, as different as the different parts of our body are, the many different parts work together so well that we usually do not notice that we are made of different parts. That is, until one part gives us trouble! Because each part is essential to the functioning of the whole, and each part is impacted by every other part of the body, there is a sense in which none of the parts are more or less important than any other.

Since each part is designed to work together, no part is endowed with perfect ability, so no part can accomplish its task without the other parts. Since each part is to fulfill only particular tasks in the whole and not others, each part is given only one or a few particular abilities. Each part is impacted by other parts, positively impacted when other parts are positively impacted and negatively impacted when other parts are negatively impacted. These, in fact, are the characteristics of a **biological system**. An increasingly popular sub-discipline of biology, called **systems biology**, studies biological systems like the human body.

system, biological—multiple parts that interact in such a way that the whole has emergent properties

systems biology—a discipline of biology that studies biological systems

emergent properties—properties of an entity unaccounted for by the entity's component parts

A biological system is composed of multiple parts that interact together in such a way that the system as a whole has **emergent properties** (properties that the system possesses that the component parts do not and cannot possess). For it to have emergent properties, a biological system must be composed of parts that 1) are greater than one in number, 2) each possess only a few imperfect properties; 3) each must work together with the other parts; and 4) each impact, and are impacted by, all other parts. When a biological system demonstrates oneness even while being composed of diverse parts, it is physical illustration of the oneness of God.

In Scripture (Rom. 12:4-9; I Cor. 12:11-31), the human body is used to explain how the church is to function. The church is to operate in the same way as a biological system: 1) A church is to be made up of multiple members; 2) God endows each member with only one or a few specific abilities, and no one is endowed with perfect ability; 3) Each member is to use their ability(ies) to work with the other members, and no member is to be considered more or less important than any other; and 4) Each member is positively impacted when other members are positively impacted, and negatively impacted when other members are negatively impacted. Just like biological systems are a physical illustration of the unseen, but real, unity of God, when the diversity in the church is unified in the bond of love, it is also an illustration of the unity of God.

Body Systems

The human body is not the only biological system. In fact, the body of *every* animal and plant is a biological system, constructed of interacting organs and organ systems, and capable of functions (*e.g.* embryological development, reproduction) impossible for the components to do apart from one another. Biological systems also abound at every level of the biological order, including systems within systems *inside* organisms. Each **organ system** in a body, for example, is created with a variety of organs and has its own emergent properties (*e.g.* the circulatory system delivering oxygen to all the cells of the body, the digestive system breaking food down into the basic building blocks of organic molecules, the nervous system processing information). Then, each organ in an organ system is a biological system made of the interaction of cells and tissues.

organ system—biological system of organs that interact with other systems in a large organism's body

Once again, organs accomplish tasks (like a stomach crushing food, a heart pumping blood, a kidney filtering blood) that cannot be performed outside of the interaction of their component parts. Organs themselves are composed of yet another biological system, namely **tissue**. Tissues are composed of suites of different kinds of interacting cells. The combination of cells in a tissue can do things together that the individual cells cannot accomplish (*e.g.* signal transmission by nervous tissue, contraction by muscle tissue, body support by connective tissue, body protection by skin tissue).

Within tissues are found cells. The cell is a biological system of molecules—and often organelles—that interact in complex ways to permit the cell to do what it does. Each cell is a biological system, whether it makes up the entire organism (*e.g.* a bacterium, ameba, or paramecium), or if it is only one cell among trillions in an organism (*e.g.* a blood cell, nerve cell or muscle cell in the human body). A cell can do things (like move, grow, reproduce) that cannot be accomplished by the parts of a cell separate from the system. And, even inside (eukaryotic) cells are found **organelles**.

Each organelle is a biological system—a set of complex molecules interacting with each other to accomplish a higher function that cannot be accomplished by the separate molecules. Examples of organelles include chloroplasts that package the energy from sunlight into monosaccharides, mitochondria that release energy from organic

tissue—biological system of cells in multi-cellular organisms that make up organs in larger organisms

organelle—membrane-bound biological system of molecules, found in eukaryotic cells; the substructure of a cell

molecules in order to power the cell, and the endoplasmic reticulum that builds organic molecules from their amino acid monomers.

Communities

Returning again to the size scale of the body, there is a second way in our body is a biological system. Humans have understood for a long time that our bodies are made of many different body parts, such as hands, feet, head, *etc.* After all, since the human body was used as an illustration in Scripture, this has been common knowledge for thousands of years. More careful studies and the invention of microscopes have added more components like tissues and cells, but until recently, our body was not thought to include *other* organisms. In fact, it has been more common to understand any other organism in or on our body is something of an invader, or at best a freeloader, associating with us for a 'free ride'.

In the last few decades, however, we have been discovering more and more organisms in our bodies that seem to be *helping* us. Our gut flora help us digest food that we cannot otherwise digest. Bacteria living on the outside of our skin and even in our noses, mouths, and so on, tend to prevent dangerous organisms from invading our bodies and causing disease.

Something similar seems to be true of non-human organisms as well. Termites and cows need their gut flora to digest plant cellulose. Plants need their mycorrhizal fungi to pull in enough water and nutrients. What this is suggesting is that perhaps *each* animal and plant in the world must be understood as a biological system of more than one species—what is known as a **biological community**. When all the right species are linked in the proper manner, the plant or animal, as well as each microorganism inside it, behave as they are designed to behave. On the other hand, if the interactions are bad, such as if something is wrong with the plant or animal, or if something is wrong with one or more of the microorganisms, or if something is wrong with how the organisms interact, then the whole community is harmed.

Problems in these relationships are known to affect overall health, behavior, and even development of the larger organism. A number of human illnesses, for example, are due to these kinds of problems[a], rather than infections by single 'invading' organisms. Curing diseases

---

community—(biological) system of plant, animal, and biomatrix species found in a particular location

---

[a] For example, unhealthy populations of gut flora have been linked to various diarrheal diseases, inflammatory bowel diseases, and irritable bowel syndrome.

of this sort will require more study of the human body as a multiple-species biological system. At least some of the emergent properties of a human are due to a complex relationship between the human body and microorganisms that live within it. Restoration to health will require restoration of the community of microorganisms. Since methods developed to kill 'invading' microorganisms (*e.g.* antibiotics) also tend to hurt the microorganisms that are *supposed* to be there, modern medicine has actually inadvertently caused some illnesses.

Furthermore, modern medicine often focuses on individual body parts while finding cures. To cure some illnesses, and to reverse some of the problems caused by modern medicine, the body needs to be studied and treated as the complex system that it is.

At a higher level of biological organization, even organisms that *look* separate and unconnected can be part of biological systems. This can occur among individuals of the same species in social networks. Some social networks are weak, and only persist while the young are being raised. During breeding season, for example, many gulls and ibises gather in breeding colonies so as to provide group protection for their young while they are raising them. Other species have stronger social networks that persist beyond the raising of young, such as in herds of deer, bison, sheep, and flocks of birds.

Other social networks are even stronger. Extraordinary social structures are found among the bees, wasps, ants, and termites. A swarm of honey bees, for example, can relocate, construct comb,

defend themselves, raise young, and store food through the interaction of nurse bees, guard bees, foragers, and the queen. Not only do bees interact for the function of the colony, but worker bees even work themselves to death in a fraction of their possible lifetimes for the benefit of the colony as a whole.

At even higher levels of biological organization, separate and distinct individuals from multiple species can form more traditionally understood biological communities. Although it has been common to claim that such biological communities are merely created by the organisms that just *happen* to be found living in a given area, evidence suggests that might not be true. In fact, biological communities have characteristics of biological systems. Not uncommonly, for example, the introduction of a species from one community into another substantially alters that community, as if the various species in that community are interacting strongly, with every species impacting every other. Furthermore, communities seem to have emergent properties, such as existing where component species cannot because the communities create their own environments.

A particular mix of organisms in a given community creates a particular distribution of nutrients, water, and light. The organisms of the community create the 'shape' of the environment—by creating its surfaces, its exposures, and its hiding places. It might be that God created organisms with certain predispositions to get together

with certain other organisms and thus generate efficient biological systems—what we now recognize as communities.

It is somewhat obvious that the communities created in the Creation Week were designed by God. But it may also be that even after Noah's Flood, when organisms had to recreate communities destroyed in the Flood, God designed organisms in such a way that they would form the kind of communities that God designed them to be in. If so, it should be easy for us to accept the fact that He has designed each of us to fit into and help construct the churches He designed us to be a part of.

The Origin of Biological Unity

No one would disagree that there is a profound amount of biological similarity in the world. However, there is much disagreement as to why that similarity exists. On the one hand, the creationist believes the similarity is there because of a common creator. This is not only because one creator would naturally tend to create things with similar characteristics, but because God desired us to *know* that there was just one creator. He placed an extra amount of similarity into the creation than might be expected as a mere side effect of a common designer.

For those who reject a creator, the similarity is thought to be due to common ancestry. From this perspective, organisms are similar to one another because they inherited those similarities from the same ancestor. Different degrees of similarity among organisms are interpreted by a branching tree pattern of relationship, where more similar organisms were generated more recently, and more dissimilar organisms were generated earlier.

Not so easy to explain is the origin of biological systems. Emergent properties do arise when items in the physical world interact spontaneously—or at least appear to interact spontaneously. Subatomic particles, for example, interact to generate a variety of atoms and ions with properties quite unlike the subatomic particles. Even atoms interact to generate a variety of molecules with properties quite unlike the atoms (such as water being very different from hydrogen gas and oxygen gas from which it is made). There is some question as to whether these interactions are actually spontaneous (after all, they likely occur because of a very particular design of the universe). But even ignoring this, such interactions involve a rather small number of

different kinds of parts, connected in rather simple ways, and result in simple and few emergent properties.

In contrast, in most biological systems, many, many different kinds of parts come together in exceedingly complex ways to generate complicated and numerous emergent properties. In fact, biological systems are not only more complex than 'spontaneously' generated systems, they are typically far more complex than even any systems created by humans. It seems far more reasonable to conclude that biological systems were designed by an intelligence far greater than humans, than it is to suggest that biological systems were generated spontaneously.

An additional problem with complex systems is **irreducible complexity**. Systems with complex properties (*e.g.* human language, the blood clotting system, the human immune system) have to have many parts to be able to accomplish the complex function. For example, human language requires an ear to translate sound waves into electrical impulses, neurons to transmit those impulses to the brain, a brain to translate the electrical impulses into a series of sounds, then translate the sounds into words, then translate the words into concepts, then determine a response, then translate the response into words, then break down words into a series of sounds, then translate the sounds into a series of instructions to the mouth, tongue, vocal cords and lungs, then translate the instructions into electrical impulses, then a suite of different neurons to translate the electrical impulses

irreducible complexity—characteristic of a system where it cannot function with <3 component parts and thus cannot function while assembled one component at time (e.g. by naturalistic evolution)

into appropriate actions of the mouth, tongue, vocal cords, and lungs to produce sounds. If there are too few parts it is not possible to accomplish the function at all.

Thus, for any biological function (emergent property), there is a minimum number of parts that the biological system must have in order to accomplish that function. And in biological systems, the minimum number of parts is usually quite large. It is impossible for all these parts to arrange themselves spontaneously, but those who advocate the worldview of naturalism suggest that naturalistic evolution works one step at a time. They argue that adding one part at a time is something that *could* happen spontaneously.

The problem with such a process when it comes to biological systems is that it seems impossible to build them one step at a time. This is because as long as one has less than the minimum number of parts (a minimum complexity), the system does not seem to work at all[a]. It looks as if biological systems have to start out with all their parts, something that cannot happen either spontaneously or by evolution… but can be accomplished by a clever designer.

9.3 | Biological Unity: Our Responsibility

Our Responsibility to God

Even in a diverse world, God created a spectrum of perfection of unity. The spectrum of biological similarity, from non-living things that lack the characteristics unique to organisms, to disparate organisms that at least share the characteristics found in all organisms, to more and more similar organisms, is an example of a spectrum of perfection of unity. Another example is the spectrum of what might be termed the tightness of biological systems. Loose biological systems (*e.g.* the social network of orangutans) exist that have very few emergent properties and contain parts (*e.g.* individual orangutans) that appear more separate than they appear one. Tighter biological systems have more emergent properties and parts that seem less capable of being independent. The

---

a A typical naturalistic response to an irreducible complexity critique is to claim that the system under consideration was derived from a system with one less part that had a very *different* function—something called 'exaptation'. Given the virtually infinite number of different functions that a complex system *might* have had, it is impossible to claim that a system with one less part could have *no* function at all. However, among the thousands of biological systems that exist, a reasonable step-by-step derivation of even one of these systems has never been proposed by evolutionists—even considering exaptation. Consequently, it seems reasonable to suppose that it is not possible to derive complex systems, even via exaptations.

tightest biological systems (*e.g.* organisms) have numerous emergent properties and are composed of parts that not only appear necessary for the system but also cannot exist without the system.

These spectra of perfection of unity point to the existence of something with perfect unity, the one true God. These spectra ought to help us understand the unity of God and ought to lift our hearts in worship of the one true God. The polytheistic religions are not worshipping the One Who created all things, for the creation confirms His word that it was brought into being by one God. The remarkable way in which the diversity and disparity of life works together so well as to create systems and their emergent properties, should cause us to worship the One Who brought it all to be.

Monomers also suggest something about God's wisdom and beauty. Building all macromolecules from simple monomers is elegant design. Complex things, such as providing individual organisms with unique macromolecules, keeping the earth clean of the bodies of organisms, repairing damaged molecules, and providing nutrients to organisms are accomplished in a very simple way—by forming macromolecules from a few dozen monomers. This is both simple and beautiful (*i.e.* elegant). This gives us insight into God's wisdom and beauty.

Biological systems further impress us with God's creativity, wisdom, and power. Just conceiving of the thousands of biological systems that exist in the biological world impresses us with God's creativity. Designing those thousands of systems with so many interacting parts required a phenomenal amount of wisdom. And finally, what extraordinary power was required to actually fashion those systems and their billions of interactions so that everything works! Contemplation of the nature of biological monomers and systems cannot help but result in worship of the God of wisdom, beauty, creativity, and power.

As astonishing as the Creator is, we should not forget that one of the reasons God created biological similarity and biological systems was to give us visible and finite illustrations of His invisible and infinite unity. The awesome Creator did it so that we could know Him! Because of this, study of biological similarity and systems should not only teach us about Him, but also draw us closer to Him. Contemplation of biological similarity should stimulate us to worship He Who created it. And, as we are drawn into a more intimate relationship with Him,

we can fill our lives with worship, and draw others into that worship. In this way we can fulfill our role as priests of the creation.

Our Responsibility to the Creation

Preserving Unity

God designed unity into the biological creation to help us understand His nature and to draw us to Him. We have a responsibility to preserve and guard that unity. When He calls us into account for how well we have kept His creation, we ought to present it to Him with at least as much biological unity as it had when He presented it to us. Unfortunately, humans have had a rather poor track record in preserving the unity of the biological creation—especially preserving the integrity of biological systems.

Antibiotics

That the human body has emergent characteristics has been understood for centuries. But the fact that the body is as well designed and highly integrated as it really is has not been appreciated for as long. When a highly integrated system is disrupted in any way, it often negatively affects the entire system. The additional fact that the body of a believer is the temple of the Holy Spirit makes it even more important that we be very careful about what we put into our bodies.

Introducing anything that compromises our ability to rule the creation is even more undesirable. For these reasons Christians should avoid tobacco, alcohol, and psychotic drugs. Furthermore, since the human body is also a unified community of organisms, we should take antibiotics only when doing so promotes life[a]—in other words when doing so kills something that is either killing the human body or reducing the diversity of organisms in the human body. For example, antibiotics should not be taken when we are well.

Exotic Introductions

Attempts to treat biological communities as systems only date back to the last half century or so. Prior to that, many communities have been altered by the introduction of **exotic species** (a species introduced from another community). As in any good biological system, all the species in a community interact best *together*. Each plant and animal in a community is provided exactly the right balance of resources it needs (water, nutrients, light, *etc.*) by the other species found in the community.

Research is suggesting that the other species in the community also affect how a particular organism develops, how it behaves, how abundant it is in the community, and how it interacts with other organisms. In any community, all the **native species** (species normally a part of a particular community) are maintained and kept in balance.

exotic species—an organism introduced into one biological community from another

native species—an organism that is normally part of a particular biological community

---

[a] An 'antibiotic' ('anti' 'bios' means 'anti life' because 'bios' means life) is designed to kill biological life—usually disease-causing organisms However, most antibiotics not only kill disease-causing organisms, they also the 'friendly' organisms that are supposed to be in our body. Consequently, taking antibiotics to 'prevent' infections is not wise. For example, one disease, antibiotic-associated diarrhea, is caused when an antibiotic kills most of the beneficial gut flora, allowing the toxin-producing bacterium *Clostridium difficile* to become common in the intestines.

It is when a species is in this balance that it is fulfilling its role in the system and is interacting as it should with other species in the community. And it is when all species are in their proper roles that all the emergent properties are displayed.

When a species is taken out of its native community and placed into another, both the introduced species and its new community are altered. The introduced species will interact with the other species in the new community and change them. The tighter the community, the more the new species alters the community and all its members. But the introduced species is also changed by the different combination of species found in that new environment.

Some exotic introductions have resulted in unexpected and undesirable changes in the community. In 1859, for example, twelve European rabbits (*Oryctolagus cuniculus*) were released in Australia for sport hunting. Within a decade, 2 million rabbits could be hunted every year without negatively impacting the rabbit population, and an estimated 600 million rabbits were living in Australia by 1950. Rabbits have altered native populations of plants and animals and are thought responsible for many or most of the recent species extinctions in Australia.

As another example, in the 1883 New Orleans Exposition, kudzu vine (*Pueraria*) was introduced from China as an ornamental plant to shade porches. In the early twentieth century, the U.S. Soil Erosion Service encouraged people to plant kudzu in order to prevent erosion on slopes (such as on the banks cut into hills along roads) and to feed livestock. Without its native controls, and in the climate of the U.S. Southeast, it very effectively covers slopes, but it also smothers out every other plant species, including trees. It even grows faster than most livestock can eat it. Even with more than 6 million dollars every year spent on mowing and poisoning it, kudzu is estimated to be spreading at 150,000 acres per year.

As a third example, in 1816, vines of Chinese wisteria (*Wisteria sinensis*) were first introduced into the United States for the beauty of its flowers. In U.S. ecosystems, Chinese and Japanese wisterias grow much more aggressively than the American wisteria, even to the point of behaving like kudzu, smothering out all other plants. As a fourth example, the beaver-like rodent, nutria (*Myocastor coypus*), was introduced into the United States from South America for its fur.

In states that border Mexico, however, nutrias have reproduced in very large numbers, reducing the populations of many of the native marsh plants and destroying flood banks (levees) into which they dig their burrows.

In an effort to control an exotic species, other species have been introduced, creating further problems. For example, in 1961, the southern green stinkbug (*Nezara viridula*) was first observed in Hawaii (probably introduced accidently in a food shipment). Because of the damage the stinkbug caused to food crops, in 1963 a parasitic wasp (*Trichopoda*) was introduced to control it. By the mid 1960's the wasp had the stinkbug under control.

Unfortunately, the wasp has also decimated the populations of native insect species, including Hawaii's largest native bug, the koa bug. Exotic introductions have altered communities all over the world. There are estimated to be over 4500 exotic species in the United States, including 25% of the species in Florida and 45% of the species in Hawaii. Many of these have altered the communities in undesirable ways, for example by reducing community diversity. Before an organism is introduced into a community, the effects of that introduction need to be studied carefully—both the effects on the exotic species and the effects on the native community. In the case of communities that have been harmed by exotic introductions (*e.g.* by reducing community diversity), we should restore those communities, only after making sure what is done to restore them does not harm the communities in other ways.

Enhancing Biological Unity

In order to fulfill our dominion roles and glorify God, we should work to enhance the unity of the biological creation. It is likely that God created much more potential for unity than is currently evident in the biological creation, and that He created humans capable of discovering and releasing that potential. There is, for example, a diversity of hidden organism designs that can be revealed by careful breeding. Within that as-yet unrealized diversity there are likely to be never-before-seen combinations of organisms that would generate new communities.

Adding such communities to the present communities would increase the overall diversity of biological organisms. It is also likely that there are organism designs that can actually increase the unity of *existing* communities and thus increase the overall unity in the biological creation. Whereas the introduction of many exotic species can *reduce* the diversity of a community, some species—known as **keystone species**—may actually *increase* the diversity of a community. Perhaps God has hidden keystone species designs in organisms that would allow an increase in diversity of existing communities.

keystone species—species that may increase the diversity in a community

Furthermore, when we design communities (such as city parks, residential and business landscaping), we can maximize the unity of these 'artificial' communities by choosing organisms that thrive in the native communities living in the region. Native organisms, for example, are more likely to be immune to local diseases, and are less likely to spread diseases to which native plants are susceptible. Native organisms are probably better designed for the local climate and soil conditions. Native organisms are also more likely to benefit from the biomatrix species that are already there.

Even if no organisms appear to be living on the site to start with, a surprising number of biomatrix species are probably already present. Native organisms are also more likely to interact well with each other and any plants and animals that might already be on site or nearby. All other things being equal, native organisms are more likely to generate a unified community.

Summary of Chapter

A. To illustrate God's invisible quality of oneness, God created physical illustrations of oneness in biology: 1) common monomers for the macromolecules of life, 2) similar structures in all organisms on earth; and 3) biological systems.

B. Though each organism contains a unique set of organic macromolecules, almost all macromolecules are built from a small number of monomers. Most of these macromolecules can be grouped into four different categories:

   a. carbohydrates (with a C:H:O ratio of 1:2:1) which are built from monosaccharide monomers like glucose, fructose, ribose, and deoxyribose. Disaccharides (2 monosaccharides) include lactose (the common sugar in milk) and sucrose (table sugar). Polysaccharides (many monosaccharides) include cellulose (the main structural molecule in plants), chitin (the external skeleton of arthropods), glycogen (a long-term energy storage molecule in animals), and starch (a long-term energy storage molecule in plants).

   b. proteins (molecules where their shape is important to what they do) are built from 20 different amino acid monomers. Proteins include molecules that define the shape of cells, molecules that create doors in and out of cells, and enzymes that speed up chemical reactions.

   c. nucleic acids (macromolecules such as DNA and RNA that carry genetic information) are built from 5 different nucleotide monomers.

   d. lipids (hydrophobic molecules) are of two different types:
      i. fats, oils, waxes, and phospholipids that contain one, two, or three fatty acid chains on 3-carbon backbones
      ii. fats and oils are used for long-term energy storage
      iii. fats are used for cushioning and insulation
      iv. waxes are used to repel water
      v. phospholipids (made of a water-loving phosphate 'head' and 2 fatty acid 'legs') make up cell membranes

   e. steroids, made from four connected rings of carbon, are used as chemical messengers in the body.

C. Building macromolecules from monomers allows (1) damaged molecules to be replaced, (2) consumers to build their own macromolecules by breaking down macromolecules of other organisms, (3) all organisms to be decomposable, and (4) decomposers to clean the earth of dead bodies and produce monomers for themselves and other organisms.

D. God created a spectrum of perfection of similar structures among organisms (some organisms with few similarities, others with more similarities, and others still with many similarities, *etc.*), from which we are to extrapolate to One with infinite unity.

E. Biological systems (systems with multiple components and emergent properties where each component has a few imperfect properties but works together with other components in a strongly integrated manner) are common in biology (*e.g.* organelles, cells, tissues, organs, organ systems, bodies, biological communities).

F. The bodies of animals and plants are actually communities of organisms.

G. Typical biological systems are not only much more complex than systems that arise spontaneously but they also seem to have too much irreducibly complexity to arise in the stepwise fashion. This suggests that they were designed by God rather than derived by naturalistic evolution.

H. As priests of the creation we should better know Him, worship Him, and bring others into that worship. This is stimulated by

a. common monomers + biological systems + spectrum of perfection of similar structures → God infinite unity & His desire that we know Him

   b. elegance of monomers → God's wisdom & beauty

   c. complexity of biological systems → God's wisdom & power

I. As kings of the creation we should try to preserve the oneness of the creation

   a. use antibiotics only when necessary to preserve the unity of the human body as a community of organisms

   b. introduce organisms into a community from another only when it has been demonstrated that such an introduction will not harm that community (*vs.* rabbits in Australia harming native marsupials; kudzu and wisteria in the Southeast harming native plant species; nutrias in the Southwest harming the native marsh organisms).

   c. and we should try to *enhance* the oneness of the creation by searching for hidden organisms designed to increase community diversity.

Test & Essay Questions

1. How does God illustrate His triune nature / oneness in the biological world?
2. How is there molecular similarity across all organisms when every organism is built of unique molecules?
3. List three / four different types of organic molecules.
4. From what kind of simple molecule is a carbohydrate / a protein / a nucleic acid / a lipid / DNA made?
5. List two / three specific, but different, complex molecules found in all organisms.
6. Define cryptic species.
7. Describe the spectrum of perfection of biological similarity / tightness of biological systems and what it demonstrates.
8. What does systems biology study?
9. Define a biological system / emergent property.
10. In what way is an atom / a molecule / an organelle / a cell / a tissue / an organ / an organ system / an animal / a plant / a human / a community a system.
11. Describe community succession. / Define a climax community / pioneer species.
12. Compare and contrast the naturalistic and creationist understanding of the origin of biological similarity.
13. Essay. Explain why biological systems are more difficult to explain in the worldview of naturalism than in a creationist worldview.
14. Describe how a biological system is irreducibly complex and what that implies about the origin of that system.
15. Compare and contrast a native and an exotic species.
16. How might an exotic introduction turn out to be a bad thing?
17. Give two / three examples of exotic introductions that turned out to be bad things.
18. How is it that an organism can also be a community of organisms?
19. Considering for a moment how your body is a community of organisms, how might antibiotics harm your body?
20. List two / three / four advantages to utilizing native organisms to build city parks and install landscaping around your home.

CHAPTER 10

GOD IS THREE

Climates, Biomes, and Biodiversity

"This is how you can recognize God's Spirit: Every person who declares that Jesus Christ has come as a human has the Spirit that is from God."
1 John 4:2, GWT

10.1 | Three Persons of the Godhead

Thus far in this text we have seen that God illustrates a number of His invisible attributes in the biological creation. These are all characteristics of God which are fully possessed by God and fully possessed by each person of the Godhead. For example, God is life. But God the Father is also life, as is God the Son and God the Holy Spirit. God is also person, but so is God the Father, God the Son, and God the Holy Spirit. The same is true of the attributes of glory, distinctness, goodness, and love that we have already discussed. This will be true of other attributes we will address in turn. But now we turn our attention to a truth about God which technically is not an attribute of God, but rather something some theologians describe as God's *mode*.

God is person (Chapter 6) and God is one (Chapter 9), but God is also three. Three persons are identified in the Bible as God. First, the Father is God (*e.g.* John 8:41-42, 54[a]) and the Father has all the attributes of God. Second, Jesus is God (*e.g.* Heb. 1:1-10[b]), and Jesus has all the attributes of God. Third, the Holy Spirit is God (Acts 5:3-4; I Cor. 3:16-20), and the Holy Spirit has all the attributes of God. Jesus is also equal to the Father (Isa. 9:6; John 5:23; 10:20; 14:9-11; 17:5, 11, 21; 20:17). The multiplicity of persons in the Godhead is why God refers to Himself as 'us' in Gen. 1:26, Gen. 11:7, and Isa. 6:8. These three persons being one God is the distinctive Christian doctrine known as the doctrine of the triune God or the trinity.

Since God wishes us to understand something about Him, He wishes us to understand something about the trinity. To illustrate the multiplicity of persons in the Godhead, God filled His creation with variety[c]. The billions and billions of stars in the universe all seem to be

---

a   Further references that indicate that God the Father or the Father in heaven is, in fact, God include Matt. 5:16, 45, 48; 6:1, 4, 6-8, 14-15, 18, 32 (compare vss. 26 and 30); 7:11, 21; 10:32-33; 11:25-27; 12:50; 13:43-44; 15:13; 16:17; 18:10, 14, 19, 35; 23:9; Mark 11:25-26; Luke 10:21-22; 11:2, 13; 12:30-32; John 4:21-24; 6:32-33; 8:41-42, 54; Rom. 1:7; 15:6; I Cor. 1:3; 8:4-6; 15:24; II Cor. 1:2-3; 11:31; Gal. 1:1-3; Eph. 1:2-3, 17; 3:14-15; 4:6; 5:20; 6:23; Phil. 2:3, 11; 4:20; Col. 1:2-3; 3:17; I Thess. 1:1-3; 3:13; II Thess. 1:1-2; 2:16; I Tim. 1:2; II Tim. 1:2; Titus 1:4; Philemon 1:3; Jam. 1:7; 3:9; I Pet. 1:2-3; II Pet. 1:17; II John 1:3; and Jude 1:1).

b   Jesus was not only declared to be God by others (*e.g.* John 20:28), Jesus was, in fact, God (*e.g.* compare John 1:1-3 and John 1:14; Philippians 2:6; Col. 1:19; Heb. 1:1-10; I Tim. 2:3). While walking this planet Jesus not only declared himself to be God (*e.g.* Mat. 26:63-64; John 8:58), but the religious leaders in His time understood Him to be declaring Himself to be God (*e.g.* Mat. 26: 65; John 5:17-18; 19:7).

c   Although the creation illustrates the multiplicity of persons in the Godhead, it does not specify the exact *number* of persons in the Godhead. This is analogous to the fact that although the multiplicity of the Godhead is revealed in the Old Testament (*e.g.* the 'us' and 'our' in Gen. 1:26 and 11:7 and Isa. 6:8), the precise number is not revealed in special revelation until the New Testament.

different, as are all the planets and moons of our Solar System. Even snowflakes may all be different. This variety gives us a physical picture of the multiple persons of the Godhead. God also placed variety in the biological world in the form of a high **biodiversity** (a large number or diversity of different life forms).

biodiversity—the number of taxa (usually species)

Climactic Variety

Diversity of Biological Communities

So as to create a diversity of organisms, God fashioned the earth with a diversity of environments. To create a variety of environments God created a diversity of **climates**[a] (the climate of a particular location is defined as the average temperature and rainfall rate at that location). The temperature of an area is mostly dependent on how much the sun warms the earth's surface at that location. The land on a spherical earth gets heated by the sun to different temperatures at different latitudes.

At the equator, the sun is positioned higher in the sky, so there is less atmosphere to obstruct the light and a given package of sunlight shines more directly on a smaller area of land. The same sized package of sunlight falling upon the poles is obstructed by more atmosphere and

climate—average temperature (and rainfall) at a given location (e.g. tropical; temperate; polar; arid)

---

a Some creationists have claimed that the original earth was created with a uniform climate, but this is not consistent with Biblical claims (*e.g.* the seasons in the original earth according to Gen. 1:15; the rain implied as coming following the creation of man in Gen. 2:5; the seasons interrupted by the Flood according to Gen. 8:22), or physical evidence (*e.g.* trees in Flood sediments having rings showing evidence of seasonality), or theoretical science (*e.g.* no theory generating uniform climates has been shown to be physically viable; the tilt of the earth and the associated seasonality seems to be essential for life).

is spread out over a larger area of land. This results in cold temperatures at the poles and warm temperatures at the equator. Merely by creating the earth in a near-spherical form, God automatically generated a variety of climates on the earth.

As sunlight warms the earth's surface at the equator, the warm ground then warms the air just above the surface. Since warm air holds more water and warm air rises, air over the warm equator rises as warm, humid air. As the air rises farther from the earth's center, gravity lessens and the air expands. As the air expands it cools. This causes it to drop its water because cooler air cannot hold as much water as warmer air. What results is very warm temperature and typically heavy rainfall within 10-15 degrees of the equator (the tropical zone).

Once the air over the equator gets high enough, it moves away from the equator to higher latitudes. The high-altitude air sinks back to the earth's surface at about 30 degrees north and south latitude. Since this air has already been dewatered, it is very dry when it reaches the surface again, producing an arid zone at those latitudes. Air that sinks at this latitude moves away from this zone, getting warmed and picking up water vapor as it moves along the earth's surface. The air that moves toward the equator will rise at the equator. The air that moves away from the equator rises at about 60 degrees north and south. Because the air is not as warm as the air at the equator, this zone is called the temperate zone.

Although the rising air cools and creates another zone of high rainfall, the lower temperatures result in lower rainfall rates in the temperate zone than are found in the tropical zone. Since the air over the temperate zones is dewatered as it rises, it is again dry when it sinks to the earth's surface. The air that moves toward the equator sinks at the low latitude arid zone. The air that moves toward the poles sinks at the poles as very cold dry air to create a cold, arid polar zone.

10.2 | Land Biome Variety

For each climate that God created, He created distinct organisms for that climate. A large area of the earth's surface characterized by a certain climate and certain type of vegetation is called a **biome**. The creation of a variety of climates permitted God to create a diversity of biomes. This, in turn, generates a high diversity of organisms.

A few of the major land biomes include the following:

biome—flora and fauna found in a particular climate and across a large area of the earth's surface

rainforest—forest land biome designed for the wetter part of the earth's tropical zone

savannah—grass land biome designed for the dryer part of the earth's tropical zone

desert—land biome designed for the earth's arid zones (less than 10 inches of rainfall per year)

- Tropical **Rainforests** are found in the wetter portion of the tropical zone (50-80 inches rain/year). Because of continuous warmth and high rainfall, tropical rain forests contain the greatest diversity and complexity of all the land biomes.

- **Savannahs** are found in the dryer portion of the tropical zone (10-40 inches rain per year), often between the rainforests and the arid zone located to the north and to the south. The dryer environment makes it prone to fires. The combination of fires and lower rainfall make it difficult for trees and shrubs to get established. This makes the savannah a grassland with only scattered trees and shrubs. As a protection against fire, plants store lots of organic nutrients underground, resulting in grassland biomes having the richest and deepest soils of all land biomes.

- **Deserts** are located in arid zones. Because they are so dry (less than 10 inches per year), deserts have the lowest diversity of organisms, and because of this, deserts have the poorest soils. Many deserts have a soil crust of blue-green bacteria, lichens, mosses, and fungi, and this crust helps hold place whatever soil is present. The lack of water in the soil also gives deserts the largest swings in temperature (very hot during the day and very cold at night). Desert plants and animals all have special designs to allow them to survive (*e.g.* ocotillo leaves come out only when it rains; poppies grow quickly from seed only after a rain; barrel cacti store water;

mesquite & creosote plants have very deep roots; kangaroo rats have efficient enough kidneys to get all the water they need from nuts; Sonoran Desert tortoises hibernate in summer; bats rest in caves during the day).

- **Temperate Grasslands**, located in the drier portion of the temperate zone (10-40 inches/year), are called *prairies* in North America, *veld*s in South Africa, *pampa*s in South America, and *steppe* in Eurasia. As in the case of the savannah, the dryer environment makes it prone to fires and therefore a difficult environment for shrubs and trees. Because temperate grasslands are cooler than savannahs trees are generally completely lacking. As in the case of Savannahs, grassland plants store a lot of organic nutrients underground, making them some of the richest and deepest soils of all the land biomes. Because of rich soils and moderate temperatures, temperate grasslands have been made into some of the richest farmlands of the world.

- **Temperate Deciduous Forests** are found in the wetter portion of the temperate zone (20-60 inches/year). To survive the coldest part of the year, most temperate zone trees drop their leaves in the winter, so temperate forests are dominated by deciduous[a] trees.

temperate grasslands—grass land biome designed for the dryer part of the earth's temperate zone

temperate deciduous forest—forest land biome found in the wetter part of the earth's temperate zone

deciduous—when a tree drops its leaves at a certain point in the year and then regrows them at another

---

a When a tree drops its leaves at a certain point in the year and then regrows them at another, that tree is considered **deciduous**.

coniferous forest—forest land biome found between the temperate and polar zones of the earth

- **Coniferous forests** are found between the temperate and polar zones. Trees can survive the cooler and drier conditions, even in the very cold winters, by means of thick, waxy coverings on their leaves. Consequently, most of the trees are evergreens (meaning they have green leaves the entire year). Since most of these evergreens produce their seeds in the form of cones, they are called conifers (meaning *cone-bearing*).

tundra—biome found in the earth's polar zone

- **Tundra** biomes are closer to the polar zone, and usually receive less than 10 inches of rain per year. Although they experience continuous sunlight during the summer, they are covered with snow and experience continuous darkness during the winter. Close to the poles (such as the Arctic tundra) the ground is permanently frozen to a rather great depth. This permanently frozen layer is called permafrost. Only a thin layer of ground above this permafrost thaws during the summer seasons. The permafrost beneath slows decomposition and holds onto water, leaving wet, rich soils during summer thaws.

10.3 | Water Biome Variety

On Day Three of the Creation Week (Gen. 1:10-11) God separated the dry land from the oceans. On the next day He created the sun. The heat of the sun evaporates water from the ocean. Since evaporation leaves salt behind, the evaporated water is fresh, not salty. In the atmosphere, water evaporated from the oceans is carried around the world, eventually condensing and falling as fresh water rain. Wherever this water accumulates on the land, lakes are formed. God created special organisms for the still, fresh water of lake environments.

The rainfall that lands on the continents, but does not accumulate in lakes, flows back to the ocean. Some flows back beneath the surface in the form of **groundwater**. The remainder flows back on top of the surface in the form of rivers and streams. Since the moving water of rivers and streams has not been heated as much by the sun, it is usually cooler than lake water. And, since cooler water can dissolve more oxygen and streams expose more surface area to the atmosphere, stream water usually contains more oxygen than lake water.

groundwater—water that flows underground on its way to the ocean

The differences between lakes and streams allowed God to create another set of special organisms for the cooler, oxygen-rich, moving waters of stream environments. Where river water enters the ocean, there is a zone where fresh and salt water mix, and this area is called an **estuary**. The intermediate salinity of the estuary requires a unique set of organisms for that environment. As consistent as the salinity is in the world's oceans, the shape of the ocean bottom creates a variety of different ocean—or marine—environments.

estuary—ocean outlet of river where water saltiness transitions from fresh to marine

On shorelines with tides (ocean levels that rise and fall inches to feet every twelve hours due, primarily, to the moon's gravity) the alternating submerged and exposed intertidal zone requires a unique set of organisms. In shallow areas within 25 degrees of the equator the sun warms ocean waters to create optimal conditions for the **coral reef** biome. Coral animals and sponges, growing atop solid surfaces and atop one another, form calcium carbonate ($CaCO_3$; limestone) structures called reefs. These reefs host the highest diversity of organisms of any marine ecosystem.

coral reef—high diversity, low-latitude marine biome living in limestone structures built by organisms

God designed yet another set of organisms for the shallow marine zone. Light can penetrate only a certain depth into ocean water—approximately 200 meters in very clear water. Most seafood is harvested from the shallow marine zone. Finally, the darkness and limited oxygen of the deep marine zone requires yet another unique set of organisms.

10.4 | Biodiversity

God created a large variety of land animals by creating a spherical earth so that there would be a variety of climates and then creating biomes for each climate. He created a similar variety of water environments by creating the water cycle as He did and by creating ocean basins with a particular shape. God then created a distinct biome for each water environment.

God compounded the diversity even more by creating multiple continents. This allows in each biome the creation of a different set of organisms for each continent. The tropical forest biome, for example, is found on the continents of South America, Africa, and Asia. Although the tropical forest in each case is broadly similar, with organisms of similar designs, the species found on each continent are different from those on other continents. Similarly, although specially designed desert organisms are found in the desert biome of South America, North America, Africa, and Australia, different species are found in the deserts of each continent. The same is true of the Savannah and temperate grasslands, the temperate and conifer forests, and even the tundra.

The same biome on different continents houses a similar, but distinct, suite of organisms. Stream, lake, estuary, reef, and intertidal biomes are found on most continents but in each biome different continents will host similar, but distinct organisms. Interestingly enough, similar biomes on different continents can contain some

astonishingly similar organisms (*e.g.* prairie dogs and meerkats in the grasslands of North America and Africa; anteating armadillos, pangolins, anteaters, and echidnas of North America, Asia, South America, and Australia, respectively; marsupial and placental tigers, cats, flying squirrels, and mice of Australia and North America).

Biological Diversity in Communities

God created an even greater diversity of organisms by dividing the labor in each community of organisms and creating a separate **niche** for each. The word niche means 'to nest' and it refers to the space or role that a species or population of organisms has in an environment. For example, each community has a **trophic structure**, which traces how energy enters and flows through the community. In every community there are **producers**, organisms that capture energy from the environment in some way and store that energy in molecules from which other organisms in the community can get energy.

In land biomes, plants are usually the producers. In all but dark places such as the deep sea, photosynthetic bacteria and algae are usually the producers in water environments. Photosynthetic plants, algae, and bacteria are specially designed to capture the sun's energy and store that energy in molecules from which other organisms can get their energy. In dark environments there are specially designed bacteria that get their energy from other sources.

In every community there are also **consumers**, organisms that get their energy by consuming producers or other consumers. Animals make up the consumers in most environments. **Decomposers** make up a third group of organisms in every community. Decomposers get their energy from breaking down dead organisms and recycling the molecules so they can be used again in the community. The decomposers also function to keep the environment clean. Decomposers are usually fungi and bacteria and make up a part of the biomatrix of every community.

God multiplied the diversity of organisms even more by creating a variety of distinct spaces in the community and then creating different organisms for those spaces. The different spaces in a community that have the required shelter, water and food, are known as **habitats**. A whole suite of different habitats in the community allows for a considerable diversity of organisms. For example, most forest environments are tiered. There are plants without wood that cannot grow more than

niche—role of an organism in a biological community

trophic structure—the flow of energy through a community (from producers through consumers)

producer—an organism that gets its energy from the physical environment (vs. consumer)

consumer—an organism that gets its energy by consuming organic molecules (vs. producer)

decomposer—organism that gets its energy from breaking down dead organisms and recycling the molecules so they can be used again in the community

habitat—space in a biological community in which species can thrive

a few feet tall and then there are bushes designed to grow only a half dozen or so feet in height. Then, there are short trees that grow a dozen or two feet tall, and tall trees that grow forty, fifty, sixty feet tall and beyond. **Tiers** like this allow multiple species to live on the same piece of land. Different soils, different amounts of shade or sunlight, different amounts of water—these are all things that can produce a variety of habitats in a given community.

tier—vertical zone of a forest characterized by a unique flora and fauna

God increased diversity even more by dividing up the resources of a community and designing different organisms to use the different resources. For example, different plants have roots at different depths allowing different plants (*e.g.* bristly foxtail vs. Indian mallow vs. smartweed) to access different sources of water, such as rain water, or stream water, or ground water at different depths. Other examples of **resource partitioning** include different sized animals consuming different sized seeds (*e.g.* medium vs. large ground finches on the Galápagos Islands), different plant parts consumed by different kinds of animals, or different depths of a plant being accessed by different birds (*e.g.* 8 different species of woodpeckers in Oregon).

resource partitioning—division of a biological community's resources among different organisms

Yet another way that God increased the diversity of organisms is by creating organisms that are active at different times. For example, different non-woody plants emerge at different times of the year, again allowing multiple species to grow on the same plot of land without competing with one another. Animals were also designed to forage at different times of the day—some animals foraging at night, others during the day, and still others at the transitions between night and day.

The consequence of all this is that there are a lot of organisms on this planet. At this point, over 1.8 million species have been identified and named. Nobody knows for sure the total number of species there are, but estimates commonly exceed ten million.

Diversity Amplified

Another characteristic of the diversity of organisms is that there seems to be more diversity than necessary. This is indicated in at least two different ways. First, different groups of organisms, even very similar groups of organisms, have very different levels of diversity. For example, there are scores of species in the deer family and scores of species in the cow family, but only one species in the pronghorn family. Neither the deer family, nor the cow family have to be as diverse as they are. They could just as easily be represented by one species as is evidenced in the pronghorn family. There are many similar examples throughout the animal and plant kingdoms to indicate that the diversity we observe is much higher than it *has* to be.

A second reason we know diversity is abundant has to do with extinction. A number of species that existed in the recent past are no longer with us. These include the passenger pigeon, the dodo bird, and quite possibly the ivory-billed woodpecker ('Woody Woodpecker' of early cartoon fame). There is even reason to believe that the loss of

tropical rainforests is causing quite a bit of extinction. No one really knows how many species are going extinct every year (partly because we have not identified every species that exists), but estimates range from several to dozens.

We have also identified and named approximately a quarter million species of fossils that are not known to exist anywhere in the present. These fossils represent species that have gone extinct. Yet, with all the extinction that has occurred, and is occurring, life on earth does not seem to be crashing into oblivion. If the diversity we had was close to what was necessary for life on earth, then the extinction of species would have catastrophic consequences.

Disparity

Creating diversity is not the only way that God can illustrate his own diversity. He also illustrates it with disparity. **Disparity** is not the same as diversity. Disparity is a measure of how *different* organisms are when compared to one another. As an illustration, if one room contains a collie, a wolf, and a coyote while another room contains a collie, a bacterium and an ivy plant, both rooms have the same diversity (namely three species) but the second room has a much higher disparity (the species are much more different from one another). The dog, wolf, and coyote are very similar, so the room containing them has a lower disparity. Disparity is a measure of how different things are. Not only is there a high diversity of organisms on the planet, there is also a

> **disparity**—measure of how different organisms are when compared with one another

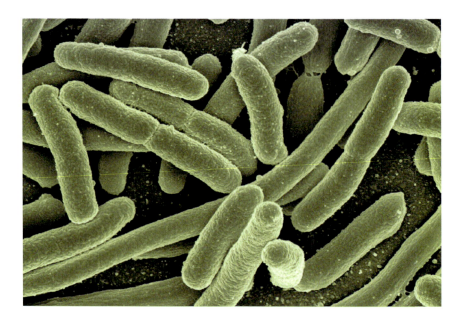

high *disparity* of organisms. Humans are extremely different from oak trees and both are extremely different from bacteria, earthworms, and sea jellies.

The Origin of Disparity

Disparity is a substantial problem in naturalism. The biological change we actually observe involves small steps. We know of no examples of rapid or large-scale changes in the present biological world. If such small changes are responsible for biological change, then to start with one organism and get another very different from it, the original organism has to get there one small step at a time. A large number of steps would be required. That many steps would result in many branches on the evolutionary tree.

Consequently, to get high disparity, diversity would have to increase first (because of the large number of branches on the evolutionary tree) and a lot of time would be required (because of the large number of steps). However, the general pattern in the fossil record is for high disparity to appear *before* high diversity. For example, the oldest rock layer that contains well-preserved arthropod fossils (the Burgess Shale) contains 21 non-trilobite arthropod species representing 20 different arthropod classes. There are only five arthropod classes in the present and more than 1.5 million species!

The time required to explain disparity is a second problem. For example, if the fossil record is interpreted in terms of millions of years, then 50 million years separate small horses that browsed on leaves to large horses that graze on grass. Using the same time scale, the first animals appear only about 500 million years ago. Biological change that produced small changes in horses over 50 million years could not possibly derive the disparity between horses and invertebrates in just nine times that time.

Even if the fossil record does record billions of years of time, there is not enough time in earth history to explain the earth's disparity by small evolutionary steps. The high disparity of biological life in the present combined with the pattern of disparity before diversity in the fossil record is much better explained by divine creation than it is by naturalistic evolution.

A third problem is how to explain how organisms from very different (disparate) groups came to be so similar. For example, as similar as meerkats in Africa are to prairie dogs in North America those similarities were not inherited from any ancestor that had those characteristics. In the naturalistic world view, organisms *without* those similarities each developed the similar characteristics separately. The same is thought to be true for the anteaters on five different continents, and for the many Australian marsupial mammals that are similar to North American placental mammals[a]. It is hard to imagine how naturalistic processes operating on different organisms on different continents could have created even two such similar organisms, let alone the hundreds of similar organisms actually found on earth.

Though these features are challenging for the naturalistic world view, they are readily explained in the creationist world view. Since God created disparity to illustrate His triune nature, and humans were present from the beginning, organisms were created showing high diversity from the very beginning. Though evolution is incapable of explaining the origin of disparity in the time available in earth's history, the God of Scripture is sufficiently powerful to make it happen.

Furthermore, fossils from any environment in creationist history should show the relatively large disparity and low diversity found in

---

[a] The embryo of a marsupial mammal is 'birthed' very early in development, then crawls to a pouch on the mother to complete its development. The embryo of a placental mammal continues to develop inside the womb of its mother (fed through a placenta) before being birthed much later in development.

modern environments. This makes the high disparity and low diversity of the fossil record fit right into a creationist understanding of earth history. Finally, since the existence of distinct but similar organisms increases the diversity on the planet, such organisms would be the expected if the Creator desired to illustrate His trinity by creating as much diversity as possible.

10.5 | Diversity: Our Responsibility

Our Responsibility to God

Life's diversity and disparity should remind us that our God exists in three persons. Note that the monotheistic religions of modern Judaism and Islam are not worshipping the Creator. In the previous chapter we saw how the monomers, biological similarity, and biological systems all point to one Creator.

But the fact that the creation also abounds with diversity and the biological creation abounds with both diversity and disparity, suggest that oneness is not the whole story. In fact, Scripture clarifies that He is three. He is three person AND one God. He is triune. The creation of all things is a consequence of *three* persons—God the Father (*elohîm*) (Gen. 1:1, 3, 6-7, 9, 11-12, 14-17, 20-21, 24-25, 26-27), God the Holy Spirit (Gen. 1:2), and God the Son (John 1:1-3, 14; Col. 1:16)—*united* as one. Only the unique triune God claim of Christianity is consistent with both the unity and diversity of biology.

We also learn from life's diversity and disparity that God loves variety. He loves variety because it is a part of His very being. For this reason, He made each of us unique and desires a unique, personal relationship with each of us. He loves all (*e.g.* John 3:16), but He loves each of us as individuals. This is further emphasized by the fact that He created diversity and disparity as an *illustration* of His nature. He wants us to know Him, so much so that he fashioned the entire creation so that we could understand His nature.

Again, we are talking about the God of Scripture. This is not the god worshipped by Muslims nor the god worshipped by modern Jews, for neither of these gods desire a relationship with humans. Rather, the biological diversity and disparity that is evident all about us is consistent with the God Who became man, dwelt among us, died on the cross for our sins, and rose again to give those who believe the assurance of eternal life with Him.

Contemplation of biological diversity and disparity ought to bring us closer to God and teach us about Him. And, as we come to know Him better, how can we respond in any other way than in worship? And, as this worship fills our lives we cannot help but draw others into that worship. In this way we fulfill our role as priests of the creation.

Our Responsibility to the Creation

Preserving Diversity

conservation biology—discipline of biology that seeks to preserve organisms and biological communities

After creating Earth's biodiversity, God handed it over to humans to 'guard and keep it'. We should at the very least work at maintaining the biodiversity of this planet—something called **conservation biology**. God created biodiversity, even abundant biodiversity, to give us a physical manifestation of His very nature. When biodiversity is lost, our understanding of God is dimmed. We lose something invaluable.

Besides that, however, we were made caretakers of His creation, to guard and keep it. We will be called into account by God for how we took care of those things He made. We were given the creation with a certain biodiversity. At the very least we should guard that biodiversity, and not 'return' it to Him with less diversity than we received it. Christians ought to be playing active and leading roles in biodiversity preservation. We have an obligation to preserve and to protect the earth's biodiversity, and we ought to do so in order to glorify God and show our love for Him.

Managing Food Species

Humans have a poor track record when it comes to preserving biodiversity. The sin of greed has repeatedly brought other species to the brink of extinction, and even to extinction itself. Although humans have the authority to use species for their own benefit, we should remember that we are also supposed to watch after those same organisms and preserve them.

We should be careful not to overharvest species we use for food, such as was done with the white abalone of California (which was harvested for food until the 1970s when its population had been reduced to only 1% of its original population), the Atlantic codfish (which was overfished from the mid-1980s to the early 1990s until its populations crashed), and the passenger pigeon (which was hunted for food until it was totally extinguished in the early 1900s). We should thoroughly study the species we harvest, and carefully manage them to assure the health of those species.

Sustainable Development

Our greed can threaten species in indirect ways as well. As we claim new land for buildings and farms we should remember that we are impacting species that depend upon that land to survive. As we cut down trees to build our buildings we should remember that this impacts species that depend upon those trees to survive. Logging in the Congo has endangered the gorilla, logging bamboo in China has

threatened the Giant Panda with extinction, and logging in the U.S. Southeast has almost (if not completely) exterminated the ivory-billed woodpecker.

As we consume water we should remember that other species need that water to survive as well. Excessive water use in Texas has reduced the size of the Edwards aquifer, which has in turn endangered the Texas blind salamander. Development of beaches has interfered with the reproduction of sea turtles that lay eggs in beach sand. Building dams in the Pacific Northwest prevents Coho salmon from getting upstream for breeding. Power plants increase the temperature of local water, threatening cold-loving trout species. Slash and burn agriculture in Indonesia has put all *Rafflesia* species (the world's largest flower) on the endangered species list. Plowing prairie into farmland has threatened the prairie fringed orchid with extinction. As we transform the environment about us we should carefully examine the impact on the organisms of that environment and devise ways to preserve the biodiversity—an activity known as **sustainable development**.

sustainable development—preserving biological communities as we transform the environment

Extinction

Although we know that a number of species have gone extinct in the last few decades, we do not have an accurate count of how many species are going extinct each year. After all, we do not even know how many species exist, and we may be familiar with only a small percentage of species that currently live on this planet. We also do not know how many extinctions are actually caused by humans[a]. Although we know more about living species, we are only slightly more capable of determining how many living species are at risk of going extinct. In 2009, the International Union for Conservation of Nature and Natural Resources (IUCN) reported that of 47,677 species they assessed, 36% were threatened or endangered. An **endangered species** is a species whose population is declining at an unhealthy rate; a **threatened species** is one that is likely to become endangered in the near future.

Although this is a census of less than 1% of the earth's diversity, it does suggest that we may have a serious task ahead of us in maintaining the earth's biodiversity. It is only very recently that humans have begun to think strategically about preserving biodiversity. One strategy is to identify and preferentially preserve biodiversity hot spots. A biodiversity **hot spot** is an area that if altered would result in the most extinction of organisms. The rationale is that investing money to save the home of a whole set of organisms is a better use of resources than using the same money to save just one species.

endangered species—a species whose population is declining at a rate that puts the species at risk

threatened species—a species that is likely to become endangered in the near future

hot spot, biodiversity—a region that, if altered, would result the greatest extinction of organisms

Preserving Unique Species

Breeding has demonstrated that many different body forms are somehow 'hidden' in the organisms God has created. In the case of dogs, for example, from only three different dog types known in Europe in the 1500s, over 250 varieties of dogs have been generated

---

a Whereas environmentalists sometimes give the impression that all extinction is caused by humans, and that we are causing the highest rate of extinction in all of earth's history, this may not be so.

(German shepherds, great Danes, Scottish terriers, Irish setters, *etc.*). It also appears that breeding merely *revealed* those body forms—that somehow the information for building the different dog varieties was already hidden inside three dog types known in the 1500s.

Similar experiences with varieties, cultivars, and breeds of plants and animals suggest that biodiversity is probably hidden within *all* organisms. And, since breeding usually generates a suite of very similar organisms (*e.g.* breeding dogs generates a wide range of different body forms, but they are all dogs), it suggests that threatened or endangered species might be recoverable from very similar species that are not in any danger of extinction. In contrast, species that lack any species similar to themselves might be unrecoverable if they go extinct. If we have insufficient resources to save all species and must make decisions of what is to be preserved and what is not, species that are unlike any others should probably be preserved before others.

Reproductive Cloning

From time to time, humans have desired to produce exact copies of organisms that already exist[a]. To that end, humans have developed reproductive cloning. A clone is an organism with the same DNA[b]. **Reproductive cloning** is when humans artificially produce a mature adult that is a clone of another organism. Three reproductive cloning methods have been developed so far. The first method is *horticultural cloning*. For most plants, it is possible to obtain a cutting from any part of the desired plant and grow an entire plant from it. Since the new plant can be planted in a very different location, it is a separate plant. However, the cutting has the same DNA as the plant from which the cutting was made. It is a clone.

Horticultural cloning is an ancient practice and may date as far back as the Garden of Eden. The other two methods were developed for cloning animals. Whereas just about any part of most plants can be fairly easily developed into a separate plant, only parts of the earliest stages of an animal embryo can be so easily developed into a

---

cloning, reproductive—producing from one organism, another organism with identical DNA. Types: horticultural cloning by rooting a plant cutting; embryonic splitting by separating the first cells of a developing animal; adult cloning by getting an adult cell to develop into a mature organism

---

a Reasons why an exact copy of an organism might be desirable include (but are not limited to): 1) duplicating organisms with desirable characteristics (such as choice food, choice athletes); 2) replacing a beloved pet or family member; and 3) creating compatible organ donors.

b DNA is discussed in more detail in Chapter 13. It should be noted that contrary to popular belief, clones are *not* identical. Because they contain the same DNA, clones are very similar, but they also differ from one another. As one example, although identical twins are clones of each other, they have distinct personalities. Consequently, clones can never be expected to replace lost loved ones.

separate animal. If an early embryo is divided in two—a process called embryonic splitting—separate organisms can develop from each half of the embryo. The resulting organisms have the same DNA, so they are clones of each other. **Identical twins**, for example, result from a natural embryonic splitting event. Identical twins have the same DNA and they are clones of each other. *Natural* clones by embryonic splitting have probably been forming from near the beginning of time.

The ability to produce *artificial* clones by embryonic splitting (by slicing up very young embryos) is only a few decades old. It was not until the 1970's that the first animals were artificially cloned by embryonic splitting. Now, artificial production of human clones by embryonic splitting is performed regularly in IVF clinics[a]. The third reproductive cloning method creates what most people think of as a clone—something often referred to as an *adult clone*. This method creates an embryo that is a clone of one cell of an adult animal and then grows that embryo to adulthood. The first successful adult clone of a vertebrate animal was 'Dolly' the sheep in 1997[b]. No matter what kind

> **identical twins**—two people with identical DNA; arise from embryo cells separated soon after fertilization

---

[a] Embryonic splitting is performed so as to reduce the costs of vitro fertilization (IVF). In a typical IVF procedure, after extracting as many eggs as possible from the woman donor and fertilizing those eggs in the lab with sperm from the male donor, the fertilized eggs are allowed to divide a few times. Then, the resultant embryos are carefully cut so as to separate the individual cells, with the intention of using each separate cell or group of cells as separate embryo. All the embryos developed from each fertilized egg are clones (identical twins).

[b] Steps of adult cloning: 1) take the nucleus out of an egg cell; 2) take the nucleus from an adult cell in the donor organism and place it into the egg cell lacking a nucleus; 3) cause the egg cell to think that it has just been fertilized; 4) place the egg cell into a surrogate mother to allow the embryo to develop and birth the baby.

of clone is produced, however—whether a horticultural or embryonic splitting or adult clone—the process makes things more similar than they would be naturally. Rather than maintaining the diversity of the creation God gave us, cloning *reduces* the creation's diversity.

Enhancing Diversity

Beyond preserving biodiversity, we ought also to be *increasing* biodiversity. From the parable of the talents, we learn that it is better to *increase* what God gave us. Because it shows us something about God, biodiversity glorifies God. Increasing biodiversity would bring God even greater glory. If it is in our power to increase biodiversity, it behooves us to do so. In fact, while humans have caused extinction, humans have also increased the biodiversity on this planet, or at least the *manifested* biodiversity.

In our breeding programs we have unveiled an astonishing array of biological variety. Breeding of orchids, for example, has increased an estimated 'natural' diversity of 25,000 species in 1940 to over 75,000 orchid types in the present. In similar manner, humans have increased the diversity of food crops, ornamental plants, pets, and agricultural animals by the thousands. When one considers that humans did not actually 'create' this diversity but brought out what was hidden

somewhere within these organisms, it is clear that the diversity was put there by God.

As humans have revealed this diversity they have brought increased glory to God. To whatever extent they acknowledge that this diversity is from God they further glorify Him. Although we can be sure that humans have caused the extinction of many species, they have also revealed thousands of different organisms that were not known naturally. Therefore, we have increased biodiversity more than we have destroyed it—a fact rarely, if ever, reported.

Summary of Chapter

A. The God described in Scripture is triune: There are three persons in the Godhead (God the Father, God the Son, and God the Holy Spirit), and the three persons are one. God physically illustrates His invisible multiplicity through high biological diversity (a large number of organisms) and high biological disparity (huge differences between and among organisms).

B. When God created a spherical earth in orbit around its heat source, the Sun, He created a diversity of climatic zones (a warm and wet tropical zone within 10-15 degrees of the equator; a hot and dry arid zone about 30 degrees north and south of the equator; a cool and wet temperate zone about 60 degrees north and south of the equator; and a cold and dry polar zone at the earth's poles). The diversity of climates permitted the creation of a diversity of land biomes—one for each climatic zone:

  a. tropical rainforest (the biome with the greatest diversity) in the wetter portions of the tropical zone;

  b. savannah (a tree-dotted grassland biome with rich, deep soils) in the drier fire-prone portions of the tropical zone;

  c. deserts (with designs for low water conditions) in the arid zone;

  d. temperate grassland (prairies, velds, pampas, and steppes with rich, deep soils) in the drier fire-prone portions of the temperate zone;

  e. temperate deciduous forest (with most trees losing their leaves in the winter) in the wetter portions of the temperate zone;

  f. coniferous forests (with most trees being evergreen) in the colder regions of the temperate zone; and

  g. tundra (with designs for cold conditions) in the polar zone.

C. When God created the water cycle and ocean basins on the earth He created a variety of water environments (fresh-water streams; fresh-water lakes; estuaries; a marine intertidal zone; a marine shallow subtidal zone; and a deep marine zone). For each water environment God created a special biome for that zone.

a. In addition, He created a very high diversity coral reef biome for the shallow marine zone within 30-35 degrees of the equator. God multiplied diversity within each biome by creating multiple continents, by dividing the labor within each community (by have separate organisms be producers, consumers, and decomposers), and by creating separate habitats within communities (by tiering, a diversity of soils and sunlight and water, a diversity of timing).

D. The fact that extinction does not destroy all life and the fact that different groups of organisms have very different diversities, suggests that there is far more diversity than there needs to be for life to persist. This is consistent with a God creating enough diversity to illustrate His trinity.

E. The naturalistic world view has difficulty explaining

   a. disparity before diversity in the fossil record. In the naturalistic world view, disparity can only increase one step at a time as a new species arises outside the range of known species. Since each new species also increases diversity, diversity must increase before disparity increases. However, in the fossil record, life's full disparity was achieved while diversity was very low.
   b. the high disparity of life in the time available. If it took 50 million years for evolution to derive large horses from small horses, it is hard to imagine how the same process could generate all animal diversity in only ten times that much time.
   c. similar organisms on different continents (*e.g.* prairie dogs and meerkats, anteaters on 5 different continents, similar marsupials and placentals in Australia and North America)
   d. whereas each are readily explained by a God Who desires us to understand His trinity.

F. The high diversity and disparity of life should remind us that God is triune and desires for us to know Him. As we better know this God, worship Him, and bring others into that worship, we fulfill our responsibilities as priests of the creation.

G. In our responsibility as kings over the creation, we should guard and maintain the high diversity and disparity of organisms. We should manage food species and cultural development in such a way that the diversity of organisms is maintained. We should prevent extinction whenever practical, and when it cannot be avoided, restrict extinction to potentially recoverable species and in places removed from biodiversity hot spots. We should also avoid reproductive cloning, as this reduces biological diversity. Finally, we should enhance biodiversity by revealing organismal forms that God hid in the initial creation.

Test & Essay Questions

1. What is the doctrine of the trinity? / What does it mean for there to be three persons in the Godhead?
2. What does diversity mean? / Define biodiversity.
3. What does biodiversity suggest about the nature of God?
4. Define climate.
5. Why did God create a diversity of climates on earth?
6. What are the four major climatic zones on earth and where are they located? / Describe the tropical / arid / temperate / polar zone and indicate where it is located.
7. What causes the earth's climatic zones / the tropical zone / the arid zone / the temperate zone / the polar zone?
8. Define biome.
9. Why did God create multiple biomes?

10. What are the seven major biomes on earth and where are they located? / Describe the tropical rain forest / savannah / desert / temperate grassland / temperate deciduous forest / coniferous forest / tundra biome and indicate where it is located.

11. What are the seven major water environments on earth? / Describe the stream / lake / estuary / intertidal / coral reef / shallow marine / deep marine environments.

12. Define trophic structure. / What are the three major types of organisms in a community's trophic structure? / Define producer / consumer / decomposer.

13. Define habitat.

14. Describe how a forest community is tiered.

15. Define resource partitioning.

16. List three/ four / five / six different ways God increased the biodiversity in His creation.

17. How does continent diversity / division of labor / trophic structure / tiering / habitat diversity / resource partitioning / timing affect biodiversity?

18. How many species have been identified and named on earth? / What is the total biodiversity on earth?

19. Define disparity. / Give an example of high disparity on earth.

20. Define extinction.

21. List two evidences that give us reason to believe there is an excess of diversity on earth.

22. Short Essay: Compare and contrast the explanation for biological diversity / disparity in the naturalistic and Christian worldviews.

23. Short Essay: How is the origin of biological diversity / disparity better explained in the Christian worldview than in the naturalistic worldview?

24. The observation of biological diversity should cause what sort of worship?

25. Short Essay: Why should we engage in conservation biology?

26. Short Essay: How do we apply conservation biology to species that we kill for food?

27. Short Essay: How does human development threaten biodiversity? / How do we apply conservation biology to human development?

28. Define endangered / threatened species.

29. Define a biodiversity hot spot.

30. Short Essay: What is the importance of a biodiversity hot spot in conservation biology?

31. Short Essay: Why might a species dissimilar from all other species be chosen for preservation over a species that is very similar to another species?

32. Short Essay: Why should we increase biodiversity?

33. How can we increase biodiversity?

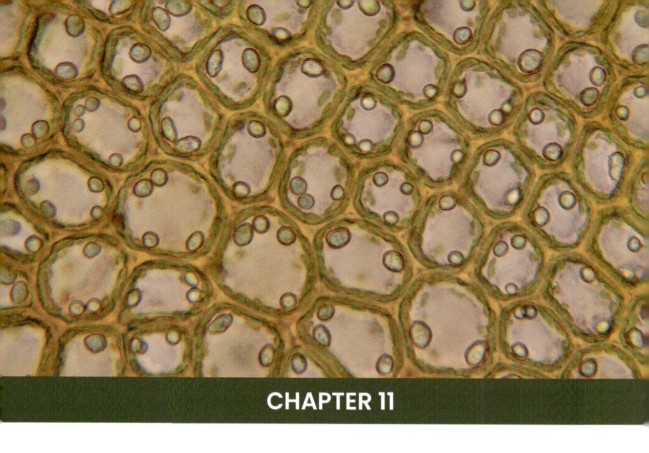

CHAPTER 11

GOD OF HIERARCHY

Molecular Biology: The Cell

"...God has given us his Spirit... and... the Father sent his Son to be the Savior of the world"
1 Jn. 4:13-14, NLT

11.1 | Divine Hierarchy

There are levels or ranks of authority in the Godhead. It is common to understand that God the Father has authority over the Son, Who jointly have authority over the Holy Spirit. After all, we read that God the Father gave life to the Son (John 5:26; 6:57), generated the Son (John 8:42; 16:28; 17:8), and sent the Son (*e.g.* John 6:38 and I John 4:14[a]). God the Son subjected Himself to the will of God the Father (John 4:34; 5:30; 6:38; Heb. 10:7, 9)—even the agonizing death of the cross (Mat. 26:39)—and He accomplished the task of God the Father (John 17:4).

As the Word (John 1:1), God the Son also spoke the word of God the Father (John 3:34; 7:16; 12:49-50; 14:24). God the Holy Spirit, in turn, was sent by God the Son (John 15:26; 16:7) and God the Father (John 14:16; 26) and the Holy Spirit does not speak of Himself but glorifies the Son (John 16:14). These different levels or ranks of authority make up a hierarchy of authority in the Godhead.

Yet, even though this simple vertical hierarchy is true, it is only *part* of the truth. It is also true that God the Holy Spirit led the Son (Luke 4:1), and Jesus will be worshipped above all, and by all (Rom. 14:13; Eph. 1:9-10; Philip. 2:9-11), suggesting that divine authority is not a simple linear hierarchy. Further insight into divine authority is provided by the oneness of God. God the Father and God the Son are one (compare John 1:1 with John 1:14[b]), God the Holy Spirit and God the Father are One (*e.g.* Acts 5:3-4[c]).

In fact, all three persons of the Godhead are equal (Mat. 28:19; II Cor. 13:14; I Pe. 1:2) and One (*e.g.* Deu. 6:4; Psa. 86:10; Rom. 3:30; I Jn. 5:7). The hierarchy of the Godhead is complex. God is not only three persons, He is also One God. The members of the Godhead not only have different roles and levels of authority, they are also fully

---

a   Other references that indicate that Jesus was sent by God the Father include Mat. 10:40, Mat. 15:24, Mark 9:37, Luke 4:43, Luke 9:48, Luke 10:16; Acts 3:26; Rom. 8:3; Gal. 4:4; I John 4:9-10, and numerous passages in the book of John (3:16-17, 34; 4:34; 5:23-24, 36-38; 6:29, 38-40, 44, 57; 7:16, 28-29, 33; 8:16-18, 26, 29, 42; 9:4; 11:42; 12:44-45; 14:24; 15:21; 16:5; 17:3, 8, 18, 21-25; and 20:21).

b   Other references for God the Son and God the Father being one include Mat. 13:41, as compared with Luke 12:8-9 and 15:10, John 10:30, John 17:11, John 17:21-23, John 20:28-29, Philippians 2:5-11, and Heb. 1:2-12.

c   I Cor. 3:16-17 indicates that God the Holy Spirit and God the Father are one. The same is indicated by a comparison of John 15:26 and John 16:7 with Luke 11:13, John 14:26, Gal. 4:6, I Thes. 4:8, and I John 4:13 and by a comparison of John 15:26 with John 8:42, John 16:28, and John 17:8.

unified hierarchy—simultaneous unity and diversity, equality and distinction of a hierarchy

equal. This simultaneous unity and diversity, equality and distinction of hierarchy is described here as a **unified hierarchy**.

It may be outside of human capacity to fully understand these things. Yet God has provided physical illustrations to aid our understanding. For example, after determining that He would create man (Gen. 1:26), "…God created man… male and female…" (Gen. 1:27). Since it was not good that man should be alone (Gen. 2:18), God created a separate person, made from man's very substance (Gen. 2:21-22), and the two were created to be one (Gen. 2:24; Mat. 19:5-6; Mark 10:7-8; Eph. 5:31).

Like the Godhead, marriage is to demonstrate both unity and diversity. At the same time, the man is to have authority over his wife (I Cor. 11:3; Eph. 5:23), the wife is to submit to her husband (Eph. 5:22-24; Col. 3:18; I Pet. 3:1-6), the husband is to serve the wife (Eph. 5:25-30; I Pet. 3:7), and the body of each is the possession of the other (I Cor. 7:3-5). Like the Godhead, the marriage is to demonstrate both diversity of authority and equality. Although it is only the union of two (rather than three in the Godhead), the marriage relationship is to be a picture of the unified hierarchy of God[a]. The marriage relationship is a challenge because we fall short of God's perfection (Rom. 3:23) and find it difficult to live as one flesh.

---

a A husband is to love his wife, even as Christ loves the church and gave Himself for it (*e.g.* Eph. 5:25), but before the Fall of man and the need for Christ's sacrifice, a man is to love his wife in such a way as to illustrate the love among the members of the Godhead.

Another helpful illustration is the church. God created great diversity among the membership of the church (I Cor. 12:4-11, 28-30), yet the members are to function as one (I Pet. 3:8ᵃ). Like the Godhead, the church is to demonstrate both unity and diversity. At the same time, while some in the church have authority over others (Acts 20:28; I Ti. 3:1-7; 5:17; I Pet. 5:1-3; Heb. 13:7, 17), these leaders are to lead by serving (Luke 22:24-27; John 13:4-17; Acts 20:28; I Pet. 5:1-4), and every member of the church is to submit to every other member (I Cor. 16:16; Eph. 5:21; I Pet. 5:5). Like the Godhead, the church is to demonstrate both diversity of authority and equality. As in the case of marriage, the complexity of God's unified hierarchy makes it difficult to maintain relationships in the church the way they should be.

11.2 | Hierarchy of Biological Organization

Since God cannot be seen, His hierarchy cannot be seen. However, as with all of His invisible attributes (Rom. 1:20), God wishes us to understand His nature by illustrating His hierarchy in the biological creation. One of those ways is by means of hierarchy of organization. This begins with the lowest organizational level, some 'unit' from which everything is constructed. Every church has church members, every factory has factory workers, and every organism has cells.

VIDEO 11.2

The Unit of Biological Life

Hierarchy is found throughout God's creation. At least three tiers of organization are found *below* biological life—subatomic particles, atoms, and molecules. Items of the first tier (subatomic particles) make up the second tier (atoms), and items of the second tier (atoms) make up items of the third tier (molecules). Systems with *only* these three tiers of organization are not biologically living things. As we have already seen in a previous chapter, for example, even macromolecules, the characteristic molecules of life, are bigger and more complex than molecules outside biologically living things. Yet even macromolecules are only components of the lowest tier of biological organization—metabolism.

hierarchy—distinct levels. Types of hierarchy: unified hierarchy has both distinct levels and equality; nested hierarchy has lower level groups fully within higher level groups; netted hierarchy is a nested hierarchy with substantial cross-hierarchy similarity

---

a That the church is to be one is indicated further by Psa. 133:1, John 17:21-23, Rom. 15:6, I Cor. 1:10, I Cor. 12:12-14, 20, 25, II Cor. 13:11, Eph. 4:3-6, Eph. 4:12-13, Philip. 1:27, and Philip. 2:2.

Metabolism

Every biologically living thing is sustained by metabolism, which is a very complex set of processes that break down and build up macromolecules and store and release energy. Metabolism is absolutely essential for biological life to thrive in a universe where the second law of thermodynamics operates to break down complex molecules. Surfaces known as biological membranes play an essential role in metabolism.

Not only is most of the machinery of metabolism mounted on such surfaces[a], but nutrient molecules must be brought into organisms through such surfaces and waste molecules must be discarded through such surfaces. Biological membranes are formed from special molecules known as **phospholipids**. In water, the hydrophobic tails of phospholipid molecules gather together on their own (spontaneously) to exclude water. When there are enough phospholipid molecules, they produce spherical surfaces of double (tail-to-tail) layers of phospholipids, called phospholipid bilayers. For the same reason that they spontaneously form, phospholipid bilayers also spontaneously repair themselves.

These membranes are also flexible. In fact, not being strongly bonded together, the phospholipid molecules are continually moving about, making them so fluid and dynamic that the phospholipid bilayers are also called plasma membranes.

The Size of the Life Unit

An organism's volume determines how much metabolism must occur, but the surface area of the organism's membranes determines the maximum amount of metabolism that can occur. Consequently, the surface areas of biological membranes place a maximum size limit on an organism.

The smallest organisms contain only a single membrane—a plasma membrane making up the outer surface of the organism. The surface area of such a surrounding membrane can only support the metabolism of an extremely small volume. Organisms whose only membrane is that surrounding the organism are called **bacteria**[b]. Because of the surface

---

phospholipid—lipid with hydrophobic and hydrophilic ends; make up biological membranes

bacterium (pl. bacteria)—single-celled organisms small enough not to need organelles (aka prokaryote)

---

a Molecular machines mounted on biological membranes are analogous to machines in factories being mounted on surfaces known as floors.

b Based on the differences in their DNA, the group of organisms that used to be called bacteria have been divided into two different groups, the monera and the archaea. Although some have redefined the old term

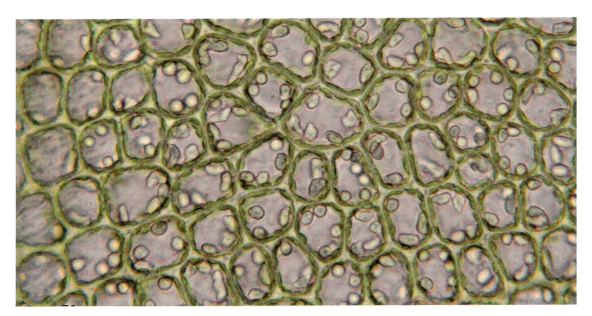

area limitations of their outer membranes, bacteria are microscopic. Spherical bacteria can only be a few microns (a few thousandths of a millimeter) in diameter. Rod-shaped bacteria, although narrower, can be a bit longer, and bacteria shaped like thin threads can be as long as a millimeter or so, but are so narrow as to be invisible without a microscope.

The volume of an organism can be substantially increased if additional membranes are included *inside* the membrane surrounding the organism. The diameter of a spherical organism can be increased around ten times the diameter of the largest bacteria. Although this increases its volume to a thousand times that of a typical bacterium, the resultant organism is still usually less than a tenth of a millimeter in size and invisible to the unaided eye.

Larger organisms must be built from more than one of these tiny membrane-bound entities. Organisms as large as humans are made of up about ten trillion of them. In 1665, Robert Hooke (1635-1703) introduced the term cell for those membrane-bound entities. The microscope was a new invention in his day. When he looked at a plant (cork) under the microscope, the tiny rectangular cubes he saw reminded him of the 'cells' found in monasteries where monks live.

'bacteria' as another name for monera, that redefinition has not been completely accepted. To refer to all organisms in the monera and archaea groups, the word 'prokaryote' is usually used. However, since the word 'prokaryote' was created for evolutionary reasons, this text will use the old colloquial word 'bacteria' to refer to all organisms of both the monera and archaea groups.

Cell Theory

Over the following decades and centuries, the microscope was used to discover all sorts of organisms make up of only one cell each. However, in spite of their first discovery as part of large organisms, it took almost two more centuries for biologists to realize how important cells were in all large organisms. In 1837, Matthias Jakob Schleiden (1804-1881) observed that plants were not only made of cells, but that plant cells seem to come from other plant cells.

Another biologist, Theodor Schwann (1810-1882) noted that animal cells also seem to come from other animal cells. After the two shared their ideas with each other, and then tested their ideas, Schwann published a book suggesting that 'all living things are composed of cells and cell products', and that many cells come from other cells. Rudolph Virchow (1821-1902) took the second claim further and suggested that all cells come from other cells. Three claims from the work of Schleiden, Schwann, and Virchow have been united to form the central tenets of **cell theory**. In their present form, these three claims are:

1. Every organism is made up of one or more cells.
2. The cell is the smallest unit of biological life. Not only is each organism biologically alive, but each individual cell within an organism is also biologically alive.
3. Every (daughter) cell comes from the division of a pre-existing (parent) cell.

In the nearly two centuries since cell theory was originally proposed, the first claim has been verified for everything except viruses. Since viruses are not made of cells, according to the terms of cell theory, viruses would be neither organisms nor biologically alive[a]. In the case of the third claim, not only has every known cell been shown to come from previous cells, but even viruses have been shown to come from cells. This has suggested to some that viruses are not necessarily organisms, but products or parts of organisms. Since the time of Schleiden, Schwann, and Virchow, further claims have been added to cell theory:

---

cell theory—1: every organism is made of one or more cells; 2: the cell is the smallest unit of biological life; 3: every cell comes from the division of a pre-existing cell; 4: all metabolism occurs within cells; 5: heredity information is located in cells; 6: every cell has the same basic chemical ingredients

---

[a] The only level of biological organization viruses share with organisms is the macromolecule DNA (or RNA). Viruses also cannot perform metabolism. Nor can they reproduce—at least by themselves. With cells, however, they *can* do these things (by taking over the machinery of cells). In fact, viruses are formed by cells, so even though viruses are not cells and are not formed of cells, they are the products of cells. Thus, rather than understand them to be organisms, viruses are probably better understood as parts or products of organisms.

4. The metabolism of an organism occurs within its cells.
5. The heredity information of an organism is in the form of DNA, DNA is found within an organism's cells, and a cell's DNA came from its parent cell.
6. Every cell has the same basic chemical ingredients.

11.3 | Traits of All Cells

Several types of molecules are found in all cells as more or less isolated molecules (part of the sixth claim of cell theory). Every cell, for example, carries its genetic information on one or more DNA molecules (part of the fifth claim of cell theory). Furthermore, every cell copies its genetic information and makes its proteins with the same molecular machines (such as ribosomes). In addition to isolated macromolecules, cells also contain systems that are constructed of many different molecules. Every cell, for example, is surrounded by a cell membrane, given shape by a cytoskeleton, and filled with cytoplasm.

Cytoplasm is water that is thick with a host of molecules dissolved in it. It is nearly motionless and molecules usually move through it by diffusion (a process driven by the second law of thermodynamics where molecules move from areas where they are in high concentration to other areas where they are in low concentration).

cytoplasm—water filling a cell, which is thick with a host of dissolved molecules

A *plasma (cell) membrane* separates the cell's cytoplasm from the environment outside the cell. Small electrically neutral molecules like water, oxygen, and carbon dioxide are allowed through the cell

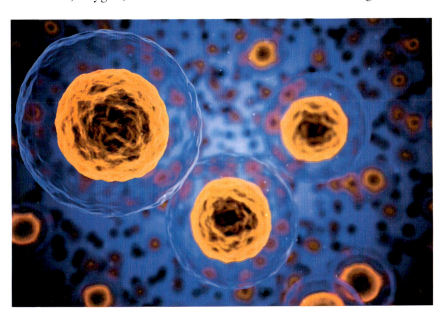

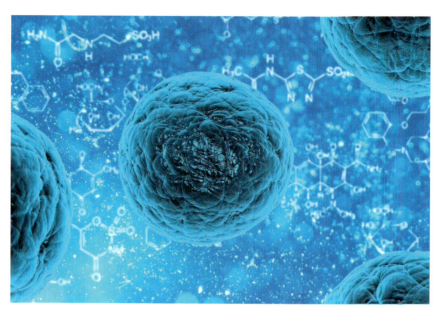

membrane, but larger molecules and ions are not[a]. Special proteins are embedded among the phospholipid molecules of the cell membrane to permit the cell membrane to perform a wide variety of functions. Membrane proteins include *transport proteins* to pass large molecules or ions across the membrane, *adhesion proteins* to fix cells together (in animals), *recognition proteins* which identify the cells that belong to the organism so that they can be protected from invading cells, *receptor proteins* to attach to molecules outside the cell so the cell can respond to molecules in the external environment, and *enzymes* that speed up the chemical reactions of the cells.

Cytoskeleton

Because of the fluid nature of the plasma membrane, if cells were only composed of cytoplasm and plasma membranes, they would be easily destroyed, have no predictable shape, and would probably not be able to move on their own power. In fact, cells are fairly stable, have standard shapes, and many can move. This is because every cell possesses a network of interconnected scaffolding, elastic cables, and pillars, known as the cytoskeleton. The cytoskeleton components are made of different types of proteins. In non-bacterial cells[b], an inter-connected mesh of pillar-like intermediate filaments, for example, gives resilience

---

a   This makes biological membranes differentially permeable.

b   Bacteria have different proteins than non-bacteria (eukaryotes). Bacteria do have proteins that make up cytoskeletons, and those cytoskeletons have structures very much like the intermediate filaments, microfilaments, and microtubules of eukaryotes, but these structures are built from different kinds of proteins.

and shape to the cell and its organelles. The intermediate filaments also determine the position of the cell components. Actin protein is used to make a microfilament, a stretchable cable-like molecule that allows cells to change shape. Tubulin protein can be quickly assembled or disassembled to create a tube-shaped microtubule, scaffolding that can drag objects (like chromosomes) around in the cell or define the direction that a cell grows. Since the filaments and tubules are found throughout the cell, connecting all its components, they make up a three-dimensional road system.

Motor proteins carry individual molecules or cell components throughout the cell, following cytoskeleton elements like trucks carry their loads on roads. Some have a motor protein called dynein that extend into hair-like cell extensions. For example, a single, long hair-like extension is called a flagellum (pl. *flagella*), and a large number of short hair-like extensions are known as cilia (sing. *cilium*). A cell with a flagellum (*e.g.* a sperm cell) can move by whipping the flagellum back and forth. With cilia, a cell can move around in fluid (*e.g.* a *Paramecium*) or move fluids on by (*e.g.* cells lining the respiratory tract) by synchronizing the waving of cilia like oars on a boat. Other cells, such as amoebas, move by extending cytoplasm portions of the cell using microfilaments and then dragging the cell along by motor proteins.

Traits of All Eukaryotic Cells

Many cells are so large as to require more surface area than is provided by the outer membrane of the cell. These cells have phospholipid bilayer membranes organized into organelles ('tiny organs') inside cells. These cells are called **eukaryotic** cells (*eu* means 'true'; *karyon* means 'nut'—referring to the nucleus that looks something like a small nut under the microscope). The major organelles possessed by all eukaryotic cells include the following:

eukaryotic—type of cell that contains organelles

Nucleus

Usually the easiest organelle to identify in a eukaryotic cell is the nucleus. The nucleus functions to store and protect the genetic material of the cell. To help understand why the nucleus is constructed the way it is, let us use the analogy of a huge construction project. Once the architectural plans are drawn up, they are usually stored in a secure file room. Rather than taking the original plans to the job site, copies of the relevant sections of the plans are made in the file room and the copies are taken to the job site.

In a similar way, the genetic material of the cell (DNA) is stored and protected in the nucleus, which is surrounded by a *double* membrane known as a **nuclear envelope**. Since the cell constructs the various molecules it needs *outside* the nucleus, copies of relevant portions of the DNA are being made in the nucleus and then taken to the construction sites. The copies are made of messenger RNA (mRNA). Even though it is much smaller than an entire DNA molecule, an mRNA molecule is very large. Consequently, the nuclear envelope has holes, called **nuclear pores**, which are large enough for RNA molecules to exit the nucleus. The nuclear envelope is composed of a double membrane so as to generate stable nuclear pores large enough to allow RNA molecules through.

Also in the nucleus is at least one irregularly-shaped dense region known as the **nucleolus** (pl. *nucleoli*). The nucleolus is where the cell builds the two components of a ribosome, a structure used by the cell to build proteins. Since ribosomes are used outside the nucleus, the ribosome components must be constructed in the nucleus and then transported out. As in the case of mRNA, the ribosome components are very large molecules, again requiring the large nuclear pores to permit the ribosome components to pass through.

nuclear envelope—double membrane surrounding a cell's nucleus

nuclear pores—holes in the nuclear envelope that allow macromolecules to pass in and out of the nucleus

nucleolus—dense region in the nucleus of a cell where ribosome components are assembled

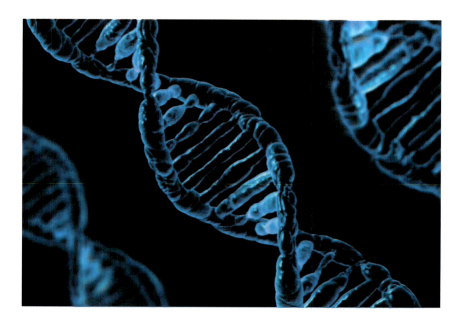

Endomembrane System

The eukaryotic cell builds the protein and lipid macromolecules it needs in the endomembrane system, a series of organelles made of membranes which are found between the nucleus and the outer membrane of the cell. Endo means 'inside' so endomembrane system refers to a system of membranes inside the cell. The series of membranes begins with an extension of the nuclear envelope known as the endoplasmic reticulum (ER). The nuclear pores open directly into what is essentially a huge surface area of membrane that is collapsed or folded (reticulated) into flattened sacs and tubes, making a maze of passages through it.

Machines for building the macromolecules are found in the membrane walls of the ER. The protein-making machines are called **ribosomes**. Since ribosomes are large molecules, they appear as bumps on the walls of the ER. The portion of the ER closest to the nucleus has so many ribosomes in its wall, that it is called the rough endoplasmic reticulum. The rough ER is where proteins are constructed. The remainder of the ER is called the smooth endoplasmic reticulum. It is where lipids are formed.

Although some of the proteins and lipids formed in the ER are used there, most of them are needed elsewhere. **Vesicles** of the endomembrane system provide a sort of mass transit system for molecules. At the end of a tube or sac of membrane, a bulb of membrane can pinch off, forming a spherical vesicle that encloses the molecules and transports them wherever needed. Vesicles from the ER typically travel to another reticulated sac of membrane known as a **Golgi body** (or **Golgi apparatus**[a]). When the vesicle makes contact, the membrane of the vesicle inserts itself into the membrane of the Golgi body and dumps its contents into the flattened sacs of the Golgi body.

The Golgi body is something like a final assembly and shipping center. Here, proteins are folded into their final complex shapes, which determine the job they do, and are then given "shipping labels" so that they get to their proper destinations. Molecules are released from the Golgi body in vesicles that pinch off the sacs at the far side of the Golgi body. A Golgi vesicle delivering molecules to another organelle fuses into that organelle's membrane and dumps its contents into

ribosomes—protein-making machines found in the membrane walls of the endoplasmic reticulum

vesicles—endomembrane system spheres that transport macromolecules among a cell's organelles

Golgi body or apparatus—reticulated sac of the endomembrane system that labels macromolecules

---

a Golgi in Golgi body and Golgi apparatus is capitalized because it is named after Camillo Golgi, who first identified the organelle in 1897.

that organelle. A Golgi vesicle delivering molecules outside the cell fuses with the cell membrane, automatically dumping its molecules outside the cell.

Mitochondria

Eukaryotic cells get most of their energy from mitochondria (sing. mitochondrion). The mitochondrion has a double membrane. An inner membrane with lots of surface area is complexly folded and included entirely within a smooth membrane that gives the organelle its kidney-bean-like shape. The mitochondria are the power plants of the cell. Organic molecules are broken down and the energy from those molecules is put into ATP molecules for use by all the cell processes that require energy.

11.4 | Specialized Cell Traits

Though the organelles described above are found in all—or nearly all—eukaryotic cells, some organelles are designed for only certain kinds of cells. The most common ones include:

Vacuoles

The cells of plants, many protozoa, and some animals have fluid-filled organelles called vacuoles. Depending on the vacuole, they may store nutrients, sugars, starch, water, or waste. In plant cells they take up most of the cell volume.

Lysosomes

A special type of vesicle found in animal cells and possibly in some plant cells is the lysosome. Lysosomes carry molecules that digest (or break down) other molecules. They can also fuse with the membrane of a damaged organelle and release its contents to break down that organelle and recycle it. Alternatively, a lysosome can fuse with the membrane of an invading organism or with a vesicle containing a dangerous organism and release its contents to kill the invader or render the danger harmless. It can also fuse with a vesicle containing food captured from outside the cell and release its contents to digest the food for nutrients that the cell needs.

Chloroplasts

A number of organisms capture the energy of sunlight and store that energy in the form of sugar molecules in a process called photosynthesis. Many of the eukaryotic organisms that perform photosynthesis, including plants, do so in specialized button-shaped organelles called chloroplasts (chloro: 'green'). Most of the photosynthesis in a chloroplast occurs on the thylakoid membrane, a membrane with a huge surface area folded into what looks like stacks of flattened disks. The thylakoid membrane is enclosed within two other membranes, giving the chloroplast a triple membrane.

Cell Walls

Aside from protozoa and animal cells, most cells also produce cell walls. These are tough rectangular boxes just outside a cell's outer plasma membrane that give the cell stability. Cell walls provide even more stability when the cell swells with water and presses against the cell walls. Cell walls are built of different molecules in different organisms (e.g. cellulose in plants, chitin in fungi).

Why the Hierarchy of Biological Organization Exists

To feed a crowd numbering in the thousands, Jesus had them sit down 'by hundreds, and by fifties' (Mark 6:40)—creating a hierarchal organization with at least two tiers (groups of 50 and groups of 100) for the purpose of efficient food distribution. To rule the people of Israel, probably numbering in the millions, Moses followed his father-in-law's advice and chose 'rulers of thousands, rulers of hundreds, rulers of fifties,

and rulers of tens' (Ex. 18:26)—creating a hierarchal organization with at least four tiers for the purpose of efficient governance.

Many organisms are composed of only one cell. This is true of the bacteria and most of the algae and protozoa. The remaining organisms are constructed of multiple cells. Since the basic units of biological life (cells) are constrained in size, the larger the organism, the more cells are needed. The more cells, the more biological organization is necessary to supply those cells with nutrients and to purge those cells of waste products. Very small organisms, for example, should be able to get along with only two tiers of biological organization (metabolism and cells). The larger the organism, the more tiers of biological organization are needed.

In the case of humans, for example, the human body is composed of trillions of cells. And, the human body has at least six levels of biological organization: macromolecules are organized into organelles; organelles are organized into cells; cells are organized into tissues; tissues are organized into organs; organs are organized into organ systems; and organ systems are organized into the human body.

11.5 | Netted Hierarchy of Biological Similarity

God illustrates something of the hierarchy of the Godhead in the organizational hierarchy of the biological world. But God also illustrates the *unified* hierarchy of the Godhead. This is most clearly evident in the pattern of similarity among organisms. That similarities and diversity exist among organisms has already been discussed in Chapters 9 and 10. It is the *pattern* of that similarity and diversity that we now examine.

Let us try a thought experiment. Imagine representing each distinct form of biological life as a separate point in a square. The number of points in that square represents the huge number of different organisms that are possible. The distance between two points represents how similar two organisms are to each other. Two points right next to each other represent two species that are nearly identical. Two points on opposite sides of the square represent two organisms as different from each other as organisms can be. A random pattern of points in the square (see Figure 11.1a) could show both high diversity (if there were many points) and high disparity (if points were found in most of the square), but there would be little evidence of similarity or design. A uniform pattern of points (where each point is located the same distance from the closest points; see Figure 11.1b) could show high diversity (if there were many points), high disparity (if points were found in most of the square), and high similarity or design.

Such a pattern, however, would not illustrate a unity or a spectrum of perfection of either disparity or diversity. Spectrum of perfection of disparity could be added if groups of close-packed bunches of points were grouped together in ever larger bundles of groups, something called a nested hierarchal pattern of points (see Figure 11.1c). If different groupings were a variety of sizes, we would also have a spectrum of perfection of diversity. If there were a variety of spacings between groupings, there would also be a spectrum of perfection of similarity.

If biological organisms were arranged in a nested hierarchal pattern as described above, it would be possible to classify organisms in that nested hierarchal pattern. Biological systematics (or **biosystematics**)

biosystematics—the science of classifying (grouping) and naming organisms

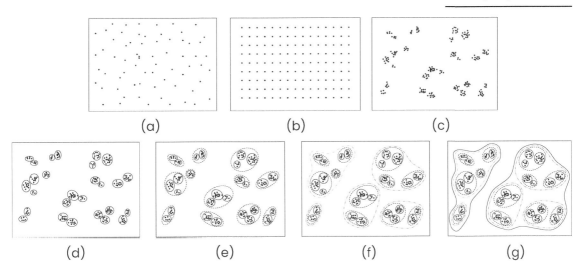

Figure 11.1
In (a) is a random distribution of points, which can demonstrate high diversity and high disparity, but neither unity nor hierarchy. In (b) is a uniform distribution of points, which can demonstrate high diversity, high disparity, and unity, but not hierarchy. In (c) is a nested hierarchal distribution of points, which can represent high disparity, high diversity, unity, and hierarchy—and even a spectrum of diversity of unity and diversity. (d) through (g) illustrates how (c) is a nested hierarchy of points. In (d) the points are grouped (as in the case of species into genera), and in (e) the groups are grouped (as in genera into families). In (f) the groups of groups are grouped (as in families into orders) and in (g) the groups of groups of groups are grouped (as in orders within classes).

taxon (pl. taxa)—group of organisms united by similarity, distinguished from other taxa by discontinuity

taxonomy—the science of naming organisms

taxonomic level—the hierarchal level of a particular taxon (e.g. species, genus, family, order...)

genus (pl. genera)—taxonomic level above species and below taxonomic family

family, taxonomic—taxonomic level above genus and below order

is the science of classifying and naming organisms, and a professional in this field is called a biological systematist or, simply, systematist.

A group of organisms created by a systematist is called a **taxon** (pl. *taxa*). The science of grouping organisms into taxa, naming taxa and identifying organisms is called **taxonomy**, and the scientist who does this is called a taxonomist. For nearly three centuries, systematists have grouped and named organisms following a nested hierarchy pattern. The taxonomist names each distinct form of organism as a separate species. Following a pattern introduced by Linnaeus, a species is given a two-word Latinized name—a capitalized genus name followed by a lower-case species epithet.

Humans, for example, have been named *Homo sapiens* (Latin for 'wise man'). *Homo* is the genus name and *sapiens* is the more specific species epithet. The systematist then groups taxa in a series of higher **taxonomic levels**:

- Similar species are grouped together into a **genus** (pl. *genera*). [For example, humans are in the genus Homo.]
- Similar genera are grouped together into a **family**. [For example, humans are grouped with ape genera in the family Hominidae.]

- Similar families are grouped together into an **order**. [For example, humans and other hominids are grouped with monkey families in the order Primata.]
- Similar orders are grouped together into a **class**. [For example, humans and other primates are grouped with carnivore, insectivore, marsupial, and other orders in the class Mammalia.]
- Similar classes are grouped together into a **phylum** (pl. *phyla*). [For example, humans and other mammals are grouped with fish, bird, and reptile classes in the phylum Chordata.]
- Similar phyla are grouped together into a **kingdom**. [For example, humans and other chordates are grouped with arthropod, mollusk, worm, and other phyla in the kingdom Animalia.]
- Similar kingdoms are grouped together into a **domain**. [For example, humans and other animals are grouped with plant, fungi, protozoa, and other kingdoms in the domain Eukarya.]

order—taxonomic level above family and below class

class—taxonomic level above order and below phylum

phylum (pl. phyla)—taxonomic level above class and below kingdom

kingdom—taxonomic level above phylum and below domain

domain—taxonomic level above kingdom

The similarities that God placed in organisms (discussed briefly in Chapter 9) are what biologists use to group organisms in this way. Organisms that share the most similarities are placed into the same species, species that share similarities are grouped into the same genus, and so it goes for each taxonomic level. The fact that a majority of organisms are readily classified into this scheme suggests that organisms really are arranged in something similar to a nested hierarchal pattern. This is confirmed by the observation that a very similar classification results regardless of what type of similarity is utilized.

The classification that is produced using only similarities found in adult organisms looks very similar to the classification produced using only similarities found in developing organisms or in how long two organisms remain similar while they develop. And both of these classifications are similar to the many classifications produced using any of the many macromolecules found in organisms.

It is reasonable to assume that if an architect wished to create two similar-looking homes that he or she would probably use similar foundations, similar floor plans, and similar building materials. The more different the architect wants the house, the more different will also be the foundation, the floor plan, and the building materials. In fact, whatever the pattern of similarity is in all the homes built by a particular architect, one might expect a similar pattern of similarity when considering only one aspect of the homes (such as just foundations or just the windows or just the roofing materials, and so on). The similar hierarchal pattern found using different types of similarity would be expected if the same architect fashioned all of life.

But this is not the end of the story. For nearly three centuries, most of the known organisms of the world have been named and classified according to a nested hierarchal pattern. However, a few organisms do not seem to fit into this type of classification. For example, even though the pronghorn (*Antilocapra americana*) is now placed into its own family, at different times it has been classified in the deer family (Cervidae) or the cow family (Bovidae)[a]. Similarly, although the lesser panda (*Ailurus fulgens*) is now placed into its own family, past taxonomists classified it at different times in the raccoon family (Procyonidae) or with the greater panda in the bear family (Ursidae).

The problem in these instances, and so many similar ones, is that in practice, scientists only use a small number of traits to classify organisms (not all the traits, as we imagined in our thought experiment). When one set of traits is used, one classification becomes clear (*e.g.* branched, temporary 'antlers' seem to place pronghorns in the deer family), but when another set of traits is used, another classification becomes clear (*e.g.* permanent bone 'horns' seem to place pronghorns in the cow family). It is also not uncommon for a different classification to result when different types of similarities are used (*e.g.* skeleton characteristics of lesser pandas place them in the bear family, behavior and coloration characteristics place them in the raccoon family, and genetic characteristics suggest they are more closely related to weasels). Although different sets of similarities produce *similar* classifications, they do not produce identical classifications.

If the character traits that produce one classification are considered along with the character traits that produce a different classification, homoplasies result. A **homoplasy** is a character trait that does not fit the pattern of classification suggested by other character traits. When a taxonomist makes a decision as to where to classify the organism,

homoplasy—character trait that does not fit the pattern of classification suggested by other characters

---

a The 'horn' of the pronghorn has a permanent core (as horns in the cow family, Bovidae, are permanent), and a sheath that is shed every year (as antlers in the deer family, Cervidae, are shed every year), but pronghorn cores are branched (as opposed to the unbranched horns of Bovidae) and pronghorn sheaths are made of keratin (as opposed to the bony antlers of Cervidae).

that homoplasy will be a character trait that organism shares with an organism in another group, but not with the other organisms in its own group. In our thought experiment, points would have to be both close together (when considering some traits) and *at the same time not* close together (when considering other traits)[a].

Described another way, points at a distance from each other on our diagram are found to have a few very profound similarities and these similarities are homoplasies. If we ignore the homoplasies, many similarities among organisms show evidence of hierarchy (species within genera, genera within families, families within orders, *etc.*). At the same time, the homoplasies provide evidence of unity (similarities between distant points on the square). Although homoplasies are not the majority of traits, they are also not rare. They usually account for about a fifth of the traits considered in classifying organisms. Such homoplasies suggest that not only is an organism similar to those with which it is classified (close points in our thought experiment), but it is also similar with organisms across the entire spectrum of life. Not only is there hierarchy among organisms (that which fits organisms into the nested hierarchal classification system), but there is also a widespread network of similarity that unites all the organisms—what might be called a **netted hierarchy** of similarity. A netted hierarchy

netted hierarchy— a widespread network of similarity that unites all the organisms

---

[a] Although this is not possible to visualize, it is mathematically possible if our square did not have just two dimensions, but had as many dimensions as we had traits. Each trait would be represented in its own dimension of our multidimensional 'square'.

is both hierarchal and unified, a physical illustration of the invisible unified hierarchy of the Godhead.

The Origin of Biological Hierarchy

In the naturalistic worldview, the diversity of biological life arises by evolution, starting with the simplest of organisms. Modern organisms with the simplest organization are those constructed of one cell, and the one-celled organisms with the simplest organization are those that lack organelles. In the worldview of naturalism, one-celled organisms whose only membrane is the cell membrane and who are lacking the complex nucleus, should have been the first organisms to evolve. Since such organisms should have evolved *before* cells with nuclei, scientists with a naturalistic worldview called them **'prokaryotes'** (*pro*: 'before' + *karyon*: 'nut', for nucleus), and sometimes use their existence as evidence for evolutionary theory.

prokaryote— evolutionary term for a cell lacking organelles (aka bacterium)

However, organisms that small have sufficient surface area on their outer cell membranes to perform all the cellular metabolism required by cells with such a small volume. It is most likely that these cells lack organelles because that is the best design, rather than because they have not yet evolved them. Besides, these cells are by no means simple, for most of them are capable of doing so much more than any eukaryote can do. For example, most of the functions of the biomatrix, such as fixing nitrogen, molybdenum, nickel, *etc.* and getting energy from hydrogen gas, sulfur compounds, *etc.* are performed by these tiny cells. It seems more likely that the absence of organelles is due to good design rather than inadequate evolution.

Cells are complex biological systems (see Chapter 9 for a discussion of biological systems). Even the cells with the simplest organization are constructed from millions of molecules, integrated so well that they function very much as units. Based on our experience, such complexity and integration could never be generated spontaneously. The fact that cells are capable of a variety of emergent properties that no human-created systems can accomplish (*e.g.* reproduction), suggests that they were designed by an intelligence exceeding that of humans.

Finally, cells are irreducibly complex. If they possess less than a certain minimum number of components they simply cannot function. They cannot be generated one step at a time, such as the naturalistic worldview requires. The complexity, emergent properties, and irreducible complexity of cells are more consistent with creation by God, than anything suggested in the naturalistic worldview.

Because of the needs of metabolism and the limitations of biological membranes, cells seem to be about as large as they can be and still not have to be divided into smaller chambers. Consequently, cells seem to be the optimal design and the fundamental construction unit of biological life. If biological life were designed, it would have been constructed from cells or something very much like them. The fact that all biological life was made this way not only suggests that life *was* designed, but that all biological life was designed and constructed by one Creator. Once the Creator chose the cell as the construction unit, large organisms would have to be made from trillions of component cells.

Among human designs, the most efficient systems are those with a hierarchy of organization which is appropriate for the size of the system (*e.g.* 2 hierarchal tiers for feeding thousands; 4 hierarchal tiers for ruling millions). The number of levels of biological organization in organisms (*e.g.* six levels for trillions of human cells) is similar to what is expected in an efficiently designed system. The similarity of organizational hierarchy in biology and well-designed human systems suggests that not only cells were well designed, but all the organisms made from those cells were also well-designed. Although this might be expected from a Creator, such efficiency is not a reasonable expectation in a naturalistic worldview.

macroevolution—the origin of taxa above species or the process that generates new taxa above species

The best explanation for the origin of organisms in the naturalistic worldview is macroevolution. According to **macroevolution**, every species is continually generating new species, distinct from the original species but similar to it. This is somewhat analogous to a human couple generating children that are distinct from the couple, but at the same time similar to them. In this manner, by reproduction, a human couple can generate a very large and diverse family tree over time. In a similar manner, by macroevolution, a species is thought to be able to generate a large and diverse evolutionary tree over time. One species becomes two (an early branching of the tree) and on each branch a species branches into two more. In such a manner one species generates the modern diversity of life.

If this is how life's diversity came to be, life should follow a simple nested hierarchal pattern, for every taxon was derived from only one previous taxon and should be classified in only one taxon of the next highest level (just as every branch grows out of only one larger branch). If life's diversity was generated by **macroevolutionary theory** (the best theory in the naturalistic worldview), then it should be possible to classify all organisms in a simple nested hierarchal pattern, and different types of similarities should generate the same pattern.

macroevolutionary theory—naturalistic evolution theory explaining the origin of taxa above species

Macroevolutionary theory has difficulty explaining the organisms that cannot be classified into a simple nested hierarchy. It also has difficulty explaining homoplasies—those traits of organisms that suggest different patterns. The widespread homoplasy of life suggests that life was not generated by macroevolution. Rather, the netted hierarchal pattern of life is more consistent with creation by a God Who desired to illustrate His own unified hierarchy in the biological creation.

Biological Hierarchy: Our Responsibility

Our Responsibility to God

The unified hierarchy of the Godhead is a challenge to comprehend. It is difficult enough to understand how three persons can be one; it is even more difficult to understand how three persons in hierarchy can be one and mutually submissive. Yet God desires more than our comprehension of His nature. He wishes us to illustrate His nature in our lives—to show unified hierarchy in our families and in our churches. Towards that end, the hierarchy of biological organization and netted

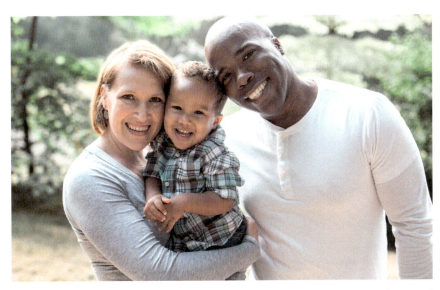

hierarchy of biological similarity can help us. As we better understand the hierarchy of biology, we can get closer to understanding the nature of God. This in turn can help us better live out unified hierarchy in our homes and churches.

The organisms which are difficult to classify, such as the lesser panda and the pronghorn, along with homoplasies, cell complexity, and the optimality of biological organization should demonstrate the inadequacy of evolutionary theory. This, in turn, should remind us of the only reasonable explanation—that God created these things. The remarkable complexity of cells, including the amazing capabilities of bacteria, ought to impress upon us the wisdom of God's design, and evoke worship of the great God Who knows what is best for us.

Since these features are best explained by a God of unified hierarchy, they should also remind us of the hierarchy of divine authority and evoke worship of the God Who has authority over all. At the same time, since God designed this hierarchy so that we could see it, they also remind us that God desires for us to know Him. That such an awesome God desires a relationship with us, ought to inspire us into even more worship.

As a consequence, a study of biological hierarchy should not only improve our understanding of God, but it should also encourage us to live a life filled with His worship. As we continue, we cannot help but share this worship with others and so glorify our God. Contemplation of biological hierarchy can thus stimulate us to better know God, worship God, glorify God, and bring others into our worship of God.

Summary of Chapter

A. The three persons of the Godhead exist in a perfect unified hierarchy (three persons, but one; having hierarchy of authority, but equal). We are called to reflect this unified hierarchy:

   a. in the relationship between the husband and wife (two persons, but one; husband in authority over wife, but equal; the body of each being the property of the other)

   b. in the church (multiple persons, but one; positions of authority, but everyone should submit to everyone else)

B. To illustrate His invisible, unified authority God created levels of organizational authority:

   a. in the entire physical creation:
      i. Hierarchal Level 1: energy organized into subatomic particles
      ii. Hierarchal Level 2: subatomic particles organized into atoms
      iii. Hierarchal Level 3: atoms organized into simple molecules

   b. in organisms (2 hierarchal levels for the smallest organisms; more and more hierarchal levels for successively larger organisms):
      i. Hierarchal Level 4 (in all organisms): monomers organized into biological macromolecules (*e.g.* DNA, ribosomes) and metabolism
      ii. Hierarchal Level 5 (in large cells): macromolecules and metabolism organized into organelles
      iii. Hierarchal Level 6 (in all organisms): macromolecules and metabolism organized into cells (= bodies of small organisms)
      iv. Hierarchal Level 7 (in large organisms): cells organized into tissues
      v. Hierarchal Level 8 (in larger organisms): tissues organized into organs
      vi. Hierarchal Level 9 (in even larger organisms): organs organized into organ systems

C. Every biologically living thing is sustained by metabolism (the breaking down and building up of biological macromolecules and the storing and releasing of energy to break down and build up macromolecules). Essential to metabolism are biological membranes

   a. that (for metabolism) are surfaces
      i. on which are mounted the molecular machines of metabolism
      ii. through which nutrient molecules must be brought in and waste molecules must be discarded

   b. composed of bilayers of phospholipid molecules (a phospholipid having a hydrophilic head and two hydrophobic tails)

   c. which (because of the interaction of weak electrical charges on water and phospholipid molecules)
      i. are flexible, fluid, dynamic (so known as plasma membranes)
      ii. assemble spontaneously into spherical surfaces
      iii. repair themselves spontaneously
      iv. grow, simply by adding phospholipid molecules
      v. bud off a separate spherical surface when pinched together
      vi. can fuse into other membranes

D. According to cell theory (developed early in the twentieth century)

   a. Every organism is made up of one or more cells.
   b. The cell is the smallest unit of biological life. Not only is each organism biologically alive, but each individual cell within an organism is also biologically alive.
   c. Every (daughter) cell comes from the division of a pre-existing (parent) cell.
   d. The metabolism of an organism occurs within its cells.

e. The heredity information of an organism is in the form of DNA, the DNA is found in cells, and that DNA of each cell came from its parent cell.

f. Every cell has the same basic chemical ingredients.

E. ALL cells have

a. a cell membrane (with phospholipids and proteins) that surrounds and protects the cell, keeping in the macromolecules of life and allowing in and out small electrically neutral molecules (like H_2O, CO_2, O_2);

b. cytoplasm (water thick with dissolved molecules) that fills the cell; and

c. a cytoskeleton (constructed of cable-like, pillar-like, and scaffolding-like proteins) that creates a road system in the cell, gives the cell its shape, and allows the cell to move.

F. Bacteria do not have organelles because they are small enough for the cell membrane around each cell to have enough surface area for the organism to perform all its metabolism. It is NOT because they have not yet evolved them (as has been suggested by evolutionary theory).

G. If a cell is more than a few microns in size, the surface area of its outer membrane is not enough to support the metabolism needed by the cell. Larger cells like this (eukaryotic cells) have membranes inside the cell organized into organelles.

a. All eukaryotic cells have

i. a nucleus (spherical, surrounded by a double membrane nuclear envelope with large nuclear pores) that contains and protects the cell's genetic material (DNA) and one or more nucleolus (where ribosomes are built);

ii. an endomembrane system

- rough endoplasmic reticulum (rough ER; complexly folded membrane with a huge surface area extending out from the nucleus and studded with ribosomes) where proteins are built;

- smooth endoplasmic reticulum (complexly folded membrane with a huge surface area extending out from the rough ER) where lipids are built;

- vesicles (spheres of membrane that bud off the smooth ER and Golgi body) that transport and deliver macromolecules; and

- Golgi body (one or more complexly folded membrane with a large surface area) where macromolecules are given their final shape and 'shipping labels'.

iii. mitochondria (kidney-bean-shaped, given shape by an outer membrane and having a complexly folded inner membrane) where organic molecules are broken down to create energy for the cell.

b. organelles found in some cells:

i. vacuole (membrane surrounding a pool of water) for the storage of molecules (in plants, many protozoa, and some animals);

ii. lysosomes (vesicles containing digestive enzymes) for breaking down organic molecules to destroy invading organisms and recycle macromolecules and organelles (in many plant and animal cells);

iii. chloroplasts (disc-shaped, two outer membranes plus an innermost membrane with a huge surface area) where photosynthesis occurs (in plants and algae); and

iv. cell wall (stiff surface outside the cell membrane) to protect the cell and prevent it from expanding too large from osmosis.

H. In the modern biological field of biological systematics, systematists have classifying organisms and taxonomists have been naming organisms for nearly three centuries. Similarities among and between organisms have been used to classify organisms in a nested hierarchal pattern (species within genera; genera within families; families within orders; orders within classes; classes within phyla; phyla within kingdoms; kingdoms within domains).

I. Some biological similarities show similarity across the nested hierarchal pattern:

   a. organisms that do not easily fit into a nested hierarchy (like pronghorns and lesser pandas); and

   b. homoplasies (similarities that suggest a different classification).

J. These similarities plus those that fit a nested hierarchal pattern make up a netted hierarchal pattern. The netted hierarchal pattern of biological similarity was created by God as an illustration of His own unified hierarchy.

K. The naturalistic worldview has difficulty explaining 1) the extreme complexity of cells (*e.g.* their emergent properties; their irreducible complexity; not constructible spontaneously or by humans); 2) the efficiency of biological hierarchy to supply the metabolism needs of organisms; and 3) the abundance of homoplasy and the existence of organisms that do not fit into a simple nested hierarchy.

L. We fulfill our responsibilities as priests of the creation as we study biological hierarchy and come to better understand God's unified hierarchy, His wisdom in creation, and His desire that we know Him.

Test & Essay Questions

1. Describe what it means for God to have the attribute of unified hierarchy.

2. Describe the unified hierarchy of the family / church.

3. How is the family / church to reflect the unified hierarchy of God?

4. What are prokaryotes / Why are bacteria so small? / How can a eukaryotic cell be so much larger than a bacterium / prokaryotic cell?

5. Why were bacteria called prokaryotes and what is another possible explanation for their structure?

6. How does a phospholipid generate a biological membrane? / Why do phospholipids produce phospholipid bilayers? / Why are biological membranes also called plasma membranes?

7. Why are cells / the fundamental biological units of life so small?

8. List two / three / four claims of cell theory.

9. List the three traits that are found in all cells / both prokaryotic cells and eukaryotic cells.

10. On a diagram of a cell, label cell membrane / cytoplasm / flagellum / cilia / nucleus / nuclear envelope / nucleolus / nuclear pores / endoplasmic reticulum / rough endoplasmic reticulum / smooth endoplasmic reticulum / ribosomes / vesicle / Golgi body / mitochondrion / cell wall / vacuole / chloroplast.

11. Describe the function of the flagellum / cilia / the nucleus / the nuclear envelope / the nucleolus / nuclear pores / the endoplasmic reticulum / the rough endoplasmic reticulum / the smooth endoplasmic reticulum / ribosomes / a vesicle / a Golgi body / a mitochondrion / the cell wall / a vacuole / a chloroplast / a lysosome. [or match any of these things with function]

12. Why are nuclear pores so large?

13. How vesicles work to move molecules from the ER to a Golgi body / from the Golgi body to the cell membrane?

14. List the five tiers / levels of biological organization.

15. How does a nested hierarchy differ from a netted hierarchy?

16. What type of hierarchy of biological classification is expected in a naturalistic / creationist worldview?

17. What type of hierarchy of biological classification is used in biology, and why does it not seem to work?

18. Define biosystematics / taxonomy / taxon.

19. List, in order, the eight taxonomic levels from the lowest level (with the least members) to the highest level (with the most members).

20. Explain how a homoplasy is a problem to evolution / naturalism / nested hierarchy.

21. Explain what homoplasies indicate.

CHAPTER 12

THE ALMIGHTY GOD

Cellular Energy Metabolism

"You are great and powerful, glorious, splendid, and majestic. Everything in heaven and earth is Yours, and You are King, supreme Ruler over all."
1 Chr. 29:11, GNT

12.1 | Almighty God

God is a powerful God (*e.g.* Job 36:5 and Rom. 13:1[a]). He does anything He pleases (Psa. 115:3), for there is no limit to his power (Gen. 18:14; Jer. 32:17, 27; Mat. 19:26; Luke 1:37)[b]. One of his names is Almighty (*e.g.* Gen. 17:1 and Rev. 21:22[c]). By His power He created the universe (Jer. 10:12; 51:15), and by His power He holds the universe together (Heb. 1:3). The Almighty God is the source of every power, whether it be a power of the physical world or a power of authority (Rom. 13:1).

Physical World Energy As Illustration

As ubiquitous as God's power is, His power is also invisible. Yet, since God wants us to understand His invisible attributes, He illustrated those attributes in the physical creation (Rom. 1:20). God illustrated His power by creating the physical world in such a way that energy is required for anything to happen in the universe. And God allowed the universe to be active or dynamic by creating an enormous amount of energy in the universe and placing it in a wide variety of forms. Energy can be put into storage, released from storage, converted from one form of energy to another form of energy, and/or used to do make things happen.

As in the remainder of the creation, energy also plays an important role in the biological world. Everything we do requires energy. We need energy to move, to grow, to speak, to breathe, to circulate our blood, to repair our molecules, to digest food, and even to use our brains. Inside cells, energy is needed for a host of cellular activities (such as building macromolecules from monomers and breaking down damaged molecules into monomers). Consequently, at every moment

---

a Further references indicating that God is a powerful God include Deu. 7:21, I Chr. 29:11-12, Job 36:5, Job. 37:23, Psa. 24:8, Psa. 50:1, Psa. 62:11, Psa. 132:2 & 5, Psa. 147:5, Neh. 9:32, Isa. 1:24, Isa. 9:6, Isa. 10:21, Isa. 30:29, Isa. 40:26, Isa. 49:26, Isa. 60:16, Jer. 32:17-19, Jer. 51:15, Nahum 1:3, Hab. 1:12, Zeph. 3:17, Matt. 28:18, Eph. 1:19, Heb. 1:3, Jude 25, and Rev. 7:12.

b There are, of course, a number of things that God cannot do and will not choose to do. He cannot do anything that is contrary to His nature. Because of His righteousness, God cannot sin. Because He is truth, He cannot lie. Because He is life, God cannot die. And so it continues. God does whatever He wishes to do, but will never choose to do anything contrary to His own nature.

c Further references to the Almighty God include Gen. 28:3, Gen. 35:11, Gen. 43:14, Gen. 48:3, Ex. 6:3, Ruth 1:20, Psa. 68:14, Psa. 91:1, Isa. 13:6, Eze. 1:24, Eze. 10:5, Joel 1:15, II Cor. 6:18, Rev. 1:8, Rev. 4:8, Rev. 11:17, Rev. 15:3, Rev. 16:7 & 14, and Rev. 19:15. Many more references are found in the book of Job (*e.g.* 5:17; 6:4 & 14; 8:3 & 5; 11:7; 13:3; 15:25; 21:15, 20; 22:3 & 17 & 23 & 25-26; 23:16; 24:1; 27:2 & 10-11 & 13; 29:5; 31:2 & 35; 32:8; 33:4; 34:10 & 12; 35:13; 37:23; and 40:2).

of every day each of our trillions of cells is storing, transforming, and/ or using energy. And what is true in each cell of our bodies is true in every organism and each cell of each organism. All cells are active at all times in storing, transforming and/or using energy.

Cellular Energy Metabolism: Biology's Power Source

Metabolism (Greek *metabolē*, meaning 'change') refers to all the chemical reactions in cells that transform molecules from one form into another (and that is every single reaction!). Besides reactions that build molecules (**anabolism**) and reactions that break molecules down (**catabolism**), metabolism also includes energy metabolism reactions that store or release energy, so that the cell can do all that a cell must do.

Different organisms get the energy they need in different ways. **Autotrophs** [Gr. *eauton*: 'self' + *trophe*: 'nutrition'] can get their energy from the non-biological world around them. Autotrophs are **producers** and include many bacteria and most of the algae and plants. The most well-known autotrophs (most plants and algae) get their energy from sunlight, so they are called **photoautotrophs**.

Many bacteria, in contrast, are **chemoautotrophs**, since they get their energy from chemistry—more specifically inorganic molecules such as hydrogen sulfide (H_2S), elemental sulfur (S), ferrous iron (Fe^{++}), hydrogen gas (H_2), or ammonia (NH_3). Chemoautotrophs play important roles in biogeochemical cycles. Autotrophs—both photoautotrophs and chemoautotrophs—produce the food ultimately consumed by all organisms. Consequently, many autotrophs are designed to take the energy they acquire and store it in organic molecules—both for themselves and other organisms.

Heterotrophs [Gr. *heteros*: 'different' + *trophe*: 'nutrition'] get their energy from organic molecules (such as the molecules created by autotrophs to store energy). Heterotrophs are consumers and include most of the animals, fungi, and protozoa. Most of these heterotrophs use oxygen to extract energy from organic molecules in an energy metabolism process called aerobic respiration.

Other heterotrophs release molecular energy without using oxygen. **Fermentation** is probably the most well-known type of oxygen-free energy metabolism. Lactic acid bacteria, for example, ferment sugars

---

anabolism—chemical reactions in cells that build molecules

catabolism—chemical reactions in cells that break down molecules

autotroph—organism that gets energy from the non-biological world. Types: photoautotroph gets energy from sunlight; chemoautotroph gets energy from inorganic molecules

producer—an organism that gets its energy from the physical environment (vs. consumer)

photoautotroph—organism that gets its energy from sunlight

chemoautotroph—organism that gets its energy from inorganic molecules

heterotroph—organism that gets its energy from organic molecules

fermentation—chemical reactions a cell uses to get energy from organic molecules without using oxygen

into lactic acid. Humans use lactic acid bacteria to produce cheese, yogurt, buttermilk, sauerkraut, corned beef, and salami. In another type of fermentation, yeast ferment sugars into carbon dioxide and ethyl alcohol. For example, a certain type of yeast, leaven, is added to bread dough, to 'raise' the dough with the carbon dioxide generated in the fermentation process. Grain or fruit is fermented with other types of yeast to create beer, wine, cider, and other alcoholic beverages. In yet another type of fermentation, acetic acid bacteria ferment ethyl alcohol into acetic acid. Acetic acid fermentation is used to produce vinegar.

Even the simplest example of energy metabolism is a very complex process. There are so many reactions involved that even well-studied examples are not completely understood. And, in the case of many an organism, its energy metabolism process is not understood at all. It is not the purpose of this text to examine energy metabolism in detail. Rather, we will choose only two well-known examples and provide only the simplest of introductions—merely enough to provide a 'taste' of what is involved in cellular metabolism.

12.2 | Photosynthesis

Most of the energy that circulates through land and water ecosystems ultimately comes from the sun through a metabolism process known as photosynthesis. In plants and green algae, photosynthesis occurs in cell organelles known as chloroplasts. Photosynthesis is extremely complicated, involving somewhere around 500 chemical steps. When

all these chemical reactions are 'added' together, the result is the overall chemical reaction of $6CO_2 + 6H_2O +$ sun energy $\rightarrow C_6H_{12}O_6 + 6O_2$ [$C_6H_{12}O_6$ is the formula for a monosaccharide—a sugar monomer].

In other words, 'photo' 'synthesis' is using light ('photons') from the sun for the 'synthesis' (or assembling) of sugar molecules. Referring back to the overall reaction, carbon dioxide from the atmosphere, water from the soil, and energy from sunlight are combined to build sugar, releasing oxygen gas in the process.

Pigments

Sunlight warms the earth. One does not have to stand in the sunlight long to realize that sunlight contains energy. And, although sunlight on a white surface looks 'white', passing that same sunlight through a prism demonstrates that 'white' light is actually made of all the colors of the rainbow, and a white surface is actually one that *reflects* all the colors. A black substance is one that *absorbs* all the colors, reflecting none of them. This is why black objects get so warm in the sun. A pigment is something that absorbs one or more of the colors and reflects the rest. A red pigment, for example, is something that reflects red and absorbs all the other colors. Similarly, a blue pigment reflects blue and absorbs the others. Plants and green algae are green because the pigments of photosynthesis (chlorophylls), absorb the red and blue ends of the spectrum and reflect green. Different sets of pigments—such as are found in different photoautotrophs—are colored differently

(thus the common names of 'red algae', 'brown algae', 'yellow-green algae', 'golden algae', and 'cyanobacteria').

Light-Dependent Reactions

In green plants, the hundreds of steps of photosynthesis can be grouped into two sets of reactions. The first set of reactions requires light, so they are called the **light-dependent reactions**. The end result of these reactions is the production of the energy and electrons that are needed for the second set of reactions. Among the light-dependent reactions, pigments capture sunlight energy and release electrons carrying that energy. Each energy-carrying electron is released into an **electron transport system** (ETS), which passes the electron from one molecule to another, releasing carefully measured amounts of energy to accomplish specific tasks. Energy from one ETS is used to break water molecules into oxygen gas, electrons, and hydrogen ions ($2H_2O \rightarrow O_2 + 4e^- + 4H^+$). The oxygen is rejected as waste.

In the case of the electrons, some are used to replace those that are released from the pigments, while others are carried to the light-independent reactions in special molecules called NADPH (Nicotinamide Adenine Dinucleotide Phosphate). The hydrogen ions are pumped into the innermost compartment of the chloroplast using energy from a second ETS. The concentration of ions is built up for a reason similar to why water is stored behind a hydroelectric dam. When the water is released from behind a dam, it flows downhill due to gravity, rotating components of an electric generator to generate electricity.

In a similar manner, when hydrogen ions are released from the innermost compartment, the hydrogen ions flow out of the compartment due to the second law of thermodynamics, rotating components of an *ATP synthase protein* so as to generate ATP[a]. Some of the ATP is used to supply energy for the cell, while the remainder is passed on to the light independent reactions to be stored in sugar molecules.

12.3 | Light-Independent Reactions

The second set of photosynthesis reactions does not require light, so they are called **light-independent reactions**. In most plants,

light-dependent reactions—photosynthesis reactions that require sunlight

electron transport system—cell membrane molecules that extract energy from an excited electron

light-independent reactions—photosynthesis reactions that do not require sunlight

---
a   All cellular machinery runs on one form of energy—the energy stored in ATP (adenosine triphosphate). The molecular machines of the cell have been built to run on ATP in a manner analogous to how many of the appliances of a home have been built to run on electricity.

these reactions occur in the same chloroplasts in the same cells. In some plants, however, the light-dependent and light-independent reactions occur in chloroplasts of different cells and/or different times of the day. Light-independent reactions collect carbon dioxide from the atmosphere and electrons (NADPH) from the light-dependent reactions so as to store energy (ATP) from the light-dependent reactions in molecules of glucose. Central to these light-independent reactions is a metabolic cycle.

Humans have discovered that machines are often most efficient when they are designed to operate in a cyclical fashion. If the machine is designed never to stop, but to continue through a series of steps to constantly return to the same first step, then the machine not only never stops, but does not have to waste energy to start up again when stopped. In an automobile engine, for example, the burning of gas is used to move a piston and turn an axle. If two or more pistons are attached to the same axle at different points in their respective cycles of motion, the burning of gas at one piston will not only turn the axle, but also reset a second piston to its beginning position. Burning of gas in the second piston will then turn the axle and reset the first piston to its first position. Although some of the energy released by burning the gas is used to reset the piston or pistons that are *not* firing, it is much less energy than would be required to get the engine going again from a stop if it had to restart every time the piston finished burning its gas.

Cycles tend to maximize the efficiency of a process. Many metabolic processes include a metabolic cycle, probably to maximize the efficiency of metabolic processes. In most metabolic cycles, an organic molecule is modified, step by step, through a series of reactions. In the last step the molecule is returned to its original form. The metabolic cycle in the light-independent reactions of photosynthesis is called the **Calvin cycle**. In the course of one 'turn' of the Calvin cycle, one carbon dioxide molecule is *fixed* (*i.e.* combined with an organic molecule so that it is useable by organisms).

Every six times around the continuously-running cycle, six carbons are fixed into (usually) a glucose or fructose molecule. Fructose and glucose are the most important sugars in most fruits. These two sugars are used to build a wide variety of carbohydrate molecules. For example, a fructose and glucose can combine to form the disaccharide

---

Calvin cycle—chemical reactions in photosynthesis' light independent reactions that fix carbon

sucrose—what is usually known as table sugar. Also, a large number of glucose molecules can be joined to produce starch or cellulose.

Tree Design

The unique characteristics of water combined with the process of photosynthesis created substantial constraints on the design of land plants. Photosynthesis requires sunlight, water, and carbon dioxide, and must get rid of oxygen. Although water is most available to the plant *beneath* the earth's surface, sunlight is only available above the earth's surface. This is why most plants draw their water from underground and transport the water above ground, where photosynthesis occurs.

An important part of the process is acquiring the necessary water. Since water enters the plant through exterior membranes, plant roots must increase their surface area as fast as the amount of photosynthesis increases in the plant. As a plant increases in size, the roots must become more branched, root cells must produce more microscopic hairs, and more symbiotic relationships must be developed with mycorrhizal fungi. Roots are specially designed to acquire the water needed for photosynthesis. Then the water needs to be transported to the leaves.

The most efficient way to transport water is in a tube with a circular cross-section. Due to *capillary action*—a unique characteristic of water—the narrower the tube, the farther up water will travel without energy needed to pump it. In plants, water is transported in extremely narrow, circular-cross-sectioned tubes, called *tracheids*. These cells are

tube-shaped and are placed end to end from the roots to the upper leaves. Water can travel several feet by capillary action. Water can travel the remainder of the way up the tracheids by taking advantage of the flexible hydrogen bonds between adjacent water molecules. It is as if water molecules were attached to each other with springs.

In a leaf, as a water molecule is taken into a cell for photosynthesis or is evaporated from the surface of the leaf, the hydrogen bond between that water molecule and the next one down is stretched. When the bond is stretched enough, it pulls up the next water molecule, stretching the bond between that molecule and the third one down the line. In this manner, each water molecule from the leaf to the root is pulled up one molecule at a time and one position at a time.

The flexibility of the hydrogen bonds between water molecules make it possible for an impossibly heavy column of water to be drawn up to the leaves of a plant[a]. In the course of a single day scores of gallons of water can be drawn up hundreds of feet from roots deep in the ground without the plant having to expend any energy in the

---

[a] This is analogous to how a locomotive can move an impossibly heavy train. If all the cars of a train were solidly connected, locomotives would not be able to get a train to move. Rather, the connections between cars are designed with flexibility, so that even when connected, train cars can move back and forth a small distance without engaging (i.e. without pushing or pulling the next car). When it is time to move a train, the locomotive in front of the train first moves in *reverse*, pushing all the cars of the train together. Then, the locomotive accelerates forward as fast as possible. The flexibility in the connection allows the locomotive to gain some momentum before it begins to pull the first car. For a moment thereafter, the locomotive is pulling only one car while that first car gains some momentum of its own. Then the first car engages with the second, and so it goes until the entire train is moving—a train which is too heavy for a single locomotive to move if all the cars had to be moved at once.

process. Trunks, stems, branches, and leaf veins are specially designed with these tracheids to transport water from the roots to the leaves. Since photosynthesis requires sunlight, it would be most efficient if it occurs in very thin (so light can penetrate), flat surfaces (for maximum surface area).

This explains why most of the green chlorophyll is found in thin, flat leaves. When leaves are surrounded by generally dry air, their large surface area should lead to rapid evaporation of the water brought up from the roots. Chloroplast-containing cells are thus protected from evaporation by being surrounded by small, tightly packed epidermal cells and a transparent, waxy lipid layer. Then, in order to allow carbon dioxide in, and oxygen out, there are special pairs of *guard cells* that can open and close around a passageway called a *stoma* (pl. *stomata*). Guard cells act as doors through the epidermis and control the entry and exit of carbon dioxide and water. Leaves also, then, are specially designed to allow as much photosynthesis to occur as possible. In short, the entire plant, from roots to stems to leaves, is specially designed for photosynthesis.

Aerobic Respiration

For the average person, the word 'respiration' is probably only used to refer to breathing. As we respire, we inhale oxygen and exhale carbon dioxide. We do this because each cell of our body is using oxygen and getting rid of carbon dioxide in the process of cellular respiration. In **cellular respiration** our cells get energy by breaking down organic molecules (like sugars, fats, and proteins).

In human cells, cellular respiration utilized the energy metabolism process known as aerobic respiration. **Aerobic respiration** uses oxygen and releases carbon dioxide as a waste product. We have to continually breathe in oxygen to supply the ten trillion cells of our bodies with the oxygen they need for aerobic respiration. The carbon dioxide produced by the aerobic respiration of our ten trillion cells is the carbon dioxide we must continually exhale. And it is not just humans that utilize aerobic respiration. All animals use aerobic respiration. Even plants use aerobic respiration when photosynthesis is not occurring.

Glycolysis

The first step in obtaining energy from sugar is not actually part of the aerobic respiration process. Because it does not require either

respiration, cellular—chemical reactions a cell uses to get energy out of organic molecules. Types: fermentation is done without using oxygen, aerobic respiration is done using oxygen

respiration, aerobic—chemical reactions a cell uses to get energy from organic molecules using oxygen

a special organelle or oxygen, it can occur in most cells—even in cells that are unable to perform aerobic respiration. This first step is called **glycolysis** (*glycose*: glucose + *lysis*: break down), and it occurs in the cytoplasm. 6-carbon sugar molecules such as glucose are broken into two 3-carbon *pyruvate* molecules and two electrons (carried in NADH). Although it takes two ATPs to break each monosaccharide, four ATPs are created in the process. The *net gain is two ATPs* (4 ATPs minus 2 ATPs). In organisms that use oxygen to get energy out of molecules, the pyruvates and electrons pass into mitochondria for aerobic respiration. Other forms of energy metabolism (like fermentation) utilize the pyruvates and electrons in other ways.

glycolysis—respiration reaction in the cytoplasm that breaks a glucose molecule in half to extract energy

Aerobic Respiration

Very little of the energy found in a monosaccharide is released in the process of glycolysis. Aerobic respiration can extract a lot more energy from the pyruvates. Aerobic respiration occurs in the mitochondrion, an organelle found in almost all eukaryotic cells. The first step occurs in the innermost compartment of the mitochondrion, where each pyruvate is broken into a carbon dioxide molecule, a 2-carbon acetyl group, and an electron (stored in NADH). The carbon dioxide is released as waste. The acetyl group is combined with coenzyme A to produce acetyl-CoA.

The acetyl-CoA then enters a cyclical metabolism process known as the **Krebs cycle** (aka **citric acid cycle**). Since each 'turn' of the Krebs cycle breaks down one acetyl group, two 'turns' of the Krebs cycle completes the breakdown of the original monosaccharide. These two turns produce four CO_2 molecules, ten electrons (stored in 6 NADHs and 2 $FADH_2$s), four hydrogen ions (stored in two $FADH_2$s), and two ATPs. The electrons are then passed on to electron transport systems in the innermost membrane and the released energy is used to pump hydrogen ions into the outer compartment of the mitochondrion. These hydrogen ions are then used to generate ATP in a similar fashion to how it was done in chloroplasts (hydrogen ions released from the outer compartment turn an ATP synthase protein). This produces an additional 32 ATP.

Krebs (or citric acid) cycle—aerobic respiration reactions that extract energy from carbon-hydrogen bonds

Considering all the steps of both glycolysis and aerobic respiration, the result is more or less the inverse of photosynthesis: $C_6H_{12}O_6 + 6O_2 \rightarrow 6CO_2 + 6H_2O + 36ATP$. Notice that carbon dioxide and water are waste products. Also notice how much more efficient aerobic respiration is than glycolysis. Aerobic respiration gets 36 ATP from a glucose molecule whereas glycolysis only gets 2 ATP.

Lipid & Protein Respiration

Energy can be derived from just about any organic molecule. If, for example, blood sugar (glucose) levels get too low, fat cells in the human body break triglycerides into glycerol and fatty acid molecules. The liver converts glycerol into a molecule that enters glycolysis of most cells. The fatty acids are taken directly into body cells and broken down into a bunch of acetyl-CoAs. If the level of both glucose and triglycerides is low, the body can also break down protein.

The Origin of Cellular Metabolism

The processes of energy metabolism have all the earmarks of having been designed by the God described in Scripture. First, energy metabolism is astonishingly complex. Photosynthesis is so complex that we're not even sure exactly how many chemical steps there are, and of the hundreds of steps that are known, only dozens are well understood. Every energy metabolism process is many times more complex than processes that occur spontaneously (or naturally).

Second, not only are there many steps in energy metabolism, but the many steps interact with extraordinary efficiency. For example, although scientists have attempted to simulate photosynthesis, the

simulated systems are nowhere near as efficient as the systems in the biological world.

Third, even some of the smaller *sub*systems of energy metabolism look very much like humanly designed systems. For example, hydrogen ions generate ATP in a fashion similar to how hydroelectric dams are designed to generate electricity from the flow of water.

Fourth, there is the matter of the extremely small size of chloroplasts. The entire process of photosynthesis occurs within an object about a millionth of a meter in diameter. It is hard enough to imagine humans generating something as efficient as is photosynthesis, it is many times harder to imagine humans doing it on the head of pin! In the worldview of naturalism, there is no natural process capable of generating the complexity of energy metabolism. Rather, it looks like energy metabolism was designed. At the same time, since the complexity, efficiency, and size of the processes of energy metabolism exceeds the human ability to understand and duplicate, the designer of energy metabolism appears to be substantially more capable than humanity. The God of Scripture has the wisdom and manipulative ability to design such systems. And, His desire to illustrate His invisible power gives Him reason to do so.

Cellular Metabolism: Our Responsibility

Our Responsibility to God

As a lawn is mowed, thousands of blades of grass are cut. Each blade contains millions of microscopic cells capturing energy from sunlight in a system so complex and efficient that it could only have been designed by the God described in Scripture. Every green plant is evidence of that Creator and ought to remind us of the God Who created it. The same is true of algae in water ecosystems. The amazing design of photosynthesis not only provides the food for the animals of the world, but provides the oxygen they need. At the same time, carbon dioxide exhaled by animals is used for photosynthesis in plants. The creation of both plants and animals creates a well-balanced system. This should evoke worship of God because of His wisdom and awesome power.

The huge variety of biological activities that occur in the world are all powered by metabolism. This power was provided by the God of

power. This should remind us how capable God is. With the Almighty, nothing He desires to do is impossible for Him, and nothing He desires us to do is impossible if we rely upon Him. Furthermore, the essential nature of energy metabolism processes in biology should remind us of how essential it is for us to derive our spiritual energy from Him.

One of the reasons God created biological metabolism was to give us a visible and finite illustration of the invisible and infinite power of God. Because He did, study of biological metabolism can improve our personal understanding of God and draw us closer to Him. To help us fulfill our priestly role, God not only created us to respond to illustrations of His nature, He also gave us the Holy Spirit to lead us into truth. Because of this, contemplation of biological metabolism can stimulate us to worship God, glorify God, and bring others into our worship of God.

Summary of Chapter

A. God illustrated His unlimited (but invisible) power by creating the universe in such a way that energy is required for everything that happens in the universe. In organisms He created energy metabolism processes in cells to supply the energy required for everything they do.

B. Autotrophs (producers) can get their energy from the non-biological world around them. Some bacteria are chemoautotrophs and get their energy from inorganic molecules. Most plants and algae are photoautotrophs and get their energy from sunlight through the process of photosynthesis.

C. Heterotrophs (consumers) get their energy from organic molecules. Most animals, fungi, and protozoa use oxygen in a process called aerobic respiration.

Some bacteria and yeasts get energy from organic molecules through fermentation without the use of oxygen.

a. Fermentation by lactic acid bacteria is used to produce cheese, yogurt, buttermilk, sauerkraut, corned beef, and salami.

b. The CO_2 generated by yeast fermentation is used to raise bread dough.

c. The ethyl alcohol generated by yeast fermentation is used to create alcoholic beverages.

d. Fermentation by acetic acid bacteria is used to create vinegar.

D. The overall chemical reaction of photosynthesis (aerobic respiration's inverse) is: sun energy + $6CO_2 + 6H_2O \rightarrow C_6H_{12}O_6 + 6O_2$. Plants are specially designed for photosynthesis:

a. Roots grow toward water, branch and grow root hairs to increase surface area to absorb more water, and enter into mutualistic relationship with mycorrhizal fungi to pull in even more water.

b. Tracheids (very narrow-diameter tubes empty of cell contents) provide a continuous channel for water to run from the roots, through the stem and branches, to the leaves. Water gets up to the leaves by capillary action of water combined with the molecule-by-molecule pulling of flexible hydrogen bonds between water molecules.

c. Leaves are
  i. thin to allow light into the interior of the leaf
  ii. broad and flat to maximize surface area for photosynthesis
  iii. sealed with tightly packed epidermis cells and a waxy coating to conserve water
  iv. covered with stomata to allow CO_2 in and O_2 out.

d. Photosynthesis occurs in chloroplasts
  i. The light-dependent reactions of photosynthesis
    ◦ Photosynthetic pigments capture sunlight energy and give it to electrons.
    ◦ Electron transport systems extract the energy from the energy-carrying electrons
      ▪ to pump H^+ ions into the innermost compartment of the chloroplast.
      ▪ to break water to create electrons and H^+ ions (and release O_2)
    ◦ Releasing the H^+ ions through an ATP synthase protein generates ATP.
  ii. The light-independent reactions of photosynthesis use ATP & electrons from the light-dependent reactions to fix CO_2 (each 'turn' of the Calvin cycle fixing one CO_2 molecule—6 'turns' making a monosaccharide).

E. Glycolysis, which can occur in the cytoplasm of most cells, does not require oxygen, and divides glucose into two 3-carbon pyruvates, 2 electrons, and 2 ATPs.

F. Aerobic respiration

a. (cumulatively) in all the cells of our body is the reason we breathe.

b. in overall reaction (photosynthesis's inverse) is: $C_6H_{12}O_6 + 6O_2 \rightarrow 6CO_2 + 6H_2O + 36\ ATP$.

c. is far more efficient at getting energy out of glucose than glycolysis (36 vs. 2 ATP)

d. can be used to break down glucose, or fat (if blood sugar is low), or protein (if blood sugar or fat is low).

e. occurs in mitochondria.
  i. pyruvate is broken into CO_2 and acetyl-CoA
  ii. each 'turn' of the Krebs cycle breaks down acetyl-CoA

G. Energy metabolism

a. is far more complex than anything formed by natural process.

b. is more efficient than anything humans have been able to engineer and more complex than scientists have so far been able to understand.

    c. has subsystems that look a lot like human designs.

    d. occurs in much smaller systems than humans can design.

H. Energy metabolism looks like it was designed by a being with more intelligence and manipulative ability than humans (like a God wishing to give us physical illustration of His power). It is very difficult to explain in naturalistic evolutionary theory.

I. A study of metabolism helps us fulfill our responsibility as priests of the creation by giving us insight into the power, control, wisdom, and providence of God, and stimulating worship of God and the inclusion of others into our worship of God.

Test & Essay Questions

1. How does the invisible God illustrate His power in the biological creation?
2. Compare and contrast heterotrophs/autotrophs.
3. Compare and contrast photoautotrophs / chemoautotrophs.
4. Give two / three examples of heterotrophs / photoautotrophs.
5. What does cellular metabolism do?
6. What is the overall chemical reaction of photosynthesis / aerobic respiration? / What is/are the waste product(s) of photosynthesis / aerobic respiration? / What gas is needed / expelled in photosynthesis / aerobic respiration?
7. Why are plant roots 'hairy'?
8. Why do plant tracheids / leaves have the shape they have?
9. What is role for wax / stomata in leaves?
10. What metabolic process occurs within the chloroplast / mitochondrion?
11. Define a pigment.
12. Why is it that red algae / brown algae / yellow-green algae / golden algae / cyanobacteria the color they are?
13. What are the stages of photosynthesis?
14. What is the general purpose of the light-independent / light-dependent reactions?
15. In general, how does an electron transport system work?
16. How does photosynthesis / respiration use hydrogen ions to generate ATP?
17. Why is the Calvin / Krebs cycle a cycle?
18. What happens in one turn of the Calvin / Krebs cycle?
19. Compare and contrast cellular respiration and breathing respiration.
20. Compare and contrast anaerobic and aerobic respiration.
21. Explain what type of respiration occurs in *all* cells.
22. In what way is aerobic respiration more efficient than anaerobic respiration?
23. What is fermentation?
24. How does a proper warm-up before strenuous exercise minimize muscle cramps?
25. What metabolic process(es) is (are) used to create alcoholic beverages / vinegar / cheese / yogurt / buttermilk / sauerkraut / corned beef / salami? / What metabolic process(es) is (are) used to raise bread?
26. Why does a person who is starving to death become very thin and weak?
27. What does the spectrum of perfection of power indicate?
28. What worldview is most capable of explaining photosynthesis, and why?

CHAPTER 13

GOD THE WORD

Biological Communication and DNA

"In the beginning was the Word, and the Word was with God, and the Word was God."
John 1:1, KJV

13.1 | The Communicating God

God speaks from the beginning of the Bible to the end. In the first chapter of the Bible 'God said' occurs ten times (verses 3, 6, 9, 11, 14, 20, 24, 26, 28, 29) and 'God called' five times (verses 5 (twice), 8, and 10 (twice)). He speaks in the first and last chapter of the Bible and hundreds of times in between. He speaks in general proclamations (Gen. 1), He speaks among members of the Godhead (*e.g.* Gen. 1:26; 3:22), and He speaks to angels and humans.

With humans, God blesses and curses, counsels and teaches, intercedes and advocates, names and speaks personally, covenants and communes[a]. In fact, the entire Bible is God's communication to man—and called God's word (*e.g.* Ex. 24:4; 34:28; Jos. 24:26-27; Isa. 30:8; Jer. 30:2; I Cor. 14:37).

Communication is such an intimate part of God that God Himself is even referred to as the Word (John 1:1)—the Word taking on human flesh in the form of Jesus (John 1:14; Rev. 19:13). God's Word accomplishes the work of God, such as creating the universe (Psa. 33:6; John 1:1-2; Heb. 11:3), holding the universe together (Heb. 1:3), casting out demons (Mat. 8:16; Luke 4:36), healing (Psa. 107:20; Mat. 8:16; Luke 7:2-10), providing salvation (I Sam. 3:7; John 5:24; 6:63; Rom. 10:17), growing believers (I Pet. 2:2), sustaining believers (Deu. 8:3, Job 23:12; Mat. 4:4; Luke 4:4), preserving believers from sin (Psa. 119:11), and providing spiritual guidance (Psa. 119:105; II Pet. 1:19).

God's Word also has the attributes of God, such as being complete (Deu. 4:2), unchanging (Mat. 5:18), living (John 6:63; Heb. 4:12; I Pet. 1:23, 25), eternal (Isa. 40:8; 45:23; Mat. 23:35; Mark 13:31; Luke 21:33), pure (Psa. 12:6; 119:140; Pr. 30:5), true (Psa. 33:4; 119:160; John 17:17; II Cor. 6:7), and unbounded (Isa. 55:11; II Tim. 2:9). God's Word is to be praised even as God is praised (Psa. 56:4)—even above the name of God (Psa. 138:2).

The importance of the Word of God to the very nature of God, makes it most amazing that God has given us His word and chosen us to be the testifiers and witnesses of God to others. At the same time,

---

a A few references for the verbal actions of God listed in the text: God blesses (*e.g.* Gen. 1:28; 9:1; 26:12; I Chr. 13:14; 26:5), curses (*e.g.* Gen. 3:14-19; 4:11-12), counsels (Isa. 9:6; John 14:16, 26; 15:26), teaches (*e.g.* Ex. 4:15; Deu. 4:36; Psa. 25:12; Isa. 2:3; Micah 4:2; John 14:26), intercedes (Isa. 53:12; Rom. 8:26-27, 34; Heb. 7:25), advocates (I John 2:1), speaks personally to man (*e.g.* 3:9-22; 6:13-21; 28:13-15; Exo. 6:1-8; 19:20-24:3), names humans (*e.g.* Gen. 17:5, 15; 35:10), covenants with humans (*e.g.* Gen. 9:9-17; 15:9-21; Exo. 2:24; Psa. 89:3; Heb. 10:16), communes with man (*e.g.* Gen. 18:33; Ex. 25:22; 31:18; Num. 12:8; Deu. 34:10).

God expects the word of believers to have something of the power and nature of God Himself. All of our words are to be with purpose (Mat. 12:36) and true (James 5:12), and with faith our words ought to move mountains (Mat. 17:20; Mark 11:23).

In Old Testament times, God pardoned the people of Israel according to the word of Moses (Num. 14:20), caused it to rain according to the word of Elijah (I Ki. 17:1), and blinded an entire army according to the word of Elisha (II Ki. 6:18). In our own times, decisions of the Church somehow have the power to change heaven (Mat. 16:19; 18:18-19). Communication is so important to God that it ought to play an important role in the life of the believer.

Biological Communication

As believers in Christ we must take into account the things unseen, especially in people and animals with "nephesh" or soul life. We know that our invisible God has created these creatures with life in ways that go beyond the physical. As part of His communication, God reveals His invisible attributes in the physical creation (Rom. 1:20).

Among His invisible attributes is His ability to communicate. He has placed a spectrum of perfection of communication in the biological creation. He has also placed evidence of language in the creation, including a language for all life. This should not be surprising, for God created everything by the word of His mouth (*e.g.* the commands of the Creation Week; Ps. 33:6-9; John 1:1-3; Rom. 4:17; Heb. 11:3).

Animal Communication

Different organisms communicate in a variety of ways. Many organisms transmit simple, one-meaning messages by stimulating one of the senses. These messages are sent by way of *smell* (*e.g.* chemicals left by honey bee guards that attract other bees to sting at that particular location; mate-attracting chemicals produced by females of many butterfly and moth species, males of many snake species, and many mammals), *sound* (*e.g.* prairie dog alarm calls that warn other prairie dogs of approaching predators; mate-attracting songs in crickets, bullfrogs, crocodiles, elk, and songbirds), *sight* (*e.g.* light flashes by lightning bugs to attract mates[1]; threat displays of baboons; courtship dances of spiders, many butterflies and moths, fruit flies, fish, chameleons, and songbirds), or *feel* (*e.g.* ants tapping the ground to lead the way for other ants; treehoppers that produce vibrations to attract nearby treehoppers; sexual stimulation in most snakes, bottlenose dolphins, and many songbirds).

Many organisms can communicate more complicated messages. The simplest of these would be organisms possessing a repertoire of single-meaning messages that collectively can pass on more than one piece of information to other members of the same species. With a variety of body movements and sounds, for example, the mother of a

number of different animals can tell her young to remain unmoving and unseen, follow her, come closer, or run. Many birds and mammals use a variety of specific sounds to lead, warn, and attract other members of their own species (*e.g.* chickens[2]). Vervet monkeys, for example, have distinct warning calls for pythons, eagles, and leopards, and different responses for each call (*e.g.* look around on the ground at the python call, hide in the underbrush at the eagle call, climb trees at the leopard call)[3].

Some animals can pass on simple messages to and from other species. California sea lions[4], bottlenose dolphins[5], and elephants[6] can each be trained to perform stunts in response to several dozen verbal signals or gestures of trainers. At a more sophisticated level, some organisms can combine simple messages into more complicated sets of instructions. The waggle dance of the honey bee, for example, communicates direction and distance to flowers for nectar or pollen, or water for cooling the hive, or a location for a new hive site[7]. Such information can even be passed between species. The bottlenose dolphins 'Akeakamai' and 'Phoenix', for example, are capable of combining actions such as toss and fetch with modifiers such as right and left, surface and bottom to obey five-unit commands such as 'place bottom pipe [in] surface hoop'[8].

A few animals are even clever enough to create their own messages. One Japanese macaque, for example, invented a rattlesnake alarm call[9] and one Hamadryas baboon created a gesture to invite an infant to be

carried[10]. One female orang-utan 'Princess' even invented a gesture and taught it to her infant by huffing and sucking on her own breast (and from there on, when she huffed, her infant moved off her back to suck)[11].

At an even greater level of sophistication, animals use communication for identification purposes. Many whales and birds, for example, create unique songs, thus developing distinct dialects in different populations and unique sounds for individuals[12]. Whale songs can even last up to ten hours[13]. Amazingly enough, some chimpanzees[14], orang-utans[15], and the gorilla 'Koko'[16] have even learned scores of human sign language gestures.

Animal communication displays a spectrum of perfection of communication, especially when considered alongside the language capacity of humans. This spectrum of communication from simple, one-meaning messages through complex communication and even language, suggests that a being exists with infinite communicative power.

13.2 | The Language of Life

The DNA Myth

It seems to be a general understanding in our current society that an organism's DNA fully determines the nature of that organism. It is commonly believed—and taught—that DNA contains all the information needed to build an organism. But this is not true. Several evidences suggest that some characteristics of an organism are not determined by its DNA. For example,

1. During **fertilization** in most sexual organisms, a sperm only gives up its DNA. The only portion of a fertilized cell that is contributed by the father is half of the cell's DNA. The entirety of the plasma membrane, the cytoplasm, and the cytoskeleton, as well as every organelle—all the cell except the DNA—is contributed by the mother.

    As that cell develops into an adult organism, many 'decisions' of development (such as on which end of the organism the head develops) are determined by the arrangement of cell components in that first cell (not by the DNA of that cell). Furthermore, many of the cell components of the final organism are determined more by that first cell than by that cell's DNA. Mitochondria and

fertilization—union of a sperm and an egg when the egg accepts the DNA of the sperm

chloroplasts, for example—including the DNA they contain—come from the egg cell and not from the sperm or the DNA contributed by the sperm.

2. When DNA is taken out of fertilized frog eggs, the cells can divide as many as ten times before finally dying. Even though copying and separating DNA is normally a part of how cells divide, it would seem that at least in some situations, DNA is not required for the complex processes involved in dividing a cell.

3. Recent research is revealing that the community of microbes found in the body of a human plays an important role in the development of the human body. For example, when the microbes in a growing human fetus are unhealthy, the digestive system of the fetus will not develop properly.

   Ongoing discoveries suggest that we have only begun to realize the importance of other organisms in not only the functioning, but also the development of our bodies. This suggests that some of the information needed to build organisms may actually be contributed by *other* organisms.

4. Identical twins are due to a single fertilized cell that divides and then separates into two separate zygotes. These two zygotes each develop into a distinct individual—identical twins. Identical twins start out with exactly the same DNA. Yet, even if they are raised in the same environment, identical twins are not identical. At the very least they have distinct personalities, suggesting a substantial contribution to at least human life comes from one or more other sources.

   Perhaps the non-physical attributes of life also contribute to the development of an organism. Scripture refers to both humans and animals as having souls (*e.g.* nephesh in Gen. 1:20-21 and 24 comp. Gen. 2:7). Not only is it improbable that DNA has anything to do with the development of the soul, but it may turn out that the soul contributes to the development of the body of a human or animal.

For reasons such as these we know that DNA does NOT determine the *full* nature of life. Exactly how important DNA actually is, is not actually known.

At the very least, DNA has been established to be something like a recipe book for the many thousands of proteins needed by an organism. But this function may account for only a small amount of DNA. A segment of DNA that contains the 'recipe' for building a protein is known as a **gene**. In humans, genes seem to account for only about 2% of the total length of human DNA. Genes have been identified for most of the proteins and most of the enzymes used to form other biological molecules.

gene—segment of DNA that codes for a protein (and some simple character traits)

Although it has not been demonstrated for every single biological molecule, it is not unreasonable to suggest that DNA carries the recipes for all the macromolecules needed by an organism. Consequently, DNA is definitely important. In fact, DNA is essential for an organism's existence. At the very minimum, DNA is a recipe book for all the biological molecules needed by an organism. But that still may account for less than 10% of DNA's information. How much more DNA is responsible for has not yet been determined.

Information Carrier

Even if DNA was only a recipe book for biological molecules, it would still have to carry an enormous amount of information. The shape, length, and coding structure of deoxyribonucleic acid (DNA) allows it to carry information, and a lot of it. Overall, the shape of a DNA molecule is somewhat hair-like; being long and thin, billions of times longer than it is wide. It is constructed of long strings of monomers

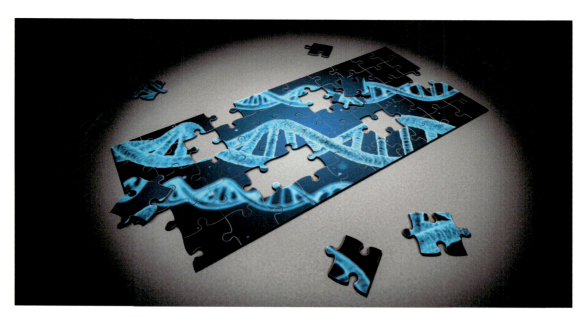

known as nucleotides. Each nucleotide is, in turn, constructed of a five-carbon sugar (deoxyribose) connected to a phosphate group and a nitrogenous base.

The sugar of one nucleotide is attached to the phosphate of the next, creating a long string of alternating sugars and phosphates with nitrogenous bases sticking out perpendicular to the string. Because four different nitrogenous bases are available to DNA nucleotides, four different nucleotides make up DNA, namely adenine, thymine, guanine, and cytosine. The different nucleotides are symbolized by A, T, G, and C, respectively. It is the particular sequence of nucleotides that carries information, and it is the extremely long nature of DNA that allows it to carry so much information.

At any particular position along the DNA there are four possible nucleotides (either A or T or C or G). If we consider two adjacent nucleotides together, there are four possible nucleotides in the first position, and for each of these there are four possible nucleotides for the second position. This gives us a total of 4 x 4 = 16 possible double-nucleotide combinations[a]. If we consider *three* adjacent nucleotides together, for each of the sixteen possible combinations in the first two

---

a No matter what nucleotide is in the first position, the second position could be either an A or a T or a C or a G. So, if the first position is an A, there are four possibilities when the two nucleotides are considered together: AA, AT, AC, or AG. If the first position is a T, there are four further possibilities when the two nucleotides are considered together: TA, TT, TC, or TG. If the first position is a C, there are four further possibilities when the two nucleotides are considered together: CA, CT, CC, or CG. If the first position is a G, there are four further possibilities when the two nucleotides are considered together: GA, GT, GC, or GG. The grand total is 16 possibilities (AA, AT, AC, AG, TA, TT, TC, TG, CA, CT, CC, CG, GA, GT, GC, or GG).

positions there are four possible nucleotides for the third position, for a total of 16 x 4 = 64 possible triple-nucleotide combinations.

If each triple-nucleotide combination was used to represent (or code for) something different, we could represent up to 64 different things with a sequence of three nucleotides. For example, sets of three nucleotides could code for the 26 letters in the English alphabet plus punctuation marks (comma, period, dash, semicolon, colon, question mark, *etc.*) and even spaces (such as between words). Once we established that code we could spell out a three-letter word with nine nucleotides. We could code for the sentence "I am that I am."—including the quotation marks—with 51 nucleotides (seventeen characters times three nucleotides per character: 17 x 3 = 51). If we had a long enough sequence of nucleotides we could code for lots of information.

In the case of humans, the DNA in each human cell is about 3.5 billion nucleotides long. That is long enough to code for 2000 books of about 500 pages each[a]! Thus, because of the extremely long sequence of nucleotides found in DNA, DNA has the potential of carrying lots of information—*if* there was some sort of coding going on.

In fact, there *is* coding going on. Triple-nucleotide sequences (called **codons**) are used to code for the twenty different amino acids that are used to build proteins[b]. The particular sequence of amino acids in a protein is what gives that protein its particular structure and function, so the sequence of codons on the DNA functions as a recipe on how to build a protein. Three of the 64 possible codons are 'start'

codon—sequence of three nucleotides on DNA or RNA that represents a 'letter' of the genetic code

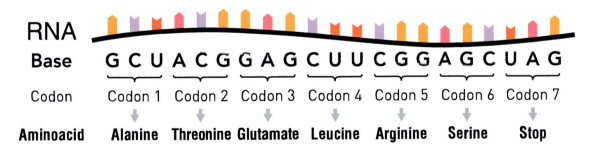

Figure 13.1
Amino acid sequence. Image by Thomas Splettstoesser via commons.wikimedia.org, used under CCby4.0.

a If each word averages about 4 letters each, each word would average a bit over 5 characters per word (4 letters per word + one space following each word + the occasional punctuation mark). Thus with about 200 words per page there would be something more than 1000 characters per page (5 x 200). At 500 pages, each book has something more than 500,000 characters (500 x 1000). 2000 books would have something more than 1 billion characters (500,000 x 2000), which could easily be coded for with 3.5 billion nucleotides.

b There are some rare amino acids that are coded for in some organisms, but with 64 possible codon sequences, it is easy to code for all of these and more.

and 'stop' codons indicating where to start or stop reading the recipe (and to mark the beginning and ending of the amino acid sequence).

The coding sequence for the amino acids preceded by a start codon and followed by a stop codon makes up a gene on the DNA. Because triple-nucleotide sequences on the DNA code for things *and* these codons are attached next to each other, one after another, it is possible for DNA to carry information. Because DNA is so very long, it can carry *lots* of information. Human DNA, for example, carries the recipes to construct over 100,000 proteins used in the human body. And all this information is carried in a molecule too small to be seen with the naked eye!

Copying Information

It is extremely important that it be easy to copy DNA and that it be copied reliably—both for forming new cells and for making protein. When one cell divides to produce another cell[a], the DNA of the first cell has to be copied so that the new cell has the DNA it needs. The DNA copy needs to be a reliable copy so that the new cell has all the right information. When a cell needs to make protein, a copy of the gene for that protein is made from the DNA and taken out of the nucleus. The copy needs to be reliable so that the protein is constructed properly. The structure of DNA not only allows it to carry a lot of information, it also allows that information to be copied easily and reliably.

A single strand of DNA carries the necessary information, but DNA is a *double* stranded molecule. The nitrogen bases of one strand of DNA are attached by hydrogen bonds to the nitrogen bases of another strand of DNA alongside it. This creates an overall structure of DNA that looks a bit like a twisted ladder. The two parallel strings of sugars and phosphates of the two strands make the 'uprights' of the ladder and each pair of nitrogenous bases connected by hydrogen bonds make up the 'rungs' of the ladder. The 'ladder' is then twisted in a structure similar to a spiral staircase. It is the double-stranded nature of DNA combined with a very simple pairing rule for the nitrogen bases that allows DNA to be copied so easily.

When it comes time to copy DNA or a portion of it, the two strands of DNA separate. Since the nitrogenous bases are connected with weak hydrogen bonds it is relatively easy to 'unzip' the DNA

---

a One cell divides to produce another cell (a) when a single-celled organism reproduces, or (b) when a cell replaces a nearby cell that is injured or dies, or (c) as a multi-cellular organism grows from one cell to many cells.

down the middle, separating the two strands. At this point, nucleotides can attach to a strand to produce the copy. Two simple pairing rules determine what nucleotides attach: guanine only pairs with cytosine (G with C) and adenine only pairs with thymine (A with T[a]). These simple rules make it very easy to make the second strand of DNA.

Because the rules are so easy, the process also tends to be very reliable. When the entire DNA is unzipped and complementary strands are made of both of the original strands, an exact copy has been made of a double-stranded DNA molecule. When just enough of the DNA is unzipped to reveal a gene, the complement of one of the strands is taken outside the nucleus to make the protein. When the complement has been taken away, the two strands of DNA are zipped back together to permit the DNA to remain safely in the nucleus.

Mutations

A **mutation** is a change in the sequence of nucleotides on DNA. The simplest type of mutation is a **point mutation** which is a simple change in one nucleotide, and is somewhat analogous to changing one alphabet letter in a sentence. A certain percentage of point mutations do not change anything because an amino acid is often coded by more than one codon. Although they are not usually fatal, many point mutations are problematic because changing an amino acid in a

---

[a] Technically, in the case of making copies of a gene to make protein, the copy is made of RNA nucleotides rather than DNA nucleotides, and pairing rule is slightly different: adenine is paired with uracil (U), rather than thymine (T).

mutation—genetic mistakes that occur as DNA is copied. Types: point mutation replaces one nucleotide with a different one; frame-shift mutation inserts or deletes a nucleotide

point mutation—a simple change in one nucleotide, analogous to changing one alphabet letter in a sentence

frame shift mutation—a nucleotide is added or omitted in the process of copying DNA

protein often changes the shape of the protein and impairs the protein's ability to accomplish its designed task. Sickle-celled anemia is caused by a point mutation. The mutation changes the shape of the protein hemoglobin, which distorts a round red blood cell into a sickle-shape that has trouble travelling through the capillaries of the body.

Another type of mutation is a **frame-shift mutation**, where a nucleotide is added or omitted in the process of copying DNA. After a frame-shift mutation, each codon is constructed of two nucleotides from one codon and one nucleotide from an adjacent codon. This changes nearly *every single codon*, so it changes nearly every single amino acid from that point forward. This would be analogous to changing every single character in a sentence. Such a sentence would be quite unreadable. So likewise, the protein that results from a frame-shift mutation would be quite unusable. This type of mutation is always problematic and very often fatal.

Mutations occur naturally all the time, but they are remarkably rare. On the average they occur at any given nucleotide only once in every million copies. This means DNA's copying process is extremely efficient (somewhat like only making one mistake in every million alphabet letters that are copied… or roughly one alphabet letter mistake in 1000 pages of a book!).

Two things seem to contribute to this efficiency in the copying of DNA. First is the ease and simplicity of copying. Second, it seems that organisms have something analogous to 'fact-checking' molecules that run along a recently copied section of DNA and make sure that the proper nucleotide pairings are in place. In some living systems there are several 'fact-checking' systems in place.

Although mutations can occur naturally, they can also be introduced artificially. What actually causes the mutations in these cases is often not understood, but it probably varies from actually kicking out the nucleotides that are supposed to be there, to preventing the 'fact-checking' processes from operating properly to preventing the operation of repair processes. The kinds of things that cause or *induce* mutations include various types of radiation (*e.g.* radioactivity, x-rays, ultraviolet light) and various types of chemicals (*e.g.* LSD, marijuana, pesticides). In most cases, the mutations have relatively little effect right away, but may result in serious consequences when combined with other experiences. This is why, for example, sunning oneself on

the beach or laying in tanning machines early in life can lead to skin cancer later in life.

Mendelian genetics

Hybrid crosses between two different organisms sometimes result in offspring that have characteristics of both parents or characteristics intermediate between the two parents. In the middle of the nineteenth century, this was commonly explained by genetic information being a kind of fluid substance that mixes together at conception. However, experiments by a monk by the name of Gregor Mendel (1822-1844) suggested that at least some genetic information was composed of (non-fluid) particles. Mendel discovered seven characteristics of pea plants (plant height, flower color, flower location, seed shape, seed color, pod shape, pod color) that could show up in two generations but not in the generation between them. Ultimately, he concluded that each of the following was true:

- Pea plants carry not one, but *two*, distinct pieces of genetic material for each characteristic, such as plant height or flower color. Each pea plant, then, is considered **diploid**, possessing two separate genes for each characteristic (a gene being a piece of genetic material).

- Each gene codes for a trait (a particular manifestation of the characteristic), such as tall or short for plant height or purple or white for flower color, and the form of a gene determining a

diploid—cell state when it contains both chromosomes of each pair of homologous chromosomes

allele—one of multiple character states for a particular gene

dominant—allele type that is fully expressed no matter the homologous allele

recessive—allele type that is not expressed at all when the homologous allele is dominant

particular trait is called an **allele**. Since pea plants are diploid, every pea plant carries two alleles.

- There are two different kinds of alleles—a **dominant** allele (that displays its trait in the pea plant no matter what kind of allele the second allele is) and a **recessive** allele (that displays its trait in the pea plant only when the second allele is also recessive). This is described as Mendel's law of dominance. The set of two alleles carried by a pea plant is called its **genotype** (*e.g.* two dominant alleles), and how those alleles show up in the plant is called its **phenotype** (*e.g.* tall plant or white flowers).

- For each characteristic, a plant only contributes one of its two alleles to the process of fertilization. For each characteristic, each pollen grain contains only one allele and each flower's egg (which when pollinated develops into a seed) contains only one allele.

Having half the genetic material of the mature plant, each **gamete** (the smallest object carrying the reproductive material for the plant—in this case the pollen grain or the flower's egg) is considered **haploid**. When combined in fertilization, then, the genetic material of two haploid gametes (one allele each) produces a diploid offspring (two alleles).

- When many of them are considered at the same time, gametes carry the same alleles in the same frequency as the plant that produced them. For each characteristic, then, if the plant's genotype is **homozygous** (the two alleles are the same), all its gametes carry that allele[a].

  - If the plant's genotype is **heterozygous** (the two alleles are different), half of its gametes carry one allele and the other half carry the other[b]. This suggests that when the plant produces its gametes, it separates its two alleles and randomly contributes one allele to each gamete. This is called Mendel's first law, or Mendel's law of segregation.

---

genotype—set of alleles possessed by an organism (2 per gene, even if the same or unexpressed)

phenotype—character(s) exhibited by an organism

gamete—haploid cell generated by meiosis in sexual organisms (aka sex cell). Types: a sperm, at fertilization, contributes only DNA; an egg cell, after fertilization, grows into an adult organism

haploid—cell state when it contains only one of each pair of homologous chromosomes

homozygous—genotype with the same allele on the two homologous chromosomes

heterozygous—genotype with different alleles on the two homologous chromosomes

---

a   Since homozygous organisms only pass on genetic information for the trait seen in the organism, these organisms are also referred to as 'true-breeding' for that trait.

b   If the trait coded for by the recessive allele involves a disease, the heterozygous individual does not display the effects of the disease, but passes the disease on to half of its offspring. Such a heterozygous individual is referred to as a 'carrier'.

- The allele carried by a particular pollen grain has no impact on which egg it fertilizes. To put it another way, gametes combine randomly—independently of the alleles they carry.

- When more than one characteristic of the plant is considered, the alleles related to one characteristic do not in any way affect what alleles are found for the other characteristics. This is called Mendel's second law, or Mendel's law of independent assortment.

The result of all this is that Mendel could not only explain why pea characteristics skipped generations but could predict what characteristics would show up in a given generation and in what percentages. Reproduction that follows these rules, no matter what the organism is, is called Mendelian genetics.

Probably the easiest way to represent Mendel's observations and to make the same predictions, is by means of **Punnett squares**, named after the method's inventor, Reginald Punnett (1875-1967). The first step in using Punnett's method is to represent each allele as if it were a particle (just as Mendel argued). This is done by using an alphabet letter[a]. Since each alphabet letter has both a lower-case and upper-case form, it is possible to use this fact to distinguish the two alleles—the dominant allele represented by the upper-case form and the recessive

Punnett square—matrix method used to calculate genotypes of offspring in Mendelian genetics

---

a   In practice, it turns out to be most useful to use alphabet letters that have lower- and upper-case forms that look very different (e.g. A & g; G & g), and *not* use alphabet letters that have lower- and upper-case forms that look similar (e.g. C & c; O & o).

allele by the lower-case form (thus representing Mendel's law of dominance).

As an example from peas, Mendel found yellow seed color to be dominant over green seed color. If we use the first letter of the alphabet to represent the alleles for pea seed color, 'A' would represent the allele for yellow seeds and 'a' would represent the allele for green seeds. Since pea plants are diploid, the genotype of each parent plant is represented by a pair of letters—a homozygous dominant plant as AA, a heterozygous plant as Aa, and a homozygous recessive plant as aa. According to Mendel's law of segregation, then, when a plant produces gametes, it separates the two alleles and puts one of the two in half of its gametes and puts the other in the other half of its gametes. Since the two alleles of a homozygous plant are the same, homozygous plants produce only one type of gamete (aa genotype plants produce only 'a' genotype gametes and AA genotype plants produce only 'A' genotype gametes). Heterozygous (Aa) plants, on the other hand, produce both gamete types (50% would be 'A' genotype gametes, and 50% would be 'a' genotype gametes).

The second step in Punnett's method is to represent Mendel's law of segregation with what is called a Punnett square. For each characteristic, each parent plant produces, at most, only two different gametes ('A' or 'a'). Since fertilization involves the combination of one gamete from each of two parents, and either gamete from each parent could just as likely combine with either gamete of the other parent, there are, at most, four equally likely combinations (the first possible allele from the male with the first possible allele from the female, or the first possible allele from the male with the second possible allele from the female, or the second possible allele from the male with the first possible allele from the female, or the second possible allele from the male with the second possible from the female).

In Punnett's method, a square is divided into two columns and two rows (representing the four possibilities). The genotypes of the two gametes from the male are represented along the top (above the two columns) and the genotypes of the two gametes from the female are represented along the left side (to the left of the two rows). The four squares, then, are filled with the genotype of the offspring that would result from the combination of the male gamete (represented above that column) and the female gamete (represented to the left of that row). The fraction of each different genotype among the four squares is the fraction of the offspring that would be expected to have that genotype (see Figure 13.1 for all the possibilities). Notice that crossing two homozygous parents—one recessive and one dominant (13.1C or 13.1D in figure 13.1)—and then crossing their offspring (13.1I in figure 13:1) explains how a phenotype (the recessive one) can skip a generation.

Beyond the seven pea characteristics that Mendel studied, Mendelian genetics seems to work quite well for a number of character traits in a number of different organisms. As a few examples, it works for seed shape and color in some plants (*e.g.* purple vs. yellow and smooth vs. wrinkled corn kernels), hair length in some mammals (*e.g.* long vs. short hair in cats), some genetic diseases (*e.g.* sickle-celled anemia), and some human traits (*e.g.* wet vs. dry ear wax).

However, compared to the list of all character traits of all organisms, Mendelian genetics works for an extremely small fraction of them. The genetics of a few more traits are explained with **codominant** alleles. An example of codominance is blood type, where each allele (one producing blood type A; one producing blood type B) shows up completely in the heterozygous genotype (AB blood type).

> codominant—allele type that is fully expressed when both homologous alleles are codominant

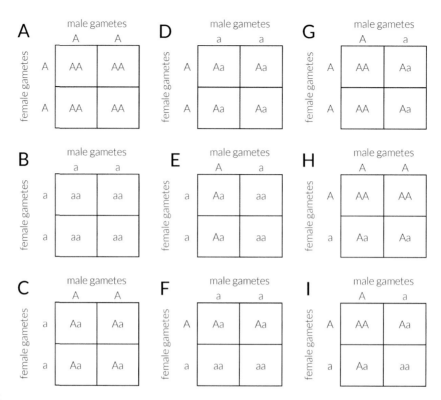

Figure 13.2
Punnett squares. A: Cross two homozygous dominant parents (AA x AA), and all the offspring will have the dominant phenotype and the homozygous dominant genotype (100% AA). B: Cross two homozygous recessive parents (aa x aa), and all the offspring will have the recessive phenotype and the homozygous recessive genotype (100% aa). C & D: Cross a homozygous recessive parent with a homozygous dominant parent (AA x aa, or aa x AA), and all the offspring will have the dominant phenotype and the heterozygous genotype (100% Aa). E & F: Cross a heterozygous parent with a homozygous recessive parent (Aa x aa, or aa x Aa), and half of the offspring should have the dominant phenotype, the other half should have the recessive phenotype, and of the genotypes 50% of the offspring should be heterozygous, and the other 50% should be homozygous recessive (50% Aa, 50% aa). G & H: Cross a heterozygous parent with a homozygous dominant parent (Aa x AA, or AA x Aa), and all of the offspring should have the dominant phenotype, 50% of the offspring should have the heterozygous genotype, and the other 50% should have the homozygous dominant genotype (50% Aa, 50% AA). I: Cross two heterozygous parents (Aa x Aa), and three quarters of the offspring should have the dominant phenotype, one quarter of the offspring should have the recessive phenotype, and of the genotypes, 25% should be homozygous dominant, 50% should be heterozygous, and 25% should be homozygous recessive (25% AA, 50% Aa, 25% aa).

incompletely dominant—allele type partially expressed when both homologous alleles are of that type

A few more traits are explained with **incompletely dominant** alleles. An example of incomplete dominance is flower color in snapdragons, where each allele (one producing white flowers; one producing red flowers) shows up incompletely in the heterozygous genotype (pink flowers). More traits can be explained by more than one allele simultaneously affecting the trait (*e.g.* human hair color).

Even more traits can be explained by combining the effects of the alleles with the effects of the environment.

Yet, even considering all this, biologists understand the genetics of only a very, very small percentage of all the character traits of organisms. The problem seems to be that although DNA carries genetic information, and does so in the form of particles (lengths of DNA known as genes) as Mendel surmised, it does not carry that information in as simple a manner as Mendel thought. Instead, DNA seems to carry genetic information in the form of a language.

13.3 | DNA as the Language of Life

DNA provides evidence of the existence of a language at the foundation of life. Note, however, that just as the words on this page are not actually language, but *evidence* of a language, so also DNA is not actually language itself, but evidence of one. The characteristics that suggest such a language include the following:

- Language and DNA both exhibit **hierarchal coding**: a few different symbols code for a larger number of other symbols, which in turn code for an even larger number of other symbols. Language codes for a seemingly infinite diversity of thoughts in a hierarchal fashion; for example, the English language involves hierarchical coding. Particular sequences of alphabet letters code for words, particular sequences of words code for sentences, and particular sequences of sentences code for paragraphs.
  - In particular, the three elements of Morse Code (dots, dashes and spaces) are used to code for twenty-six letters of the English alphabet, and the 26 letters are used to code for over 100,000 English words. This hierarchal type of coding is characteristic of all language. Similarly, in the case of DNA, the four nucleotides (A, T, C, G) are used to code for twenty amino acids, and the 20 amino acids are used to code for over 120,000 proteins in the human body.

    3 Morse Code elements → 26 English letters → >400,000 English words

    4 nucleotides → 20 amino acids → >120,000 proteins

- Lower-level randomness and higher-level order are also important in both language and DNA. Statisticians cannot distinguish a

hierarchal coding— a few symbols code for a larger number of other symbols, which in turn code for an even larger number of other symbols

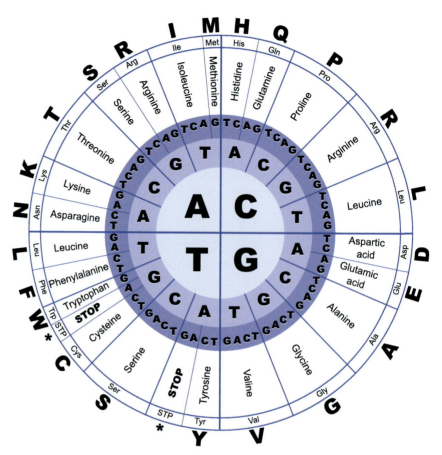

sequence of alphabet letters in a typical communication (*e.g.* a book) from a random sequence of letters. By this is meant that if you look at any given letter on a page, you cannot predict the next letter. It is possible for one letter to be followed by almost any letter of the alphabet. In a similar manner, the sequence of nucleotides in DNA cannot be distinguished from a random sequence of nucleotides (*i.e.* knowing any one nucleotide, it is not possible to predict the next nucleotide). At the same time, at a *higher* level of order, the sequence of letters is important (and quite clearly not random).

Even though the sequence of letters *seems* random, a random change of one letter to another will lead to a *loss* of information. And this was what was discovered to be true of DNA when mutations were induced randomly. Mutated DNA did not improve the organism. In fact, if it changed the organism at all, it almost always injured the organism. Although the sequence of letters, words or sounds *appears* random in language, the sequence is both non-random and important.

The same is true in DNA. Although the sequence of nucleotides *appears* random, the sequence of nucleotides is both non-random and important.

- In human language, linguistic rules determine how different types of words are used and in what sequence those different types of words are placed. Words can be nouns, verbs, adjectives, adverbs, prepositions, conjunctions, *etc*. Linguistic rules indicate the necessity and order of nouns and verbs and the proper association of those types of words with adjectives, nouns, adverbs, prepositions, *etc*.
  - In the case of DNA there does seem to be different types of genes. Some genes, for example, determine whether or not another gene is copied. Other genes determine how many copies are to be made of another gene. These genes seem to analogous to 'adjective genes' or 'adverb genes'. There are also suggestions that it matters what order the different types of genes are 'read'. Even the sequence of genes is important. Gene cascades are active during the construction of many complex organs. In a gene cascade, one gene activates several others, and each of those activates many more. By the time the cascade is over, and organs are made, thousands of genes have been activated. This creates at least a hint that linguistic-like rules exist in the processing of DNA information.

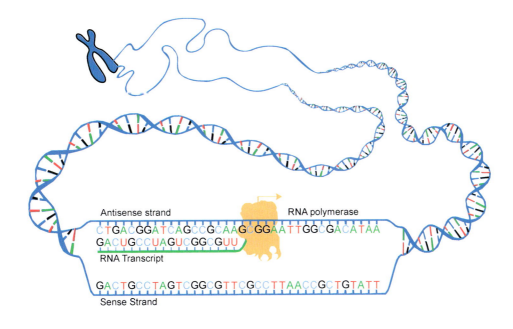

In human language, the alphabet letters on a page are read and understood (or the sounds are heard and understood) only in a larger system. Written words must be seen by eyes and spoken words must be heard by ears. The eyes or ears must translate what is seen and heard into electrical signals and those signals must be sent to the brain and processed. Human language is an emergent property which arises only from a very complex biological system. The sequences of letters in a book are useless and convey no information at all outside of that system.

- In like manner, the sequences of nucleotides in DNA are useless and convey no information at all outside of an incredibly complex system found in the cell. The process of **transcription**, for example, makes copies of genes that make up DNA. A special enzyme untwists the DNA, another unzips the two strands, while another connects nucleotides together into the new sequence while yet another checks the nucleotides to make sure they are properly paired. Still another enzyme detaches the copy. Another process in the cell cuts out certain non-coding portions of the copy.

transcription—cell system that makes an RNA copy of genetic information on DNA

translation—cell system that makes an amino acid sequence (for protein) from a nucleotide sequence

Another even more complex process occurs at the ribosomes. Called **translation**, this process uses the information on the copy to make a sequence of amino acids. Other processes fold, label, and deliver the protein. Just as language involves complex processes, so also complex processes are involved with DNA. Not only does the existence of complex interpretive systems around DNA suggest DNA is part of a language, but a couple of these systems (transcription and translation) are named because of their similarity with human-language-related processes!

The hierarchal coding, gene order, the hint of linguistic rules, and the context of DNA together seem to suggest that DNA is evidence of a language. Since every cell depends on DNA in at least one stage of its life, that language is found in every cell. Since every organism is made of at least one cell, that language is part of every organism, and it is the same language for all organisms. DNA bears evidence that there is a language of life.

The Origin of Biological Communication

In several different ways, DNA provides a substantial challenge to the worldview of naturalism. First, if the physical is all that exists, a complex entity should be understandable from its parts. A living organism should be understandable from its physical parts. Since DNA is the smallest component unique to each unique organism and not found in non-living things, a full understanding of DNA *ought* to lead to a full understanding of life. In the naturalistic worldview DNA *ought* to fully determine the nature of an organism. In naturalism, DNA *ought* to contain all the information needed to build an organism.

It is probably because of the prevalence of naturalism that these **DNA myths** are so pervasive in our culture. The fact that DNA does *not*, in fact, contain all the information needed to build an organism is a challenge to the naturalistic worldview.

A second challenge to the naturalistic worldview is DNA's efficient design. Not only does the structure of DNA allow it to carry information and be readily copied, it also seems to be *optimally* designed to do these things. There is no better length for a codon than three nucleotides and there is no simpler way to copy DNA than the way it is done. No human designs for similar processes are more efficient and no human system stores as much information in as small an object as a cell nucleus.

DNA myth—common (false) belief that DNA fully determines the nature of an organism

Such efficiency of design suggests a designer—a designer more capable, even, than humans. Since DNA is found in all organisms, there is no designer in the naturalistic worldview capable of designing such an efficient system. In contrast, the God of Scripture has the intelligence and manipulative ability capable of designing and constructing such a system.

A third challenge concerns the irreducible complexity of communication. Language and communication require complex biological systems. Numerous components must interact in specific and complex ways for language and communication to take place. Communication cannot occur with less than a minimum number of components in the system and it is not apparent how the system can be constructed one component at a time.

Naturalism's best explanation for the origin of things—evolutionary theory—lacks a means for innovation to increase the complexity of things. And, even if evolution were innovative, evolution would construct things one step at a time. Whereas naturalism provides inadequate explanation for the irreducible complexity of language and communication, the God of Scripture could create systems capable of language and communication in a single step—in fact, with a single word.

A fourth challenge is the origin of the language of life. Nothing as complex as language has ever been observed to arise spontaneously in the natural world. In our experience, language is only generated

by intelligent beings who can communicate. The existence of a biological language suggests an origin by an intelligent being who communicates. God is just the kind of being Who is not only capable of creating the language of life, but desirous of doing so in order to teach us about Himself.

Biological Communication: Our Responsibility

Our Responsibility to God

God has placed a spectrum of perfection of communication in the biological world. Some entities in His creation are completely inert, being associated with no sound or action, and no communication of any sort. The DNA of all organisms is evidence of a biological language which cannot be seen or heard. Some of these organisms are active and even interact with each other, but exhibit no further evidence of communication.

Other creatures transmit elementary information to others by means of simple cues. Various insects, birds, and mammals use a combination of body motions and sounds to guide, warn and attract other members of their own species. A variety of birds and mammals arrange a repertoire of sounds in particular ways to communicate orally, and some of them can even be creative in the process. Some animals can be trained to perform rather complex tricks, and some apes can even learn several hundred signs of human sign language.

Finally, there is the complex language of humans, including hundreds of thousands of words arranged according to very particular linguistic rules. This spectrum of perfection of communication in the biological creation should lift our spirits in worship of the infinite communicator—the communicating God described in Scripture.

DNA should remind us of the language of life and the One Who created everything by the word of His mouth. Since DNA is best explained by the God of Scripture, DNA can be understood to be an identity tag, indicating that the possessor has been created by God. Since every organism and every cell in every organism had or has DNA, every cell of every organism has a 'made by God' tag. In an adult human there are trillions of such identity tags.

On the one hand, consider how presumptuous this makes a person who thinks of himself or herself as their own master. On the other

hand, consider how comforting it is for a believer to know that he or she is God's possession. The fact that God so desires to know us that He communicates with us, and makes a way to communicate with Him, ought to bring us into spontaneous worship of Him.

One of the reasons God created animal communication and DNA was to give us visible and finite illustrations of the invisible and infinite communicating nature of God. Because He did, study of animal communication and DNA can improve our personal understanding of God and draw us closer to Him. To help us fulfill our priestly role, God not only created us to respond to illustrations of His nature, He also gave us the Holy Spirit to lead us into truth. Because of this, contemplation of animal communication and DNA can stimulate us to worship God, glorify God, and bring others into our worship of God.

Our Responsibility To The Creation

Preservation of the Language of Life

What we currently <u>know</u> about the complexity of DNA is already a challenge to naturalism. It does not seem to be possible to create such complexity by any naturalistic process. If DNA is part of a language, it is many times more complex than even what we currently know. That is probably why naturalists have been reluctant to accept that DNA really is part of a language. This means that naturalists have consistently underestimated the complexity of DNA. For example, for about forty years it was thought that each gene contains the recipe for only one protein. Since there are something in excess of 120,000 proteins in the human body, it was assumed—again for four decades—that there must be something in excess of 120,000 genes.

However, when the early results of the human genome project were announced in 2000, it was announced that there were no more than 40,000 genes, and probably more like 25,000. This would suggest that on the average, every gene contains the recipes for three to four proteins. This indicated that even the best understood aspect of DNA was much more complicated than expected. As a second example, for more than a quarter century naturalistic biologists believed that up to 98% of human DNA was 'junk'—not actually containing information. Then, beginning in 2012, this 'junk' was found to be full of biological information. Such lessons should have humbled biologists, convincing them to proceed cautiously in modifying DNA. It has not.

A capable writer may be able to improve something written by a writer with less ability, but would probably struggle to improve the text of a writer with greater ability. And, in the case of a text written in an unfamiliar language, he would not be able to improve it at all. In fact, any alterations he made to the text would almost certainly damage the text. This is analogous to the danger of humans altering DNA. DNA was created by God. The DNA text was authored by a far more capable writer than the most capable human, and it was written in a language no human can yet understand.

Many of the alterations that humans make to DNA would be expected to damage the DNA, not improve it. An early example of this would be human attempts to *induce mutations*. In the 1920's scientists began subjecting organisms to all sorts of radiation and chemicals to cause (induce) mutations and see what sort of traits might result. Since evolutionists thought organisms were improved by mutations, they thought all sorts of desirable traits would appear. Not only did none of the mutations produce desirable results, but most of the mutations were quite damaging to organisms.

Although modern biologists are now employing far more sophisticated methods of altering DNA, there are still serious risks involved in altering a language that is not yet understood. Many adult clones created thus far, for example, do not live as long as the one from whom they were cloned. Why this is the case is unknown, but it is thought to be due to something not yet understood about DNA.

Though this should act as a warning, considerable ongoing research involves the modification of DNA. In gene transplantation research, for example, genes are being transplanted from one organism to another with little to no understanding of how those genes will interact with all the other genes of that organism[a]. In gene therapy research, genes are being corrected that are believed to be damaged, even when it is not known how it is that the 'damage' causes the problem.

Until we understand much more about DNA and the language of life, we should be very hesitant to modify DNA. As stewards of God's creation, much more care should be taken in genetic research.

Summary of Chapter

A. God is a communicating God—so much so that He can create by the mere mention of a word. He illustrates His communicating character by both animal communication and the language of life found in all organisms.

B. God created a spectrum of perfection of animal communication to help us understand God's infinite ability to communicate:

  a. Many organisms seem unable to communicate in any way;

  b. Many organisms send simple messages by means of smell, sound, sight, or feel;

  c. Some organisms convey more complicated information with a variety of movements or sounds (*e.g.* the honey bee waggle dance indicating distance and direction to nectar sources; different vervet monkey warnings for different predators);

  d. Some organisms use songs to convey individual identity and group membership;

  e. Some apes can learn sign language; and

  f. Humans communicate by means of full-scale language.

C. Several things indicate that DNA does not contain all the information needed to build an organism (*i.e.* the 'myth of DNA' is wrong):

  a. Most cellular components are inherited from the egg cell, not DNA;

  b. When DNA is taken out of fertilized frog cells, the cells can divide as many as ten times before finally dying;

  c. Microbes in organisms play important roles in the development of those organisms; and

---

[a] The first two transplantations of a gene from one organism to another were performed in 1979. To provide inexpensive human insulin for diabetics, the human gene for insulin was transplanted into bacteria. To provide inexpensive human somatotropin (growth hormone) to prevent dwarfism, the human gene for somatotropin was transplanted into bacteria. In 1994, the first food approved by the FDA that contained a transplanted gene was a tomato containing a gene from another plant that allowed the tomato to stay firmer longer. In 1993, the first *artificial* (human-fashioned) gene approved by the FDA was bovine somatotropin (a growth hormone gene for cows).

d. Identical twins have the same DNA yet have distinct personalities.
D. Although DNA is a recipe book for the formation of an organism's macromolecules, it is not known how much more DNA might be responsible for.
E. DNA is designed to
   a. carry large amounts of information
      i. by hierarchal coding:
         - a sequence of 3 nucleotides (a codon) built from 4 different nucleotide types (A, T, C, and G: adenine, thymine, cytosine, and guanine)) can code for one of the 20 amino acids
         - a sequence of amino acids (codons) can code for one of 100's of thousands of proteins needed to build biological molecules
      ii. a long enough string of nucleotides (3 billion in humans) to carry all the information necessary to build all the molecules an organism needs
   b. readily copy information
      i. by possessing a double strand that can be 'unzipped'; and
      ii. by being able to create a complimentary strand by pairing nucleotides according to the rule of G with C (guanine with cytosine) and A with T (adenine with thymine)
F. As evidence of a 'language of life' at the foundation of all life, both written language and DNA
   a. exhibit hierarchal coding. For example:
      3 Morse Code elements → 26 English letters → >400,000 English words
      4 nucleotides → 20 amino acids → >120,000 proteins
   b. are random at a lower level (of nucleotides and alphabet letters) and ordered at a higher level (of genes and words).
   c. are ruled by linguistic rules:
      i. there are different types of genes and words with different functions (some seem to be modifiers, some seem to be commands, some seem to be objects, *etc.*)
      ii. different types of genes and words must be 'read' in the right sequence
   d. are read and understood only in a larger system
      i. the process of making copies of genes is so similar to humans making copies of written language is called transcription, and
      ii. the process of transforming the sequence of nucleotides into a sequence of amino acids is so similar to humans transforming one language into another is called translation.
G. Mutations are changes in the sequence of DNA nucleotides, usually occurring as rare mistakes in copying DNA (usually only about once in a million copies).
   a. A point mutation is one nucleotide switched with another, thus changing one amino acid in a protein. Many point mutations have no effect, many others have some negative effect, and a few have serious negative consequences.
   b. A frame-shift mutation is the loss or insertion of a nucleotide, thus changing every subsequent amino acid in a protein. Frame-shift mutations almost always destroy the function of a protein and are very often fatal.
H. DNA is a substantial challenge to naturalism:
   a. If naturalism were true, DNA would contain all the information necessary to build an organism (because reductionism should be true and DNA is the smallest component unique to a particular organism). However, the 'myth of DNA' false.
   b. DNA seems to be optimally designed to carry and copy information—more efficiently than is expected in naturalism without a designer.
   c. The language of life requires a complex biological system—*e.g.* molecular machines for copying, transcribing, and translating DNA—with an irreducible complexity that makes it

impossible to construct one step at a time (as naturalism's evolution requires).

d. In our experience, language only arises when designed by a communicating being. It does not arise by natural process.

I. The complexity, efficiency, and language nature of DNA are better explained by creation by a communicating God—especially One Who created the language of life as a physical illustration of His communicating nature.

J. As priests, we should be led to worship God as we contemplate the complexity, efficiency, and language characteristics of DNA. Furthermore, the DNA found in every cell of the human body should impress us that we are God's possession.

K. If DNA really is part of the language of life, then scientists are a long way from understanding that language. Consequently, we should be hesitant to modify DNA unless we fully understand those modifications and their effects.

a. Inducing mutations (causing random changes in the DNA) is not appropriate.

b. Most examples of adult cloning, gene transplantation, and gene therapy are probably premature.

Test & Essay Questions

1. List three biblical reasons why God is to be understood as a communicating God.

2. List two different ways that God physically illustrates His communicating nature.

3. What is the myth of DNA and how do we know it is a myth?

4. Present and explain three / four different evidences for why DNA does not contain all the information of life.

5. What is it about the structure of DNA that allows it to carry information / to be easily copied?

6. Describe the structure of DNA and how nucleotides connect to each other to produce that structure.

7. What is a codon and what does it do?

8. Present and explain two / three / four evidences that DNA is a part of the language of life.

9. Explain what insight Morse Code provides into the structure and function of DNA.

10. Explain what the spectrum of perfection of communication is and what it indicates.

11. Explain how the myth of DNA / efficient design of DNA / irreducible complexity of communication / language of life is a challenge to the naturalistic worldview / an indication of creation by God.

12. Explain why great care should be taken in DNA research.

References

1 Lloyd, J. E., 1966, Studies on the flash communication system in Photinus fireflies, Miscellaneous Publications of the Museum of Zoology, University of Michigan 130:1-95.

2 Konishi, M., 1963, The role of auditory feedback in the vocal repertoire of the domestic fowl, Zietschrift fur Tierpsychologie 20:349-67.

3 Seyfarth, Robert. M., Dorothy L. Cheney, and Peter Marler, 1980, Vervet monkey alarm calls: Evidence for predator classification and semantic communication, Animal Behavior 28:1070-94.

4 Schusterman, R. J. and K. Krieger, 1984, California sea lions are capable of semantic comprehension, Psychol. Rec. 34:3-23.

5 Herman, L. M., 1987, Receptive Competencies of Language-Trained Animals [Advances in The Study of Behavior, Vol. 17], Academic, New York, NY.

6 Alexander, Shana, 2000, The Astonishing Elephant, Random House, New York, NY, 300 p. (p.58).

7 Gould, J. L., and W. F. Towne, 1988, Honey bee learning, Advances in Insect Physiology 20:55-86; Couvillon, P. A., T. G. Leiato, and M. E. Bitterman, 1991, Learning by honeybees (Apis mellifera)

on arrival at and on departure from a feeding place, Journal of Comparative Physiology 105:177-84; Towne, W. F., and J. L. Gould, 1988, The spatial precision of the honey bees-dance communication, Journal of Insect Behavior 1:129-55.

8   Herman, L. M., 1986, Cognition and language competencies of bottlenosed dolphins, in Schusterman, R. J., A. J. Thomas, and F. G. Wood, editors, Dolphin Cognition and Behavior: A Comparative Approach, Lawrence Erlbaum, Hillsdale, NJ.

9   Rowe, N., 1966, The Pictorial Guide to the Living Primates, Pogonias, NY.

10 Kummer, H., and J. Goodall, 1985, Conditions of innovative behavior in primates, Philosophical Transactions of the Royal Society of London Series B 308:203-214.

11 Russon, Anne E., 2003, Innovation and creativity in forest-living rehabilitant orang-utans, pp. 279-306 in Reader, Simon M., and Kevin N. Laland, editors, Animal Innovation, Oxford University, New York, NY, 344 p.

12 Boran, J. R., and S. L. Heimlich, 1999, social learning in cetaceans: Hunting, hearing, and hierarchies, pp. 282-307 in Box, H. O., and K. L. Gibson, editors, Mammalian Social Learning, Cambridge University, New York, NY; Slater, Peter J. B., and Robert F. Lachlan, 2003, Is innovation in bird song adaptive?, pp. 117-135 in Reader, Simon M., and Kevin N. Laland, editors, Animal Innovation, Oxford University, New York, NY, 344 p.

13 Payne, K., P. Tyack, and R. Payne, 1984, Progressive changes in the songs of humpback whales (Megaptera novaeangliae): A detailed analysis of two seasons in Hawaii, pp. 9-57, in Payne, R., editor, Communication and Behavior of Whales, Westview, Boulder, CO; Payne, K. B., and R. S. Payne, 1985, Large scale changes over 19 years in songs of humpback whales in Bermuda, Zeitschrift fur Tierpsychologie 68:89-114.

14 Gardner, R. A., and B. T. Gardner, 1969, Teaching sign language to a chimpanzee, Science 165:664-672; Premack, D., 1971, Language in chimpanzee?, Science 171:808-822; Fouts, R. S., 1973, Acquisition and testing of gestural signs in four young chimpanzees, Science 180:978-980; Rumbaugh, D. M., and T. V. Gill, 1976, Language and the acquisition of language-type skills by a chimpanzee (Pan), Annals of the New York Academy of Sciences 270:90-123.

15 Miles, H. L. W., 1990, The cognitive foundations for reference in a signing orangutan, in Parker, S. T., and K. R. Gibson, editors, 'Language' and Intelligence in Monkeys and Apes, Cambridge University, New York, NY.

16 Patterson, F. G., and E. Linden, 1981, The Education of Koko, Holt, Rinehart & Winston, New York, NY.

CHAPTER 14

FULLNESS OF GOD

Biological Reproduction

"'I fill all of heaven and earth', says the Lord"
Jer. 23:24, NCV

Christ "...fills everything in every way."
Eph. 1:23, NCV

14.1 | Fullness of God

God is an infinite God. He has every good attribute and He possesses each attribute fully, completely, and perfectly. He is everywhere, *filling* the heavens, the earth, and all things (Jer. 23:23-24; Eph. 4:10). Consequently, he wishes to *fill* our mouths (Psa. 81:10), *fill* the land with fruit (Isa.27:6), and provide *fullness* of joy (Psa. 16:11). God is *abundant* in goodness (Exo. 34:6), truth (Ex. 34:6), forgiveness (Isa. 55:7; Titus 3:5), grace (Rom. 5:17), and mercy (I Pe. 1:3). He *abounds* with blessings and grace (Pr. 28:20; II Cor. 9:8). He gives *abundantly* (Deu. 28:47; Psa. 36:8; 132:15) and not just more life, but more *abundant* life (John 10:10). He is *rich* in mercy (Eph. 2:4) and gives *richly* (I Tim. 6:17). God is *perfect* (Mat. 5:48) and desires us to be *perfect* (Deu. 18:13; Mat. 5:48; II Cor. 13:11) and to be *perfected* (Eph. 4:12-13; Col. 1:28; II Tim. 3:17; Heb. 6:1; 13:21; James 1:4).

To emphasize His fullness, completeness, and perfection, God delights in giving a lot, then giving even more, so that He has given abundantly, and then when we think no more can be given, He gives again (*e.g.* Mal. 3:10).

The biological creation also illustrates his fullness. When God created the animals of the sea and the air, He did so *abundantly* (Gen. 1:20-21), and then He commanded them to 'be fruitful and multiply'! God intends for organisms to be found everywhere on this planet, and everywhere they are found they are to abound.

God's blessing and presence is demonstrated by the abundance of organisms, and His judgment and absence by the paucity of organisms. But God's mercy often replaces judgment with blessings, so that even after the earth's greatest judgment of life (the Flood), God blessed Noah's family, and the animals, to multiply abundantly and fruitfully (Gen. 8:17).

"Be Fruitful"

One of the most impressive characteristics of the biological world is its ability to reproduce. Non-living materials cannot reproduce. What we observe across the creation is that even though energy can take on different forms, the total amount remains the same (the First Law of Thermodynamics). Atoms and molecules cannot reproduce either. Their components can be broken down or built up as they rearrange

themselves into different atoms, molecules, and compounds, but this is not reproduction.

Organisms are unique in being able to increase the numbers of their *kind*—one bacterium becomes two bacteria, two foxes become three, *etc.* Reproduction is producing another of the same kind. In reproducing, organisms are uniquely designed to illustrate God's fullness and completeness, because by reproducing, organisms alone can *abound*.

Genesis 1:21 indicates that when created, the sea creatures were abundant. Being abundant, animals were a physical illustration that God is an abundant God. He abounds in each of His attributes. He abounds in His ability, and He abounds in demonstration of that ability. The abundance of life points to God. The abundance of life glorifies God. Even so, the next verse implies that although abundant, the sea creatures had not yet filled the seas, so He commanded the sea creatures to 'be fruitful and multiply and fill the waters in the seas' (Gen. 1:22).

Increasing in numbers suggests God is vibrant and active. He is not asleep, unconcerned, bored, or boring. It suggests that God is more than abundant. Whatever amazes us, He exceeds. When we think that He can do no more, He does even more. Increasing beyond abundance physically illustrates the infinity of God. God does not just abound in His attributes. He is perfect, complete, and infinite in each and all of His attributes. Animals were commanded to reproduce and multiply and fill the earth in order to bring glory and pleasure to the God Who created them.

Cellular Reproduction

Cell theory claims that organisms are composed of cells. Some organisms spend their entire lives as single cells. Such organisms are referred to as **unicellular**. The remaining organisms are **multi-cellular**. Yet, even multi-cellular organisms begin their life as a single cell. Even organisms *capable* of developing from multi-cellular fragments, like a rose generated from a cutting or a sea star developing from an amputated arm, do so as an alternative form of reproduction. They usually develop from a single cell. Ever since the Creation Week, multi-cellular organisms begin their life as a single cell. And, since each of these cells was generated from another cell, the reproduction of every organism begins with **cellular reproduction**. Whether speaking of the reproduction of bacteria, redwoods, or humans, they all begin with cellular reproduction.

Cellular reproduction usually involves one ('parent') cell dividing into two smaller ('daughter') cells. This involves four steps:

1. Getting two full sets of DNA into two different parts of the parent cell and—in the case of eukaryotes, developing a nucleus around each set (**karyokinesis**);
2. Getting two sets of the remaining cell components in two different parts of the parent cell;
3. Dividing the parent cell into two daughter cells (**cytokinesis**); and
4. Growing each daughter cell to full size and restoring a full complement of cell components.

In eukaryotes there is often more than one organelle in the cell (*e.g.* mitochondria and chloroplasts). For these organelles, some end up in one daughter cell and the rest end up in the other. Less well-defined plasma membrane organelles, like the endoplasmic reticulum and Golgi bodies, are divided along with the cytoplasm. A portion ends up in one daughter cell while the remaining portion ends up in the other. In all cases, it is *after* cytokinesis that the organelles are restored to the normal cell quantity in each daughter cell.

---

unicellular—organism that is made up of only one cell

multi-cellular—organism made up of more than one cell

cellular reproduction—cell process that divides a (parent) cell into two or more daughter cells.

karyokinesis—nuclear division; eukaryotic cell process that divides a cell nucleus into two nuclei

cytokinesis—cell process that divides a cell in two

In the cellular reproduction of eukaryotes, the most significant challenge is the dividing of the DNA and nucleus in such a way that each daughter cell ends up with both a nucleus and a complete copy of the DNA. This process is called karyokinesis or nuclear division, and the process occurs *before* cytokinesis. Nuclear division itself involves three steps:

replication—cell process that makes a copy of the cell's DNA

sister chromatids—(following replication) two identical DNA molecules attached at their centromeres

centromere—section of a chromosome containing proteins for attachment

condensation—the dense packing of a chromosome that occurs at the beginning of mitosis or meiosis

chromosome—(in a eukaryotic cell) a linear segment of DNA and its associated proteins

1. **Replication** makes a copy of the DNA so that each daughter cell will have a complete copy of its own. In replication, each piece of DNA is unzipped and a complement is built off each half. For each piece of DNA, this generates two identical DNA molecules (called **sister chromatids**). As the cell enters the next stage, each pair of sister chromatids remain attached to each other at a site known as a **centromere**.

2. **Condensation** packages DNA in such a way as to make it easy to separate into different parts of the parent cell. When it is functioning normally, DNA has to be fully extended, allowing for portions to be untwisted, unzipped and copied. But in its extended form it is an extremely long molecule (a total of about a yard long in the case of the DNA in single human cell). As such, it is extremely difficult to move around (rather like trying to separate or move several very long strands of yarn). To make it easier, each DNA molecule is 'condensed'—that is, carefully wrapped up in a compact form known as a **chromosome** (rather like wrapping up

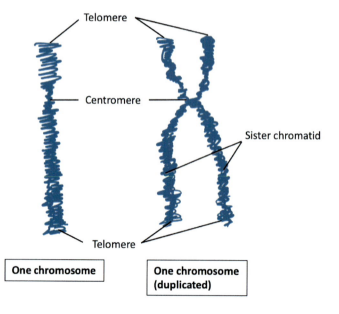

yarn in a skein to allow it to be transported, sold, and used for knitting).

Whereas DNA cannot be seen under a normal microscope when it is in its extended form, condensed DNA is stubby enough to show up. And, since the proteins that help form the chromosome stain rather easily, condensed DNA shows up under the microscope as a 'colored body' ('chromo': colored; 'some': body).

3. When chromosomes first show up under the microscope, the cells are said to enter mitosis or meiosis, depending upon the type of reproduction that is occurring. If the parent cell is making an exact copy (clone) of itself, the cell nucleus goes through mitosis. This type of division occurs when a unicellular organism (*e.g.* a yeast) reproduces by asexual reproduction. It also takes place in multicellular organisms when they grow by increasing cell numbers or when organisms repairing themselves by replacing damaged cells.

The result of mitosis is two daughter cells with DNA identical to each other and to the DNA of the parent cell. If, however, the parent cell is producing sperm or egg cells for sexual reproduction, the cell nucleus goes through meiosis (pronounced 'my-ōsis'). Meiosis yields four daughter cells, each with half the amount of DNA the parent has, and each with unique DNA.

As impressive as the process of cellular reproduction is, it is not the 'normal' part of cell life. Through the entire process of mitosis or meiosis, DNA is condensed, and while it is, the information on the DNA cannot be accessed. Furthermore, until the quantity of cell components reaches that of mature cells, a cell usually cannot fulfill the optimum functions for which it was designed. Consequently, most of the lifetime of a given cell is spent *between* divisions — in other words after the cell has fully grown from the previous division and before it begins replicating its DNA for the next division.

It has become traditional in biology to label the stages in the life of a cell as the **cell cycle**. The **interphase** is that portion of the cell cycle when the nucleus is not dividing. It includes three substages. The first substage of the interstage is labeled G1, which includes the growth of a cell following the previous division as well as the 'normal' part of the cell's life when the cell carries on its normal operations. The second and third substages are short in duration and prepare the cell for dividing

cell cycle—stages and substages in the life of a cell. Stages: interphase when the nucleus is not dividing; mitosis or meiosis when the nucleus is dividing

interphase—stage in the life of a cell when the cell is not dividing

the nucleus. The second substage is the S (Synthesis) substage when replication occurs. During the third substage, G2, the DNA is double-checked for copy errors and the DNA begins to condense.

Mitosis

When an exact copy of a eukaryotic cell is needed (such as when a unicellular organism reproduces asexually, or as a multi-cellular organism grows, or replaces cells) mitosis is how a cell divides its nucleus and DNA. There are four substages to mitosis:

prophase—first mitosis substage, when chromosomes, centrioles, and spindle fibers appear and nuclear envelope and nucleoli disappear; see prometaphase

metaphase—mitosis substage when chromosomes align and attach to centrioles via spindle fibers

anaphase—mitosis substage when sister chromatids separate to opposite ends of the cell

telophase—last mitosis substage, when, in both cells, chromosomes and centrioles disappear, and nuclear envelope and nucleoli appear

prometaphase—mitosis substage some place after the prophase, when the nuclear envelope dissolves

1. **prophase** ('pro' or *pre* phase), when chromosomes complete their condensation, the nucleolus and the nuclear envelope dissolve, centrioles (aka centrosomes) develop on opposites ends of the cell (toward which cell components will be pulled for each daughter cell), and spindle fibers begin forming (microtubule molecules that function as ropes to pull cell components to each centriole);

2. **metaphase**[a] ('meta' or *change* phase), when the chromosomes line up in a plane halfway between the centrioles, and the sister chromatids of each chromosome attach by spindle fibers to opposite centrioles;

3. **anaphase** ('ana' or *separation* phase), when the sister chromatids are separated from each other and pulled to centrioles at opposite ends of the cell; and

---

a   Many modern biologists see enough significance to the interesting process of dissolving the nuclear envelope to insert the **prometaphase** between prophase and metaphase (containing the end of the old prophase and the beginning of the old metaphase). Mitosis is described in this text with four stages (without the prometaphase) simply to make the mitosis stages easier to remember.

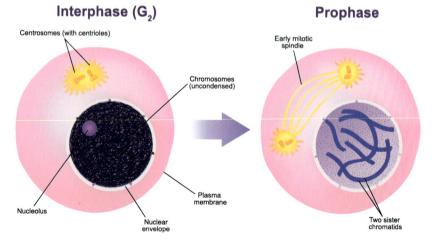

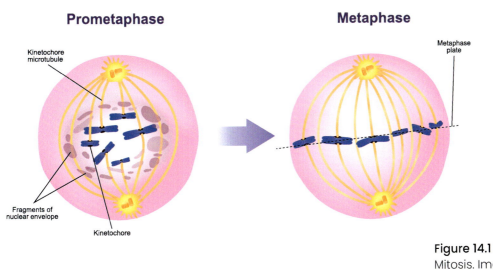

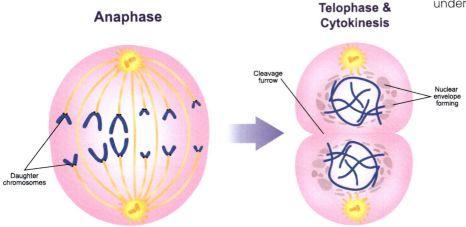

Figure 14.1
Mitosis. Image by Ali Zifac via commons.wikimedia.org, used under CCby4.0.

4. **telophase** ('telo' or *purpose* phase), when spindle fibers dissolve, centrioles disappear, chromosomes begin decondensing, a nuclear envelope begins forming around each set of chromosomes, and nucleoli begin reforming.

In the process of mitosis, a cell with one nucleus containing two identical copies of the DNA becomes a cell with two nuclei—one on each end of the cell—each nucleus containing an identical set of DNA to the parent cell before replication. Following mitosis, cytokinesis divides the cell into two daughter cells.

14.2 | Meiosis

God desired to demonstrate the diversity of His own being by creating a high diversity of species. He also created great variation *within* each species. Genes found in the same species can assume different forms. To increase this diversity even more, God designed many organisms requiring *two* copies of every gene, not just one. By the many different combinations that are possible, this diploid condition, squares the number of different traits. For an organism to be diploid, every chromosome has a **homologous chromosome** of the same length with similar sets of genes in the same sequence.

Even more variety is introduced by sexual reproduction, when offspring are created by combining DNA from two parent organisms. To make sure that this results in the proper amount of DNA in the offspring, special **gamete** cells (also known as sex cells) are produced. These are called haploid cells because each carries only half the DNA of the parent.

The gamete carrying half the male parent's DNA is a **sperm cell**. The gamete carrying half of the female parent's DNA is an **egg cell**. Fertilization occurs when the DNA of the sperm cell is injected into the egg cell. In most cases, the fertilized egg that results contains the same amount of DNA as each parent—two haploid sets of DNA make up a full diploid set of DNA. Meiosis is the type of nuclear division needed to produce egg cells or sperm cells.

Since the process of mitosis divides twice the original amount of DNA into two cells, each with the original amount of DNA, if mitosis were performed twice in a row, the result will create four cells, each with half the original amount of DNA. That is roughly what meiosis

homologous chromosomes—two chromosomes containing the same genes in the same sequence

gamete—haploid cell generated by meiosis in sexual organisms (aka sex cell). Types: a sperm, at fertilization, contributes only DNA; an egg cell, after fertilization, grows into an adult organism

sperm—(haploid) gamete of a sexual (usually male) organism that contributes only its DNA at fertilization

egg cell—(haploid) gamete of a sexual (usually female) organism that grows into an adult after fertilization

involves—a cell going through something like mitosis twice in succession, thus creating four gamete cells. The substages of meiosis are as follows:

1. Meiosis I
   a. **prophase I** is identical to prophase in mitosis (with the completion of condensation, the disappearance of the nucleolus and nuclear envelope, and the appearance of centrioles and spindle fibers) *and* homologous chromosomes connect at their centromeres and **crossing over** occurs (equivalent sections of chromosomes are traded from one chromosome to another);
   b. **metaphase I**[a], when pairs of homologous chromosomes line up in a plane halfway between the centrioles and spindle fibers connect the centromeres of each homologous chromosome to opposite centrioles.

prophase I—first meiosis substage: chromosomes, centrioles, & spindle fibers appear, nuclear envelope & nucleoli disappear, homologous chromosomes pair, and crossing over occurs; see prometaphase I

crossing over—process in prophase I when homologous chromosomes exchange genetic material

metaphase I—meiosis substage when homologous chromosome pairs align and attach to spindle fibers

prometaphase I—meiosis substage some place after prophase I, when the nuclear envelope dissolves

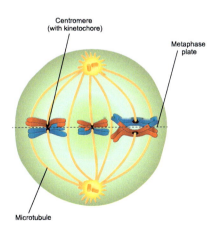

Figure 14.2a
Meiosis I & II. Image by Ali Zifac via commons.wikimedia.org, used under CCby4.0.

The chromosomes condense, and the nuclear envelope breaks down. Crossing-over occurs.

Pairs of homologous chromosomes move to the equator of the cell.

---

a As in the case of mitosis, meiosis is often described now with an additional phase, **prometaphase I**.

anaphase I—meiosis substage when homologous chromosomes separate to opposite ends of the cell

telophase I—meiosis substage, after homologous chromosomes separate, when centrioles disappear

cytokinesis I—meiosis substage, after homologous chromosomes separate, when the cell first divides

c. **anaphase I**, when the homologous chromosomes are separated from each other and pulled to centrioles at opposite ends of the cell;

d. **telophase I**, when spindles dissolve, releasing a set of chromosomes at each end of the cell;

e. **cytokinesis I**, when the cell makes its first division;

2. Meiosis II

   a. **prophase II**, when centrioles develop on opposites ends of each cell, and spindle fibers begin forming;

   b. **metaphase II**, when the chromosomes in each cell line up in a plane halfway between the centrioles in that cell, and the sister chromatids of each chromosome attach by spindle fibers to opposite centrioles;

   c. **anaphase II**, when the sister chromatids in each cell are separated from each other and pulled to centrioles at opposite ends of the cell; and

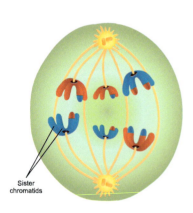

Homologous chromosomes move to the opposite poles of the cell.

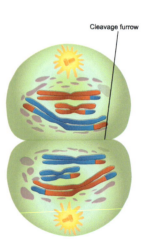

Chromosomes gather at the poles of the cells. The cytoplasm divides.

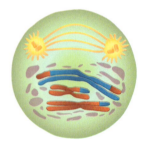

A new spindle forms around the chromosomes.

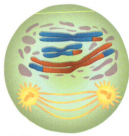

Figure 14.2b
Meiosis I & II. Image by Ali Zifac via commons.wikimedia.org, used under CCby4.0.

d. **telophase II** (the reverse of prophase I), when spindle fibers dissolve, centrioles disappear, chromosomes begin decondensing, a nuclear envelope begins forming around each set of chromosomes, and nucleoli begin reforming.

Following meiosis II, both cells divide by **cytokinesis II**, creating a total of four cells. In the process of **meiosis**, a cell with one nucleus containing two identical copies of the DNA becomes four cells containing only half the DNA of the parent cell before replication. As a result of the crossing over, each sperm and each egg carries a unique combination of traits. Depending on which sperm fertilizes which egg, even more variety can be produced.

telophase II—last meiosis substage, when, in all four cells, chromosomes and centrioles disappear, and nuclear envelopes and nucleoli reappear

cytokinesis II—meiosis substage after the separation of sister chromatids, when the two cells divide

meiosis—(in generating sex cells) cell cycle stage yielding four nuclei with unique, haploid DNA sequences

prophase II—meiosis substage, after cytokinesis I, when centrioles and spindle fibers appear in each cell

metaphase II—meiosis substage after cytokinesis I: chromosomes align & attach to spindles in each cell

anaphase II—meiosis substage after cytokinesis I, when sister chromatids separate in each cell

Metaphase II

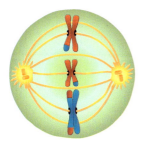

Metaphase II chromosomes line up at the equator.

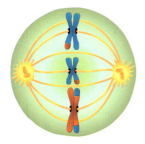

Anaphase II

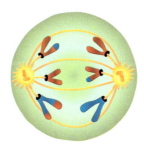

Centromeres divide. Chromatids move to the opposite poles of the cells.

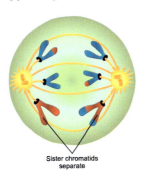

Telophase II & cytokinesis

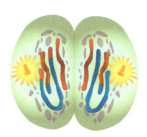

A nuclear envelope forms around each set of chromosomes. The cytoplasm divides.

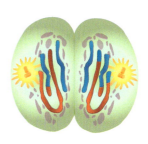

Non-Physical Reproduction

There was some discussion in Chapter Two about the non-physical aspects of biological life. Aside from its physical components, a living organism might actually possess some non-physical components or substance. 'Life', 'soul', and 'spirit' may well be the non-physical components found in different organisms.

Although biology focuses on the physical aspects and components of organisms, the subject of reproduction raises an interesting question that must not be ignored, and that is 'How are the non-physical aspects of life produced in organisms?' Cellular reproduction describes the reproduction of the physical body, but how are non-physical aspects of organisms transferred from one generation to the next? Can they be copied? Can they be divided? Do they need to grow? Can two of them combine to produce a third?

The short answer to these questions is 'We do not know'. We cannot see the non-physical in order to observe the answers for ourselves. And Scripture does not give us direct answers either. However, test tube babies and identical twins may some shed light on the issue. First, human **in-vitro fertilization (IVF)** involves taking sperm from a man and eggs from a woman and using the sperm to fertilize the eggs in a lab. Once fertilized, the egg is then implanted in a woman's uterus to be carried to birth.

---

in-vitro fertilization— medical procedure fertilizing an egg cell and inserting it into a womb to develop

Since the birth of the first IVF baby (early on called a 'test tube baby') in 1978, many have been conceived by IVF. These people possess non-physical soul/spirits just like everyone else. This indicates that the reproduction of the non-physical does not require the direct contact of parents. Second, identical twins develop from just one fertilized egg. One sperm fertilizes one egg, then that fertilized egg divides into two cells and then four cells by mitosis. If the first two cells are separated before the second mitosis, or if one or two of the first four cells are separated from the others before the third mitosis, each cell or group of cells can develop into separate babies.

The two babies arose from one fertilization event—from the same sperm and the same egg—so they have the same DNA[a]. Since each twin possesses his or her own soul/spirit, reproduction of the non-physical does not require a separate fertilization event. Between IVF and identical twins, it would seem that the reproduction of the non-physical is not directly linked to any particular step in cellular reproduction.

Although Scripture does not say so explicitly, and we cannot directly observe it, it seems most reasonable to conclude that God directly creates the non-physical components of life. It is likely that the biogenic law can be extrapolated to the non-physical—namely, that non-physical aspects of life can only be generated by non-physical life. It would seem that the non-physical God, the origin of life itself, is the most logical cause for the non-physical aspects of life. This would also be consistent with the biblical claims that God forms us in the womb (Job 31:15; Isa. 44:2, 24) and that His Spirit goes about on the earth creating its nephesh life (Psa. 104:30).

If direct creation by God is our best guess on *how* non-physical reproduction occurs, the next question is *when* does God create the non-physical? According to Psalm 51:5, some aspect of David's sin was present at David's conception[b]. Since David's physical body at the time of conception (a fertilized egg) was hardly capable of sin, it is David's non-physical being that sinned or carried the sin nature. This in turn suggests that David's non-physical being existed at his conception.

---

a   Incidentally, cloning is similar. When the cell of an adult sheep was made into an embryo to generate Dolly, the first animal cloned from adult cells, that embryo did not arise from a fertilization event, though it had to have the appropriate genetic material.

b   Some believe this to be sin committed by David himself; others believe this refers to the sin nature inherited from Adam. Either way, something non-physical about David existed at his conception.

Creation of the non-physical somewhere around the time of conception would also be consistent with consciousness from the womb ('You [God] covered me [David] in my mother's womb': Psa. 139:13), spiritual responses from the womb (Luke 1:41), and references to divine callings of Samson, Isaiah, Jeremiah, and John the Baptist from the womb[a]. It is also consistent with the death-for-death punishment required of a man who kills a child in the womb—at any point in the pregnancy (Ex. 21:22-25).

Conception also seems like a reasonable time for the creation of the human soul-spirit, for then it would be able to contribute to the development of the physical body throughout its entire development. Most probably God creates the non-physical aspect of an organism at or near conception[b]. This suggests that conception is when life begins for each individual organism, and conception is when God creates the soul of every animal and the soul/spirit of every human. This means each human is specially created, the essence of being human begins at conception, and from conception each human is fully alive and fully the image of God.

14.3 | "Multiply" and "Breed"

The commands 'be fruitful and multiply' (Gen. 1:22, 28; 8:17; 9:7), 'bring forth abundantly and multiply' (Gen. 9:7), 'multiply' (Gen. 1:22), and 'breed abundantly' (Gen. 8:17) *could* conceivably refer only to the process of increasing numbers by the process of reproduction. However, the variety of phrases employed in the text suggests that more than one process is actually involved. It is likely that at least one of those other processes is **diversification** (an increase in species diversity).

diversification—increase in number of species

---

a 'the child [Samson] shall be a Nazarite unto God from the womb': Jud. 13:5-7; 'The Lord has called me [Isaiah] from the womb': Isa. 49:1; 'Before I [God] knew you [Jeremiah] in the belly I knew you': Jer. 1:5; 'he [John the Baptist] shall be filled with the Holy Ghost, even from his mother's womb': Luke 1:15.

b   For a vast percentage of people, conception is the union of a sperm and an egg. This process involves several steps occurring over several minutes of time, including the entry of the sperm into the egg, the acceptance of the sperm DNA by the egg, and the egg continuing its development. To make sure that all human life is protected, conception should be defined at the very *earliest* step in this process—the entry of the sperm into the egg. For those who do not begin by the union of a sperm and an egg (e.g. the embryos that begin by embryonic splitting and the embryos that might arise by adult cloning) an analogous step should be selected. In the case of embryonic splitting it should be probably be the point that the embryo is split. In the case of adult cloning it should probably be the point when the nucleus is placed within the egg without a nucleus.

Each kind (baramin) of land animal was represented on the ark by two individuals—or fourteen[a] in the case of clean animals (Gen. 6:18-7:3). If a species is defined as a breeding population, each unclean animal baramin at the end of the Flood would have contained only a single species[b]. In the case of clean animals, there could have been as many as seven species, if there were seven pairs and each pair bred separately from all the other pairs. In the present, however, most baramins contain many species. Some even contain thousands of species.

Following the Flood, there must have been a substantial increase in species diversity within baramins (**intra-baraminic diversification**). It is likely that highly specialized species that show up late in the fossil record are species that developed late in this process of diversification. Two such species are mentioned in the Bible as existing within a few centuries of the Flood: lions (Job 4:10-11, 38:39-40) and camels (Gen.

intra-baraminic diversification— proliferation of species within baramins, especially soon after the Flood

---

a It is not known for sure whether seven or fourteen of each kind of clean animal was to be taken onto the ark. The animals were to be taken by 'sevens', but the Hebrew word translated 'sevens' is a dualistic form of the word, suggesting a double form of something (e.g. the Hebrew words for eyes and ears are dualistic words, suggesting the two eyes and two ears normally found on a human. So it could be that seven 'twos' (i.e. 14) of each clean animal kind were to be taken onto the ark.

b It is possible that this would have been true for a while following the Flood even if the two individuals on the ark were taken from different (interbreedable) species before the Flood. Since they would have been the only representatives of the baramin at the end of the Flood, they would have had to mate with each other, so their offspring would have been inter-specific hybrids. Those offspring in turn would have to interbreed with other hybrids (their siblings) and/or with their parents. In many cases this may have blurred the distinction between the parents, creating a single inter-specific species with a form lying between and perhaps even including the form of the parents.

12:16; 24:10-22; Job 1:3, 17). This suggests that the command to 'multiply' was fulfilled by very rapid intra-baraminic diversification.

Human breeders are continually active at working with species that already exist in order to develop new varieties, cultivars, and breeds (*e.g.* scores of horse breeds, hundreds of dog breeds)—or even new species (as in the case of thousands of orchid species). Many of the new forms are developed by artificial selection (choosing young with the desirable form of a particular characteristic and breeding them generation after generation to enhance that characteristic). Many of the remaining forms were developed by crossing different breeds, varieties, cultivars, or species to create a unique combination of characteristics.

Artificial selection reveals characteristics that were already there, but hidden. Crossing combines revealed characteristics in new ways. In each case, humans did not create something entirely new, but merely revealed more of the variety already present in these organisms. This is why such forms can be generated so quickly. It is likely that this revealed variety is a small sample of the hidden variety God placed in all organisms at the time of the creation. Some of that variety was placed there so that humans could glorify God by revealing more diversity. But most of the variety was probably placed there so as to maintain the diversity that God created to illustrate His plurality (Chapter 10).

Since at the time of the creation God knew that diversity would be decimated at the time of the Flood, He hid enough diversity within organisms to restore that diversity following the Flood. And, since God did not want even one person to miss the message, God also created a process capable of quickly expressing that diversity—in the course of the first human generation following the Flood. Since Arphaxad, the first birth listed in Scripture following the flood, lived for 438 years beginning two years after the Flood (Gen. 11:10-13), the process of restoring biodiversity must have occurred before 440 years had transpired following the Flood.

Biologists still do not know where or how the variety was hidden, nor do they know how that variety was revealed. However, evidence for the hidden seems fairly strong:

- The rapid increase in land animal diversity between the Flood and the present suggests land animals possess a huge amount of hidden variety and a process of rapidly revealing it.

- The rapid and extensive development of thousands of artificial breeds of plants and animals suggests many different organisms possess a huge amount of hidden variety.

- Since it is rather easy to lose the ability to hybridize, inter-specific hybridization among a majority of species in families of all types of organisms suggests that a large amount of **speciation** has occurred very recently within many families of organisms.

- The occasional appearance in a modern organism of a complex trait known in fossil forms (**genetic throwback**) suggests hidden variety in that organism. An example would be the occasional birth of a 3-toed horse, reflective of fossil horses having three toes. Since mutations should destroy unused genes in thousands or tens of thousands of years, genetic throwbacks also suggest that the modern species probably arose from the fossil forms within the last thousands or tens of thousands of years.

- **Vestigial organs** are those with reduced function in descendant species. Although it is difficult to demonstrate that *partially* vestigial organs ever had greater function in the past, those with no function at all in the present (fully vestigial organs) were most probably derived from organs with function in the past.

speciation—origin of a new species or the process that generates a new species

genetic throwback—occasional organism manifesting a complex trait otherwise only known in fossils

vestigial organ—organ in a modern species having a reduced function from that in an ancestral species

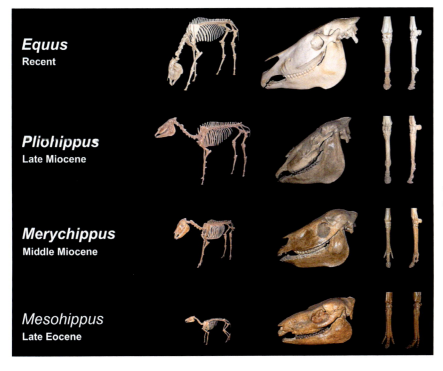

Figure 14.3
Horse series.
Image by H. Zell via commons.wikimedia.org, used under CCby3.0.

- An example would be hip and leg bones in fetal whales that disappear before birth. And, again, since mutations should destroy unused genes in thousands or tens of thousands of years, fully vestigial organs that correspond to functional organs in fossil forms suggest the modern species probably arose from the fossil forms within the last thousands or tens of thousands of years.

• The fossil species of a number of land mammal families are arranged in a stratomorphic series—that is, a steady succession or trajectory of different *morphologies* (body forms) in a corresponding steady succession of *strata* (rock layers). The most famous example is the horse series, which is a sequence of horse fossils from small bodied, three-toed horses that browse on broad leaved plants in Lower Tertiary rocks to the large-bodied, one-toed horses that graze on grass in the present.

• Evidences that indicate new species have arisen in the past are also consistent with intra-baraminic diversification. Such evidences include:

- In the case of many species on earth, the most similar species to it has an adjacent geographic range.

- In the case of many island species, the most similar species to them are found on the nearest mainland.

- In the case of many species with limited geographic ranges, the most similar fossil species has a similar restricted geographic range.

- A **ring species** (a circular arrangement of populations around an uninhabitable region where all but one pair of neighbor populations can interbreed successfully) suggests the following historical scenario: 1) a species migrated to the border of the uninhabitable region at the point opposite the current location of the non-interbreeding neighbor populations; 2) the species split into two separate lineages living side-by-side; 3) each lineage spread along one side of the uninhabitable region, changing as it went; and 4) when the two lineages met again on the opposite side of the uninhabitable region, they could not interbreed with each other. Examples include fruit flies

---

ring species—circle of populations where all but adjacent population pair can interbreed successfully

around mountains in Hawaii, salamanders around the dry Central Valley in California, and frogs around the perimeters of large lakes.

14.4 | "Fill"

As suggested above, the command to 'fill' from the complex phrase 'be fruitful and multiply and fill the earth' *could* conceivably refer only to abundance. If the environment contains fewer organisms than can be supported by that environment, it is not 'full'. So, the command to 'fill' the earth might have been obeyed by simple reproduction—in other words by increasing the *number* of individuals in order to 'fill' all the environments of the planet. However, the most natural meaning of 'fill the earth' involves something much more than that.

The study of where organisms live and how they got there is called **biogeography**. The area of the earth where an organism lives is called its **geographic range**. At their creation, humans lived in the Garden of Eden. From there, humans were to spread out and fill all the environments across the planet where humans could possibly live. To fill the earth, humans were to expand their geographic range from Eden to the entire planet. Although animals were probably supposed to do the same thing, we know too little about the geographic ranges of animals at creation to verify this.

biogeography—discipline of biology that studies where organisms live and how they got there

geographic range—the area of the earth where an organism lives

What we do know, however, is that geographic ranges changed radically at time of the Flood. In the days of Noah, a global Flood destroyed all land animals and humans, except those preserved on the ark (Gen. 7:21-23). For much of the duration of the Flood, the full geographic range of land animals and humans was the ark.

At the moment organisms disembarked from the ark, the full geographic range of land animals and humans was a single mountain in the region of Ararat (Gen. 8:4). From there, land organisms were to increase their geographic range from that mountain to the entire earth (Gen. 8:17). Since the Flood was global, it is likely that the geographic range of every organism on the planet was changed by the Flood, whether the organisms lived in water, on land, or in the soil.

In obedience to God's command, organisms now live just about everywhere. Even floating in the air, one can find plant pollen, spores from plants, fungi, protozoa, and bacteria, and even bacteria themselves. This is not just true of the air near the earth's surface, as asthma sufferers well know, but it is also true miles above the earth's surface. Although their numbers decrease with altitude, they appear to play an important role in cloud formation, as water and ice condenses on such particles. On the earth's surface, organisms are found in every known land environment. In many places, the presence of organisms is obvious, such as where plants thrive enough to produce a mat, grassland, or forest of green.

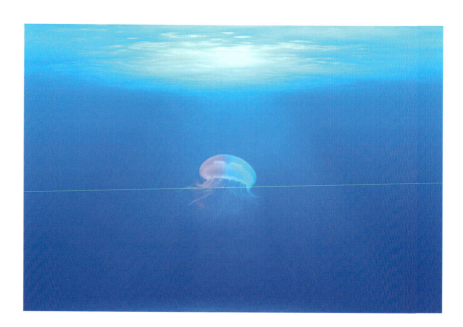

Even, however, in places where most organisms cannot live, and evidence of biological life might be uncommon—or at least not so obvious—a close examination will reveal organisms living there. Deserts, salt flats, mountain glaciers and ice fields are all homes to organisms specially designed for these harsh environments.

This is also true of water environments. Organisms are not just found in lakes and ponds where we can see lots of algae or plants growing in them. Nor are ocean organisms only living in coral reefs. Organisms have been found in every water environment we know about. Not even sea ice, volcanic hot springs, and hot spring vents at the bottom of the ocean are free of organisms.

Finally, organisms abound in the soil and rocks of the earth as well. Recently, in fact, bacteria have been found miles beneath the earth, abounding in greater numbers there than they are known to exist on the earth's surface. It would seem that organisms inhabit every possible environment on this planet—even a number of environments biologists once thought impossible for organisms.

The Origin Of Biological Reproduction

The process of reproduction is so incredibly complicated as to be impossible to generate naturalistically. Nothing as complex as the reproduction of even the simplest of organisms is known to form by natural process. Cellular reproduction is an incredibly complex biological system, and as such there seems to be a large number of components that must all work together to make it possible. It does not seem to be possible to create something like mitosis in even the simplest cells in some sort of stepwise process.

Yet the only explanation of the origin of things available in the worldview of naturalism is naturalistic evolution, which is thought to create things one step at a time. Humans have never designed anything as complex as the process of reproduction, so it would seem that only a being with the wisdom and power of the God of Scripture is capable of creating this process.

The origin of sexual reproduction is another challenge to the naturalistic worldview. Evolutionary theory is the best theory available to the naturalist. In its best conceived form, evolution prefers the organisms that do the best at passing on their DNA to the next generation. An asexual organism passes on 100% of its DNA to the next generation; a sexual organism passes on only 50% of its DNA to the next generation.

Asexual organisms should succeed over sexual organisms and asexual reproduction should never have evolved in the first place. The greater complexity involved in meiosis only reinforces the claim. Whereas a God desiring to infuse His creation with physical illustration of His internal diversity can explain how sexual reproduction came to be, the naturalistic worldview can provide no explanation for its origin.

The Origin of Modern Diversity

Both the creationist and the naturalistic worldviews agree that the present diversity of organisms has increased from a lower diversity in the past. The disagreement is over how that diversification occurred and how fast it occurred. According to the naturalistic worldview a new species arises by the **natural selection** of undirected mutations over hundreds to thousands of generations, and modern diversity arose over tens of millions of years.

> **selection, natural**—natural process resulting in death of (the less capable) part of a population; caused by overproduction and designed and introduced by God as an evil-minimizing effect of the curse

The creationist suggests that a new species arises by the expression of hidden variety in one or a few generations, and modern diversity arose in only centuries just several thousand years ago. Biologists have observed the origin of many new varieties, cultivars, and breeds. But none of them are known to arise by the natural selection of mutations as the naturalist would expect. All were generated from hidden variety expressed by intelligent intervention.

Furthermore, although inter-specific hybridization, genetic throwbacks, and vestigial organs do suggest diversification occurred, the same evidences suggest that diversification occurred recently—within the thousands of years expected by creationists rather than the millions of years expected by the naturalist.

In the latter half of the nineteenth century, there was a form of evolutionary theory that was very popular among advocates of the naturalistic worldview. According to this theory, a new species arose when an organism did not stop developing when it reached its adult form but continued to develop into a new form. Consequently, as an organism stepped through its stages of development it was thought to pass through the adult forms of its ancestors. Because development (also known as embryology) was thought to quickly pass through (or recapitulate) ancestry (also known as phylogeny), the process was summarized by the statement 'embryology recapitulates phylogeny' and sometimes referred to as **embryological recapitulation**. It was thought that this process lengthened the overall development of an organism, because even when organisms shortened the time they spent in each stage of development, each new species added another stage to development.

embryological recapitulation—developmental stages that look like the stages of that species' evolution

Although evidence of this process was pointed out in other organisms, it was most popular to find proof in human development. Advocates of embryological recapitulation would point to the long development of humans and the fact that various stages in human development *looked* like the ancestors from which humans were thought to be descended. For example, each human being begins development as a single cell, and humans are thought to have evolved (with all other organisms) from a single cell way back in the past. Then, according to embryological recapitulation, a human develops through a series of stages that look like subsequent ancestors: a spherical ball of cells (like spherically symmetric invertebrates), an elongate mass of cells (like worms), a stage with gill-like slits (like fish), a stage with a tail (like monkeys), and so on.

Although popular at one time, this particular theory of evolution has been all but abandoned. One reason is because many organisms do not develop through stages that look anything like their ancestors must have looked (for example, the hundreds of thousands of species

of organisms that undergo indirect development like starfish, frogs, and butterflies). Another reason was that which *looked* similar was not actually the same. A human develops from a *human* cell, not a single cell ancestral to humans, and the baby does not breathe through what looks like gill slits, and what looks like a tail does not have the additional vertebrae of tails, and so on.

The ultimate demise of the theory came in being unable to explain *how* it was that a new developmental stage could actually be added and the former stages sped up. Nonetheless, the similarity between developmental stages and reputed ancestors continues to be used as evidence for evolution. The reason is that these things really do *look* uncannily similar. Creationists would suggest that creation explains why. Consider first the development of a human. The stages of human development were designed in the mind of God as the most efficient means of developing a human from a single cell.

Now consider the theory of human evolution. The steps of human evolution were imagined in the mind of the naturalist as the most straightforward way of evolving a human from a single cell. The mind of the naturalist was created by the mind of God in such a way to think God's thoughts after Him. It is no wonder that similar minds would create compellingly similar processes to connect similar beginning points and similar endpoints. The differences in detail would be expected both because they are actually generated by different minds and because each process works under different conditions. The creationist worldview explains both the differences that suggest it is not explained by evolution *and* the compelling similarity that cause evolutionists to continue to use it as evidence for evolution.

Biological Reproduction: Our Responsibility

Our Responsibility to God

The incredible process of reproduction ought to invoke worship of the God Who created it. The complexity of even the simplest unicellular organism is comparable to the complexity of a city orbiting the earth in space. The cell has to accomplish similar feats of construction and repair, importing and exporting, energy production and dissemination, protection, propulsion, *etc.* And then, to top it off, the cell can reproduce itself… in a matter of minutes!

Imagine a city being able to reproduce itself and all its functions. Even taking years this would be an amazing accomplishment. Yet, in your body, just to maintain the cells in your blood thousands of cells are reproducing every second. The process of cellular reproduction ought to invoke awe towards the Designer Who created it. Such wisdom puts human wisdom to shame. Contemplation of the reproduction of any organism should give us reason to worship He who made it.

On a larger scale, the abundance of organisms ought to remind us of God's abundance. Ecosystems full of organisms ought to remind us of God's completeness. An abundance of organisms working together in community, ought to remind us of the perfection of God. As we see organisms *abounding* and *filling*, we remember that such things have been created to be continuous physical illustrations of God's fullness, completeness, and perfection.

One of the reasons God created biological reproduction was to give us a visible and finite illustration of the invisible and infinite abundance of God. Because He did, study of biological reproduction can improve our personal understanding of God and draw us closer to Him. To help us fulfill our priestly role, God not only created us to respond to illustrations of His nature, He also gave us the Holy Spirit to lead us into truth. Because of this, contemplation of biological reproduction can stimulate us to worship God, glorify God, and bring others into our worship of God.

Obeying God's Command to Multiply and Fill

Cherish Human Reproduction

Humans were commanded to reproduce and multiply and fill the earth (Gen. 1:28)—even after God had to destroy the earth and most of humanity because of human sin (Gen. 9:1). Human reproduction glorifies God in a way similar to the way animals glorify God in their reproduction. If humans are abundant, it is illustration of God's abundance. If man continues to increase in number, it is illustration of God's activity, infinity, fullness, and completeness.

But human population growth does much more than that. Humans are the image of God (Gen. 1:26-28). Since nothing else is mentioned in all of Scripture as being or having the image of God, it is likely that humans are unique in this regard. Humans—and *only* humans—possess the image of God. Thus, humans represent God in a way that no other part of God's creation can.

Just as the image of a pharaoh set up at the outermost boundary of the pharaoh's kingdom indicated that the pharaoh ruled there, or just as the flag of a European explorer planted on the soil of the New World indicated ownership, so also, wherever humans are found they are physical indications that God owns that location and rules there. As humans reproduce and multiply and *fill* the earth, they indicate that God owns and rules the *entire* earth. When humans landed on the

Moon, they indicated in a very special way that God owns and rules the Moon as well.

Since God created a man and a woman and commanded *that union* to reproduce and multiply, humans were to reproduce and multiply in the context of the family. And, since the husband-wife relationship is to illustrate the relationship among the members of the Godhead, if humans multiply and fill the earth, the earth is full of physical illustrations of relationships among the members of the Godhead. Even with a full earth, after the Fall of man, reproduction has a place in replacing those that die. God has provided human reproduction as an exceptional way of glorifying Him.

Not only should we cherish human reproduction, but we should value children as much as Scripture values children (*e.g.* Psa. 127:4-5). Even choosing a child for adoption into our family is a beautiful picture of God choosing us for adoption into His family. The general perspective of our current society is antagonistic to human reproduction and the children that result. Birth control (typically meaning the prevention of children) is thought to be a normal part of marriage, and gynecologists generally assume a couple would rather not have children[a]. Pregnancy is understood by many as a burden[b], and most insurance policies cover pregnancy as if it were a disease. The millions of babies aborted in the last few decades evidence how our culture sees no value in the youngest babies[c]. Finally, the more children that parents bear, the more our culture shuns them. Unfortunately, these perspectives are even common in Christian households. They ought not to be. Children ought to be cherished as a heritage from the Lord.

---

[a] Examples of anti-pregnancy perspectives: Menstrual cycle abnormalities are routinely treated with birth control pills when other methods that do not prevent pregnancy could be used. Complete hysterectomies are performed even when modern surgical techniques exist that could save the uterus and permit future pregnancies.

[b] God designed a woman's body to provide for the needs of a baby in her womb, even if it means taking what is needed from the body of the mother. If the mother does not eat everything needed by both her body and her baby, then her own body sacrifices for the baby. Under these conditions, the baby more or less parasitizes her body. For reasons such as these some believe a baby in the womb to be a parasite, making pregnancy a disease.

[c] This is further evidenced by 'birth control' methods that abort fertilized eggs (*e.g.* Intra-uterine devices (IUDs) that prevent implantation of the fertilized egg), birth enhancement methods that can result in the abortion of fertilized eggs (*e.g.* fertility drugs that can result in more fertilized eggs than will be allowed to develop in the womb; in-vitro fertilization (IVF) that typically fertilizes more eggs than will be implanted and implants more eggs than will be allowed to develop), and on-going research into therapeutic cloning which creates a fertilized cell so as to kill the embryo in order to produce tissue for implantation.

Fill the Earth

An important part of God's command to reproduce and multiply is that it be done until the earth is full. It is the filling that illustrates God's fullness and completeness. It is not just indiscriminate or uncontrolled population growth that is condoned, but an increase in population that leads to a *full* earth. Any particular region is capable of supporting a certain population of organisms.

To properly rule over this region, we should determine what that **carrying capacity** is and make sure that it is not exceeded. However, as long as the population is below the carrying capacity, population growth should be encouraged. When carrying capacity is approached or exceeded, we are to restore that population to proper levels, but do so while acting as shepherd kings and honoring humans as God's image bearers. We have the power and right to kill plants or animals for food. We even have the power and right to bring a population down to the region's carrying capacity, but this must be done in such a way as to serve the organisms over which we have authority. And, if we must kill animals in the process, we should do so mercifully, so that they do not suffer at our hands.

In the case of humans, we do *not* have the authority to kill humans to reduce population levels[a]. Since humans are the image of God, killing a human is murder and murder is wrong. It is not right to kill humans

---

carrying capacity—the number of organisms a particular environment can support over the long term

---

[a] Rather, we would have the authority to relocate humans so as to bring population levels down to the carrying capacity in a particular region.

to improve the health of a population, or even to save the lives of other humans. And, since humans are the image of God from the moment of conception, killing a child in the womb is also murder. Abortion is murder and wrong. In fact, it is wrong to kill a human embryo no matter what stage of development it is in, or where it is located and is as much murder as shooting a human in cold blood. Examples of procedures that are wrong because they cause abortion include:

1. Anything designed to kill an unborn child is abortion. This includes traditionally understood abortion procedures.

2. Anything that prevents the implantation of a fertilized human egg in the uterus, such as an <u>i</u>ntra-<u>u</u>terine <u>d</u>evice (IUD), directly results in the death of the fetus, so it is abortion.

3. Anything that destroys fertilized human eggs or human embryos, such as

    a. an in-vitro fertilization (IVF) clinic discarding fertilized eggs that the parents no longer need; and

    b. human therapeutic cloning, where a human embryo is created for the express purpose of destroying the embryo so as to transplant embryonic cells or tissues or organs into another person to heal that person.

4. Anything that increases the likelihood of subsequent abortions, such as

    a. fertility pills, that increase the likelihood of so many children in the womb that some are aborted ('fetal reduction') so that the remaining child or children can survive; and

    b. typical in vitro fertilization (IVF) techniques that generate more fertilized eggs than will be used and implant more embryos in a womb than will be allowed to develop.

Our Responsibility to the Creation

Enhancing Carrying Capacity

Bringing a population up to its carrying capacity is certainly part of our responsibility to God. But we have a responsibility beyond that as well. As part of our stewardship responsibility we are to enhance those aspects of the creation that bring glory to God. Since bringing a population up to its carrying capacity glorifies God, raising carrying capacities is a way of enhancing the creation. Since more of His character is illustrated, God receives more glory.

In our experience, in a given region under most situations, the total carrying capacity (the sum of the carrying capacities of all the species) is greater when there is a greater diversity of interacting organisms. It increases even more when more species interact constructively in the form of a community. Consequently, we can increase total carrying capacity in a particular region by improving ecosystems in order to enhance the diversity of organisms so that they interact constructively. With the proper understanding of genetics and ecology, it may even be possible to discover new species that God hid in His creation, and add those that interact well with that community, with the goal of further enhancing the system.

Summary of Chapter

A. God physically illustrates His (invisible) fullness, completeness, and perfection through biological reproduction and (immediately after the Flood) intra-baraminic diversification.

B. Only biological organisms can reproduce. Though some organisms *can* develop from multi-cellular fragments, *normally*, every individual organism begins life as a single cell. Consequently, biological reproduction revolves about cellular reproduction (parent cell dividing into daughter cells):

  a. DNA replication followed by (in eukaryotes) karyokinesis (nuclear division):
   i. mitosis ($\rightarrow$ 2 nuclei, each with identical DNA to the original cell) for reproduction of asexual unicellular organisms and growth and repair of multi-cellular organisms. Stages observed under the microscope:
     ◦ prophase: chromosomes & centrioles & spindle fibers appear; nuclear envelope & nucleoli disappear
     ◦ metaphase: chromosomes line up on equator; spindle fibers connect centrioles & centromere
     ◦ anaphase: sister chromatids separate
     ◦ telophase: 2 nuclear envelopes & nucleoli appear; spindle fibers & centrioles & chromosomes disappear
   ii. meiosis ($\rightarrow$ 4 nuclei, each haploid and with a unique combination of DNA) to create gametes (sperm or eggs) for sexual reproduction
     ◦ prophase I: chromosomes & centrioles & spindle fibers appear; nuclear envelope & nucleoli disappear; homologous chromosomes pair up; crossing over

- metaphase I: analogous chromosomes line up on equator; spindle fibers connect centrioles & centromeres
- anaphase I: analogous chromosomes separate
- telophase I: 2 nuclear envelopes appear; spindle fibers & centrioles disappear
- cytokinesis I: cell divides
- prophase II: centrioles appear; spindle fibers appear
- metaphase II: chromosomes line up on equator; spindle fibers connect centrioles & centromeres
- anaphase II: sister chromatids separate
- telophase II: 2 nuclear envelopes appear; spindle fibers & centrioles & chromosomes disappear

b. separating cell components;

c. cytokinesis; and

d. growing daughter cells.

C. Cellular reproduction does not reproduce the non-physical component of life (*e.g.* IVF and identical twins). The non-physical component of an organism's life is most probably created by God when that organism begins developing (*e.g.* fertilization of egg by sperm in sexual organisms—'conception' in humans).

D. >99% of the species on earth went extinct in Noah's Flood. Intra-baraminic diversification restored the species diversity within a couple centuries of the Flood. Evidence that intra-baraminic diversification occurred by revelation of hidden species designs:

a. rapid intra-baraminic diversification (→ probably hidden species designs)

b. many artificial breeds in many baramins (→ hidden species designs);

c. inter-specific hybridization in many baramins (→ recent origin of many species)

d. genetic throwbacks and fully vestigial organs (→ recent descent from other species)

e. speciation (*e.g.* stratomorphic trajectories; geographic proximity of similar species; ring species)

E. Biological reproduction and diversification are substantial challenges to evolutionary theory

a. reproduction is too complicated to form naturalistically; reproduction is so complicated to require an intelligence comparable to that of God

b. natural selection would not choose sexual reproduction over asexual reproduction; God would design sexual reproduction to generate diversity so as to illustrate His diversity and personality

c. inter-specific hybridization, genetic throwbacks, and vestigial organs suggest that species diversified on a biblical time-scale of thousands of years (too rapidly for naturalistic evolution)

F. The elegance and complexity of biological reproduction ought to lead us into worship of God—the only explanation of its origin.

G. God commanded humans and animals to 'reproduce and multiply and fill the earth':

a. Reproduction rates should be adjusted to reach and maintain carrying capacities.

b. To enhance the glory of God, we should reveal latent diversity by breeding.

c. Human reproduction glorifies God

  i. as it is to occur within the family which illustrates God.

  ii. as it requires divine intervention (to create the non-physical).

  iii. as each child is image of God (from conception to death). Abortion is murder:
   - Abortion (anything designed to kill an unborn child—fertilized egg on) is sin.
   - Anything designed to prevent implantation of a fertilized cell (*e.g.* RU 486) is sin.

- Anything designed to destroy fertilized human eggs (*e.g.* therapeutic cloning; traditional IVF) is sin.
- Anything that increases the likelihood of subsequent abortions (*e.g.* fertility pills; traditional IVF) is sin.

Test & Essay Questions

1. How does God physically illustrate His invisible attributes of fullness, completeness and perfection?
2. Why does a chapter on the reproduction of organisms focus most on *cellular* reproduction?
3. Define cytokinesis.
4. Why is nuclear division the most difficult part of cellular reproduction?
5. What is replication / condensation and why does a cell do it?
6. Compare and contrast sister chromatids and homologous chromosomes / mitosis and meiosis / sexual and asexual reproduction.
7. List the stages of the cell cycle in order beginning with the cell in its 'normal' state.
8. What type of nuclear division does the cell employ when cells need to be replaced? / when cells are needed to grow an organism? / when sperm are needed? / when egg cells are needed? / when gametes are needed?
9. What happens in each of the four phases of mitosis? / Match the stage of mitosis with what happens during that stage.
10. During what stage(s) of mitosis and/or meiosis do sister chromatids line up in the middle of the cell? / do homologous chromosomes line up in the middle of the cell? / do sister chromatids separate? / do homologous chromosomes separate? / does crossing over occur? / does the nuclear envelope disappear? / do the nucleoli disappear? / do nuclei reappear? / do nuclear envelopes reappear? / do spindles appear? / do spindles disappear? / do spindle fibers appear? / do spindle fibers disappear? / do spindle fibers connect centrioles and centromeres? / do chromosomes begin condensing? / do chromosomes complete condensation? / do chromosomes begin uncondensing?
11. Arrange the stages of mitosis / meiosis in proper order.
12. What happens during crossing over?
13. What does it mean for an organism to be diploid / haploid?
14. Compare and contrast a 'normal' body cell with a sex / gamete / sperm / egg cell.
15. For what purpose do sex / gamete / sperm / egg cells have half the amount of DNA found in any other cell in an organism's body?
16. In what ways has God created organisms to permit variety?
17. How do organisms reproduce their non-physical components?
18. How does an IVF baby suggest that a human soul can arise without the presence of parents?
19. How do identical twins suggest that a human soul can arise without a separate fertilization event?
20. When does a person's life begin? / When does a human soul come to be?
21. Why is sexual reproduction difficult for a naturalistic worldview?
22. In what way(s) is biological reproduction better explained by creation than evolution / better explained in a Christian worldview rather than a naturalistic worldview?
23. In what way(s) ought biological reproduction invoke worship?
24. Why did God create an abundance of animals and then command them to reproduce and multiply?
25. Define carrying capacity.

26. When should population growth be encouraged / stopped?

27. Why is killing a human wrong? / Why is abortion wrong? / Why is it wrong to kill a human baby in the womb? / Why is killing a human embryo wrong? / Why is killing a fertilized human egg wrong? / Why is the morning-after pill wrong? / Why is an intra-uterine device (IUD) wrong?

28. Why should we try to increase carrying capacity?

29. What can be done to increase carrying capacity?

CHAPTER 15

HISTORY OF LIFE

The Natural History of Living Organisms

"'I am the Alpha and the Omega, the Beginning and the End', says the Lord, Who is and Who was and Who is to come, the Almighty."
Rev. 1:8, NKJV

15.1 | The Communicating God

Because He desires to be known, God reveals Himself to every person who has ever lived—or will live. As indicated in Rom. 1:18-20, and in examples throughout this volume, one way He does this is through the creation that each person observes during his or her lifetime. God maintains the creation (Neh. 9:6, Psa. 36:6; Col. 1:17; Heb. 1:3) so that, among other things, it will at any given time in earth history provide physical illustration of His nature.

But God also reveals Himself *through* history. He orchestrates the flow of events so as to bring glory to Himself and fulfill His purposes. The history of organisms is part of that plan, so we would expect an examination of the flow of life to also teach us about the One Who makes it all possible.

The overall plan for the history of organisms is closely tied to the history of the rulers God placed over them—the relationship between God and man. As can be seen over and over again in the history of Israel and Judah, the obedience of a king results in blessing on that nation's people, and the disobedience of a king results in judgment on the people of that nation. Likewise, human obedience results in blessings on organisms placed in human charge, and human disobedience results in judgment on those organisms.

In creation, when man was blessed, animals were also blessed (Gen. 1:22, 28). Adam's disobedience brought a curse on all mankind and a curse on all organisms (Gen. 3:14-19; Rom. 8:21-22). When Cain is cursed, the organisms he encountered were cursed as well (Gen. 4:12). When all but a remnant of humanity was destroyed in the Flood, the same Flood destroyed all but a remnant of organisms (Gen. 7:21-23).

At the end of the Flood, God covenanted with both man *and* animals never to destroy the earth again in a Flood (Gen. 9:9-16). Organisms will even be released from the curse when the curse is lifted from man (Rom. 8:18-23). The glorious creation/fall/redemption story of God's dealings with man is also the same creation/fall/redemption story of biology. Consequently, as we learn about the history of organisms, we cannot help but learn not only about God, but also about God's relationship to man.

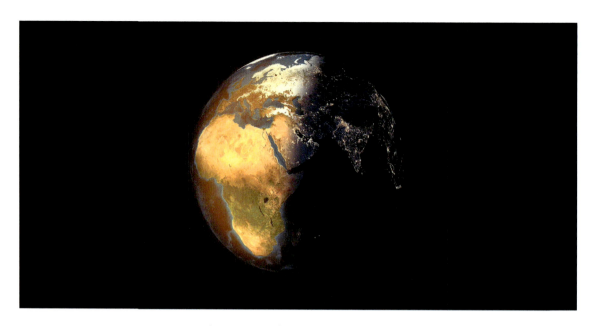

The History of Organisms

Biblical Time

If we only had the New Testament and our experience since its authorship, we might think that time on earth did not matter much to God. We are not given the time between the birth of Jesus and His public ministry. We are not given a clear timeline on the birth of the Church, the life of the apostles, or the writing of the New Testament books. And we certainly would not anticipate nearly two thousand years of time that have elapsed between the first and second coming of Christ. We are even reminded on New Testament pages of that well-known phrase that "one day is with the Lord as a thousand years, and a thousand years as one day" (II Pet. 3:8).

Biblical Chronology

The Old Testament, in contrast, gives a wholly different impression of time. Genesis opens with a notation about time ('in the beginning') and immediately describes 'days' of creation. The sun, moon, and stars are even created for the purpose of marking out time (Gen. 1:14). In Chapters 5 and 11, Genesis continues with genealogies unique in the Ancient Near Eastern world—and even in the rest of the Bible—including the age of the fathers at the birth of their sons.

The only known purpose of such information is the construction of precise chronologies. There are precise time statements, such as 'In the six hundredth year of Noah's life, in the second month, the seventeenth day of the month' (Gen. 7:11) and 'in the six hundred and first year, in the first month, the first day of the month' (Gen. 8:13). And there are summary statements about time, such as 'at the end of four hundred and thirty years' (Exo. 12:41) and 'in the four hundred and eightieth year' (I Kings 6:1).

Not only are there many statements about time, but it is fairly easy to string them together to develop a fairly precise chronology from the beginning of the creation to the time of Christ. According to this chronology, the creation occurred approximately 4000 B.C., the Flood occurred about 1656 years later, Abraham lived about 2000 B.C., and Solomon lived about 1000 B.C.

Biblical Chronology Accepted

Scripture makes claims about time. It tells us how much time elapsed in the Creation Week, how long each of the patriarchs lived, how old the patriarchs were when their sons were born, and so on. In contrast, the physical world—including the biological world—makes no statement at all about its age. If we wish to infer the history of the physical world using only evidence from the physical world, then the actual physical evidence we have is what we see in the *present*. What happened in the past and when it happened must be *inferred*. To do this, we must first infer what course history actually took, then we must infer how fast that history elapsed. If we are wrong about what actually happened, or if we are wrong about how fast it happened, we will likely arrive at incorrect conclusions about that history.

Inferring history by using the physical world alone is similar to a prosecuting attorney constructing a case against a defendant from circumstantial evidence. From the physical evidence observed in the present (in the courtroom) the prosecutor infers what happened in the past. Although this is a perfectly respectable way of determining history, such an account of history is never considered as reliable as the testimony of a good eyewitness to the event—even if the two involve radically different accounts of what happened. In the case of the history of the physical world, it is perfectly respectable to infer a history based upon data from the physical world alone, but it can never be as reliable as the testimony of a reliable eyewitness to the history of the world. And Scripture gives us that eyewitness account. God was not only present for the history of the world—and at one point the *only* one present—but He is also the perfect eyewitness. He saw everything that happened, His perspective was perfect, His understanding was perfect, His ability to communicate with us is perfect, and He cannot lie.

If God saw fit to give an account of the history of the world it should have priority over *any* inference of history made by humans, no matter how reasonable that inference. It turns out that God did see fit to give an account of the history of the world—at least from the beginning of the world to the time of Christ. The first part of that history is given in Genesis and it is presented to us in the literary genre of historical narrative—the very genre that would be expected from an eyewitness account.

Half of the biblical chronology—the two millennia or so from the time of Abraham to the time of Christ—is not controversial. Most everyone agrees on the timing of the events on this part of the chronology, including Abraham living about 2000 B.C. The earlier portion of the chronology however—the two millennia or so from the beginning of the creation to the birth of Abraham—relates a radically different history of the creation than is inferred from using the physical creation alone. However, it is the precision and clarity of this earlier chronology, combined with the fact that it is an inerrant eyewitness account of a perfect eyewitness, that leads us in this text to follow the biblical chronology rather than the history deduced from the physical world alone.

Earth Altering Events

Scripture not only provides a reliable eyewitness account of the origin of things, it also suggests a reason why history based upon physical world data alone would lead to radically different scenarios of earth history. In II Peter 3:3-4 we are told that there will be people in the 'last days' who will challenge God's promise of returning, using the argument that 'all things continue as they always have from the beginning of the creation'. These people have apparently decided that the way things are happening in the present world is the way things have always happened, and the way they will continue to happen. From this they incorrectly deduce that Christ's return is impossible.

The passage continues (verses 5-7) by listing three events in the overall history of the world that these people are ignoring—the creation, the Flood, and the judgment to come. The passage implies that these events involve processes that are not occurring in the present. Scripture confirms that this is true for the creation event, for after finishing the creation, God *ended* His creating (Gen. 2:1-3), and this rest from creation continues to this day (*e.g.* Heb. 4). During the Creation Week events occurred that are not occurring in the present, including acts of **special creation** (the direct action of God to bring something into being). If a person assumes that things came into being by means of processes operating in the present, they will deduce an incorrect history.

> **special creation**—direct action of God to bring something into being

Scripture also suggests that the same is true of the Flood account. Following the Flood, God promised that He would never again curse the ground for man's sake (Gen. 8:21), never again disrupt the day/night and seasonal cycles (Gen. 8:22), and never again bring a global flood on the planet (Gen. 8:21; 9:9-17). As in the case of the Creation Week, during the Great Flood, processes occurred that are not occurring in the present. If a person assumes that modern processes were the only ones operating in the past, they will deduce an incorrect history. When a single command can immediately cause the earth to appear (Psa. 33:6-9), or a storm to still (Mark 4:39), a child to heal (Mat. 8:13), or a body to come to life (John 11:43-44), it is clear that God is not restricted to the usual processes of the present world to accomplish His tasks.

Not only is He capable of utilizing unusual process, but He has actually employed such processes in the past. If any such process impacts the entire earth or most of it, it can throw off historical inferences. Scripture suggests that such earth altering events will occur in the judgment to come and occurred in the past at the times of the Creation, the Fall, and the Flood. If only data from the physical world is used to infer history before the time of Abraham, that history is likely to be very wrong, and to be more and more wrong the farther back that history is inferred. This explains why the biblical chronology before the time of Abraham differs so radically from the history of life inferred from the data of the physical world alone.

Radiometric Dating

One of the most significant examples of this principle is the case of radiometric dating. Using radiometric dating, a chronology of earth history has been developed that corresponds to biblical history back to about 1000 B.C., but diverges from it more and more the farther back in time we go. By the early days of Abraham, the two are off by centuries; by the end of the Flood, the radiometric dates are thousands of times greater than biblical dates; by the time of creation the radiometric dates are millions of times greater.

The only data used in radiometric dating is data from the physical world—namely the number of extremely rare radioactive isotopes in the world. Isotopes are atoms with the same number of protons but a different number of neutrons. For example, Carbon 12 and Carbon 14 are isotopes of the element Carbon. Carbon 12 has 6 protons and 6 neutrons and Carbon 14 has 6 protons and 8 neutrons. Carbon 14 is a radioactive isotope of Carbon 12 because it is unstable and breaks down (decays) over time at a specific rate that can be measured. **Radiometric dates** are calculated assuming radiometric decay rates have operated in the past in the same way as they operate now. However, there is evidence that radiometric decay rates have been different in the past. Different dating methods applied to the same object generate different dates in a consistent manner.

It is most likely that earth altering events in history, such as at the Creation, the Fall, and the Flood, changed some of the rates that researchers assume have never changed. Although the dates computed from any one radiometric dating method probably put the events in the correct order, we will use the thousands of years of biblical chronology rather than the millions and billions of years of radiometric chronology.

radiometric date—age calculated from amount and rate of decay of radioactive elements

15.2 | Creation Week

The entire physical creation came to be in the first week of time by a series of special creation events. As He created, God kept humans in mind. He created the physical world and man in such a way that humans could learn about Him (the anthropic principle characteristics *not* essential for life). He filled the creation with physical illustrations of His invisible attributes so that man could understand His nature. And He created a host of spectra of perfections so that we could begin to understand His infinity.

God created the earth in wisdom (Prov. 3:19; Jer. 10:12; 51:15), in such a way that it would be inhabited (Isa. 45:18). Thus, even when God was fashioning the non-biological world, He was doing so in such a way that life could thrive—both human and non-human life. Beginning about six thousand years ago, time as we know it began when God spoke the earth into existence (Gen. 1:1; Exo. 20:11; Psa. 33:8-9; II Pet. 3:5). It was probably at this time that the basic structure of the universe was fashioned (*e.g.* its three-dimensionality), and it was fashioned with anthropic principle characteristics that would permit life to thrive.

On Day Two of the Creation Week was the special creation of the atmosphere (Gen. 1:6-8). By the end of that day the atmosphere and the water cycle was in perfect running order, fully capable of protecting and supporting the organisms that would soon appear.

On Day Three of the Creation Week the special creation of God brought the land into existence (Gen. 1:9-10, 13). Not only was this land ready to support life, but God probably created a variety of different continents so that He could later create a greater variety of life. It is also likely that somewhere in the course of the first three days God specially created the structure and chemistry of the oceans—again, for the purpose of supporting the life that was about to appear within them.

The Origin of the Biomatrix

Archaea, bacteria, protozoa, algae, and fungi are not specifically mentioned in Genesis (or anywhere else in the biblical account). Yet, they must have been created during the Creation Week, because the biomatrix is necessary for the survival of other organisms and special creation is the only way these organisms could have arisen (Gen. 2:1-2). In fact, given its holistic flavor, Scripture probably includes the organisms of the biomatrix as part of those things with which they are most intimately associated. For example, it is most likely that the biomatrix organisms now found in the earth's rocks and oceans were created on Day One as part of the earth and its initial covering of water. Likewise, the biomatrix organisms now found in the soils and waters of the continents were likely created on Day Three as part of the continents. With the creation of these biomatrix organisms would have been the beginning of life-supporting biogeochemical cycles.

In the same manner, biomatrix organisms in plants were created on Day Three with plants. Biomatrix organisms in animals and humans were created on Day Five and Day Six with the organisms in which they lived. Since biomatrix organisms are not mentioned in Scripture, we're not told if they were created 'according to their kind', but since their classification structure seems similar to that of other organisms, they were probably created in baramins ("created kinds"). Since God ended His creation at the end of the Creation Week (Gen. 2:1-2), all the present baramins of the biomatrix were created in the Creation Week, as well as additional baramins that have since gone extinct. The ability we see in modern biomatrix organisms to change suggests that God also placed hidden information in them. Some of this was probably in anticipation of the many changes God knew the earth and its organisms would experience in the course of history, and some of it was included so that humans could glorify God by revealing its potential.

The Precambrian Fossil Record

Of the oldest (**Precambrian**) sediments on the earth thousands of feet contain fossils of bacteria, but no fossils of plants or animals. The fossils suggest the rocks were probably not directly created. The lack of plant and animal fossils suggests that the sediments were formed before the creation of plants on Day Three. These sediments might be explained if the creation of the continents included raising them out of the oceans.

Precambrian—oldest sediments on the earth; fossils in all but the uppermost layers are only bacteria

On Day Three, God commanded the water covering the earth 'to be gathered together into one place', so allowing the dry land to appear (Gen. 1:9-10). We are not told in this passage exactly how the continents were formed, but water running off the rising continents, would have formed sediments along the edge of continents and would have buried some of the biomatrix organisms that had already been created. This **Day Three regression** (of water) might have generated at least a portion of the lower Precambrian fossil record. Since it is a part of the Creation Week, the Day Three regression may well have involved some type of special creation. After all, it does not seem possible for modern processes to raise continents and deposit that much sediment in a single day.

Day Three regression—theory that water running off rising continents on Day Three of the Creation Week formed most of the Precambrian rocks and fossils

The Origin of Plants and Animals

Aside from the plants and animals of the Garden (Gen. 2:8-9, 19), all the land plant baramins were specially created on Day Three (Gen. 1:11-13), all the animal baramins of the air and water were specially created on Day Five (Gen. 1:20-23), and all the animal baramins of the land were specially created on the first part of Day Six (Gen. 1:24-25, 31). In each case all the baramins of the present world were created plus the baramins that have gone extinct since the creation (*e.g.* lycopod trees, trilobites, and dinosaurs), so life was created with high disparity from the beginning. Organismal forms were also chosen by God to display a nested hierarchy of similarity.

Organisms were also created with great diversity and the potential for more diversity. Although creatures were 'abundant' from the beginning (*e.g.* Gen. 1:21), air and land animals were commanded to 'fill the earth' (Gen. 1:22), so the earth was not created at its carrying capacity. There was an abundance of resources and room for growth, probably both in numbers and diversity within baramins (intra-baraminic diversity). God also placed a considerable amount of hidden information in each organism—some so the baramin could fill the earth, some so the baramin could survive the curse, some so that the baramin could restore its diversity after the Flood, and some so that humans could reveal that diversity to the glory of God.

Finally, the organisms of the earth were created without any biological evil (no suffering, nephesh death, **carnivory**, or disease) in mutual symbiotic relationships (see discussion of the Edenian epoch), and complex integrated communities. Considering the extinction of baramins that have occurred since then and considering the communities we have been able to infer from the fossil record (see discussion of the Antediluvian World), a greater diversity of communities was created than is even seen in the present.

carnivory—the eating of animals by other animals (aka predation), an evil-minimizing effect of the curse

The Origin of Man and the Garden

In the description of the new heaven and new earth given at the end of Revelation, God lives among humans, but chooses to manifest himself on a throne at a particular location on the earth. Something similar was apparently also the case in the first heaven and earth. God designed the earth so that He would walk with humans in a specially designed garden at a particular location. It may have been over that location that the Holy Spirit hovered on Day One of creation (Gen. 1:2), and it may have been with respect to that location that the days and nights of the Creation Week were defined.

It also seems that as God created plants on Day Three, and animals on Day Five and Day Six, that He left an area in Eden without plants or animals (Gen. 2:5). On Day Six, after completing the remainder of the creation, God turned His focus to the Garden where He planned to walk and talk with man. In a string of special creation events God fills the garden with plants and animals and creates the rulers He sets over His entire creation. In sequence God (a) watered the garden area (Gen. 2:5-6); (b) formed Adam's body and breathed life into it (Gen. 2:7)[a]; (c) planted the Garden of Eden with plants (Gen. 2:8-9); (d) placed Adam into the garden and commissioned him (Gen. 2:15-17); (e) created garden animals (Gen. 2:19) and had Adam name them (Gen. 2:20); and (f) created Eve from the side of Adam and presented her to him (Gen. 2:21-22).

The Sabbath

Once the sixth night ended (Gen. 1:31), God ended His creation work (Gen. 2:1-2). God's Sabbath, which was the termination of His work of creation (Gen. 2:3), continues to this day (*e.g.* Heb. 4). God

---

a Note that Genesis 2:7 reads that after creating Adam, 'God breathed into his nostrils the breath of life and man became a *nephesh hayim*'. In this verse *nephesh hayim* is translated 'living soul'. In the previous chapter the same phrase (in Gen. 1:20-21 and 24-25) is translated 'living creature'. Therefore Genesis 2:7 states that man became a living creature; a living creature did not become a man. Thus, any idea that man is somehow derived from animals is explicitly denied by this passage.

has apparently chosen to never again repeat the special creations of the Creation Week. Consequently, modern processes had nothing to do with the origin of the universe, the earth, or the earth's organisms.

Since the naturalistic worldview cannot appeal to special creation events, it is no wonder that naturalism comes to radically different conclusions about the origin of things. For example, since naturalistic processes typically operate many, many times more slowly than those where God is involved, it is not surprising that naturalistic explanations for the origin of organisms involve chronologies thousands to millions of times longer than the biblical chronology.

The Frontiers of Research

Of all the time periods of history, the Creation Week is the most difficult to understand. Because special creation events do not occur in the present we have no means of directly studying them. It may turn out that the human mind cannot begin to understand them. However, there are things in our world today (such as light from stars, the sun, and the oldest rocks of the earth) that came to be during the Creation Week and may not have changed much since the beginning. Studying them may give us insight into the events of this incredible week.

15.3 | Edenian Epoch

At their origin, organisms were subject to God's acts of special creation. Once created, however, organisms cannot thrive without God's direct intervention. Every organism owes its origin to God, and every organism owes its continued existence to God (Neh. 9:6, Psa. 36:6; Col. 1:17; Heb. 1:3). God did not altogether cease from interacting with an organism as soon as it was created, He merely ceased creating and began interacting with it in a different fashion—as its Sustainer. Once Day Seven began, God was now relating to His *entire* creation as Sustainer, rather than Creator. Though the Creation Week was short in duration, the end of it was a substantial event in earth history and marks the beginning of a new era.

Six times in the course of the Creation Week, God labeled His creation work 'good': light on Day One (Gen. 1:4), continents and oceans during the earlier part of Day Three (Gen. 1:10), plants during the later part of Day Three (Gen. 1:12), the sun, moon and stars on Day Four (Gen. 1:18), animals of the sea and air on Day Five (Gen. 1:21), and animals of the land on Day Six (Gen. 1:25).

At the end of the week, He evaluated everything He had done and for His seventh evaluation, pronounced everything He had fashioned 'very good' (Gen. 1:31). At that time, the biological world was exactly as God intended. Organisms were living in mutualism, and there was no biological evil. Animals and humans did not die, animals and humans ate plants and not animals, there was no disease, there was no degenerative aging, and there was no suffering. The biological world was in a state of *shalom* (peace), just as God intended it and just as it will be after the curse is finally eradicated. Because this biological optimum persisted for as long as Adam and Eve lived in the Garden of Eden, the period of shalom is called the **Edenian epoch**.

Scripture does not indicate how long the Edenian epoch lasted. At best, we can place very rough maximum and minimum values on the length of the Edenian epoch. For example, when Adam and Eve 'heard the voice of the Lord God walking in the garden in the cool of the day' (Gen. 3:8), the text intimates that they had heard the voice of God before, and that it had become a daily ritual. This suggests that the Edenian epoch lasted for a least a few days. A rough maximum can be calculated from the birth of Adam and Eve's children. Scripture indicates that Seth was born to Adam when Adam was 130 years old (Gen. 5:3), and Eve considered Seth a replacement for her son Abel (Gen. 4:25), who had been killed by her firstborn son Cain (Gen. 4:8). Cain, in turn was not conceived until after Adam and Eve were kicked out of the Garden of Eden (Gen. 3:23-4:1). So, the Edenian epoch

Edenian epoch—period of earth history between the Creation Week and the curse (<100 years)

lasted for 130 years minus at least the age of Cain when he killed his brother Abel. The length of the Edenian epoch was something between a few days and a century or so.

Compared to the entire course of earth history, the Edenian epoch did not last long. However, since it was the creation as God desired it to be, the Edenian epoch is important in Earth history. It provides, for example, an ethical norm to guide us as we determine how best to care for and restore the creation. It is also the state to which the entire creation yearns to return (Rom. 8:19-21).

The Frontiers of Research

Even though the Edenian epoch ended only thousands of years ago, it is a difficult time period for us to understand, for it functioned in a very different manner than does our modern world. For example, it is nearly impossible to imagine a world without any biological evil (no degenerative aging, no disease, no nephesh death, no natural selection, no carnivory). Furthermore, several world-changing events occurred between then and now, including a global flood that would have destroyed most of the evidence from this time period that might still have been present at the time the Flood began.

Because so little is known about the Edenian epoch, we have much to learn. Exciting research topics include such things as how communities function without animal death, how accidental animal death is avoided, how population levels are controlled without death, and what baramins, species, and communities existed across the planet during this time period.

Antediluvian Epoch

The Origin of Biological Evil

The Edenian epoch ended quickly and tragically. When Adam ate of the tree of the knowledge of good and evil (Gen. 3:6), he disobeyed God's explicit command (Gen. 2:17). Satan must have disobeyed God before this, for he was condemned while God was cursing the serpent (Gen. 3:15), but the biological world was not altered at the fall of Satan. Nor was Eve's act of eating of the tree responsible for the world's alteration (I Tim. 2:14). It was the disobedience of the appointed king of creation, Adam, which led to a catastrophic change in the history of the entire creation (*e.g.* Rom. 5:12-19; Rom. 8:19-23).

But it was not Adam's sin itself that changed God's creation; it was God's curse that altered everything. It was another change in the way God interacted with His own creation. If man had never fallen, the universe would have persisted forever, just as will be the case in the new heavens and the new earth to come. But it would have persisted forever not because of some innate ability that the universe has to exist forever, but because God would have sustained it forever. The present universe waxes old (Heb. 1:10-12) because God is sustaining it in a different fashion than He sustained it before the Fall.

Before man's sin, animals and humans did not die—and would not have died—not because we or animals have some biological potential in us to live forever, but because God sustained us in a different fashion before the disobedience of Adam. Before the Fall there was no disease, no degenerative aging, and no suffering because God sustained the creation in such a way that these things did not occur, just as He will sustain things in heaven to come (Rev. 21:4). Therefore, the Edenian epoch ended and a new era in earth history began, when God changed His relationship to the creation.

Before the curse, biological repair systems were apparently capable of complete biological repair. Ever since the curse, this has not been the case, and a host of biological evils have entered the creation as a consequence. This would include DNA information loss every generation, a progressive weakening of lineages, and degenerative aging experienced by individual organisms. The same degeneration

has introduced disease into the world and transformed mutualistic organisms into parasites and pathogens. The number of diseases and pathogens has been rising ever since, as has the animal suffering that comes as a result.

In spite of all of these negative effects, God did not wish for the curse to destroy His creation nor to destroy its ability to illustrate His nature. Man's fall also did not come as a surprise to Him, for during the Creation Week He had already designed information in organisms that would allow His creation to persist in spite of the curse.

Thus, as the negative effects of the curse increased, death of animals and humans was introduced to relieve the suffering. Biological overproduction was introduced in all organisms to replace the dead. The overproduction, in turn, led to competition and natural selection, and natural selection tends to take out weaker organisms, thus relieving suffering and minimizing the spread of disease. The introduction of predation further relieved suffering and minimized the impact of disease. Finally, plants and prey animals were provided with defenses to preserve them in spite of overgrazing and predation. Although the negative effects of the curse have been prevented from destroying creation's ability to illustrate God's nature and bring glory to God, they have been an important part of the biological world ever since the Fall of man.

The World That Then Was

The curse marked the beginning of a new era in the history of organisms, one which has much more in common with the present era than either the Creation Week or the Edenian epoch. However, the first sixteen and half centuries or so following the curse were different enough from the present to be considered as a distinct epoch in biological history. In fact, Peter refers to this period in almost an otherworldly sense, as the 'world that then was' (II Pet. 3:6). Since many of the changes to the modern era occurred at the Flood, this period of the history of life is referred to as the antediluvian epoch [*ante* meaning 'on the other side of' + *diluvium* meaning 'flood'].

One type of difference in the antediluvian epoch was that baramins and communities destroyed in the Flood never developed again. The diversity and disparity of organisms before the Flood must have been greater than is observed in the present. In terms of communities, research has so far revealed only a few of the many communities that were never seen again after the Flood. It is even difficult to reconstruct these communities from fossils.

The early stages of the Flood were so destructive that very few communities were preserved intact. Many communities have to be pieced together from widely scattered evidences. Dinosaurs and gymnosperms (plants that bear seeds on naked stalks or cone-like structures) probably dominated communities on some of the continents. Humans, mammals, and angiosperms (flowering plants) probably dominated communities on other continents. Offshore of the continents, in ocean hot springs, another antediluvian community involved bacteria that constructed forests of mushroom-shaped stromatolites, which were built of alternating layers of sediment and organic compounds. One of the most spectacular communities proposed for the antediluvian epoch was a continent-sized floating forest. Plants from this forest buried during the Flood formed the vast Carboniferous coal deposits of the northern hemisphere.

The antediluvian world also seems to have differed in its climate. Suggestions from the fossil record suggest that the world before the Flood was warmer. Tropical, subtropical, and temperate zones were probably wider than they are at present, and the poles were probably no colder than cold temperate. Ice sheets and inhospitably cold polar regions probably did not exist before the Flood.

Human Diet and Longevity

At least two aspects of the human experience were different in the antediluvian epoch. First, humans were not permitted to eat meat until after the Flood (compare Gen. 3:18-19 and Gen. 9:3). Second, if the men listed in Gen. 5 are characteristic of all humans during this time period, the average antediluvian human lived more than nine centuries.

Because humans have long been considerably interested in living longer, extensive research has been invested in finding anything that might increase human lifespans. The result of all this research is that nothing in the physical environment is known to significantly change human lifespans. Even the increase in human longevity that has come about as a result of modern medicine is not so much due to a change in lifespan as it is permitting humans to live a greater percentage of their lifespan. The long lifespans of the antediluvian epoch are probably not at all due to a different diet or different environmental conditions in antediluvian times. Rather, the longer lifespan is due to different internal programming or a change in God's sustenance. It is also not known whether any organism other than humans lived longer during the antediluvian epoch.

The Spread of Biological Evil

Based on fossils in Flood sediments, by the time of the Flood biological evil had become widespread and common in the animal world. Predatory species were as common in each environment as they

are today. Fossils also indicate that genetic diseases, pathogen-caused diseases, and parasites had spread throughout the entire biological world. Towards the end of the antediluvian period, God determined to destroy all humans and all animals (Gen. 6:7). He desired to destroy humans because of their wickedness (Gen. 6:5-6). He desired to destroy animals because 'all flesh' was corrupted and 'filled with violence' (Gen. 6:11-13). This suggests that at least by the end of the antediluvian period, there was more violence in the animal world than is found in the present world.

After the Flood, apparently so as to prevent the world from returning to this level of violence, God places the fear of humans into animals and orders that any animal that kills a human must be slain (Gen. 9:2, 5). This suggests that that violence at the end of the antediluvian world included animals across the planet killing humans with impunity.

Cursing of the Soil

Following the Flood, God established the principle of capital punishment—that whatever killed a human, whether human or animal, should be put to death (Gen. 9:5-6). Since it was first established *after* the Flood, capital punishment was not in effect during the antediluvian epoch. The only punishment for murder recorded before the Flood was Cain's, at the murder of his brother (Gen. 4:11-15). In that case, the ground was not to produce for Cain what it would for anyone else, and Cain would be a wanderer as a result. Since a general curse had already been placed on the ground in response to Adam's sin (Gen. 3:17-19), Cain's punishment seems to have been an *additional* or second curse placed on whatever soil Cain would try to farm.

Centuries later, Noah's father chose his son's name because Noah would somehow bring relief from the work that humans had to endure as a result of the ground that God cursed (Gen. 5:29). Six centuries later, in response to Noah's sacrifice to God after the Flood, God claimed that He would 'not again curse the ground any more for man's sake' (Gen. 8:21). A couple things suggest that this latter reference is not to the Adamic curse. When God cursed the ground in response to Adam's sin, man was to experience difficulty in tilling it, thorns and thistles were to grow out of it, and in death humans were to return to it (Gen. 3:17-19). All of these still occur, so they did not cease after the Flood.

Furthermore, Scripture not only indicates that the creation is still subject to 'the bondage of corruption' brought upon the creation by the sin of humans (Rom. 8:18-22), but that the curse will not be suspended until the establishment of the new heaven and the new earth (Rev. 22:3). Since Adam's curse fell upon all humans in response to the sin of the representative head of all of humanity, it makes sense that the Adamic curse should apply until humanity is delivered from sin. What this suggests in turn, is that Cain's punishment was the divine punishment for all murders in the antediluvian world. Since the earth came to be 'filled with violence' (Gen. 6:11) and man's heart was evil (Gen. 8:21), it is likely that the toil to which Noah's father was referring was due to the terrible condition of the soil after the collective effect of secondary curses from countless human murders. This in turn suggests that in the course of the antediluvian epoch the earth's soil gradually deteriorated, increasingly generating an ecological crisis. Saving organisms from this crisis may have been a secondary reason for God destroying the organisms of the planet. In fact, restoring depleted soil may have been a primary reason for God choosing to destroy the organisms of the planet with a flood.

The Frontiers of Research

The antediluvian world is much more accessible to us than the previous two periods of earth history. However, the incredible cataclysm of the Flood did wreak considerable havoc on the evidence remaining from this time period, so reconstructing the antediluvian world is still challenging and we have much more to learn.

Exciting research questions for the future include such things as how much diversification occurred between the creation and the Flood, what communities existed in this period, how many humans existed at the time of the Flood, what technology had they developed, and what was the nature of the violence of 'all flesh' that led up to the Flood.

15.4 | Arphaxadian Epoch

The Flood

The Flood in the days of Noah was unlike any other event in earth history, before or after. The biblical account suggests that it lasted for a year and ten days (compare Gen. 7:11 and Gen. 8:13-14) and that it was violent, breaking up 'all the fountains of the great deep' (Gen. 7:11), and destroying all land organisms on the entire planet except those on the ark (Gen. 6:17; 7:21-23). The evidence left over from it, however, suggests it was even more violent than might be deduced from Scripture alone.

Packages of sediment hundreds of feet thick are sitting hundreds of feet above present sea level, draped across entire continents, sourced thousands of miles distant, laid down by east-to-west flowing currents, and containing vast numbers of bodies of mostly marine organisms. Such sediments seem to be explainable only by an ocean with sea level thousands of feet higher than present flowing rapidly across all the continents of the world. Noah's Flood is the only reasonable explanation for these things. If so, the same rocks indicate what else happened in the Flood.

Evidence exists that continents on crustal plates moved thousands of miles across the earth's surface, crashing together to form mountain chains, uniting to form a supercontinent (Pangaea), and then separating again to get to their present locations—all under the waters of a global Flood. This moving ocean scoured tens of thousands of feet off the tops of mountains and thousands of feet off the tops of continents and redeposited that sediment thousands of miles away. Supervolcanoes, superquakes, and meteorite impacts rocked the planet with unimaginable destructive force. Seventy percent of the crust of the earth sank deep in the earth's interior and hot magma in its place

contacted global ocean waters to generate a line of tens of thousands of miles of geysers that spewed water miles into the atmosphere and dramatically heated the oceans of the world.

As the Flood advanced from the ocean onto the land, the organisms of the antediluvian world were swept up and buried to produce the Paleozoic and Mesozoic fossil sequence. In the first moments of the Flood, chunks of the earth's crust that dropped quickly enough and deep enough to be beneath the scouring of the Flood, preserved the only known fragments of the antediluvian surface. In the uppermost Precambrian sediments of these blocks we find the only known direct evidences of what the antediluvian world was like in the centuries or years preceding the Flood. In Cambrian rocks—the lowest rocks of the Paleozoic, we find marine organisms from the edges of antediluvian continents. Atop these rocks, organisms swept off the floors of the shallow continental oceans are preserved in the remainder of the Paleozoic rocks.

As this was occurring *under* water, the forest floating atop the ocean was also being destroyed, ripped apart from the outside in, and its components buried with the sea creatures below. The core of this forest was buried in upper Paleozoic sediments to form the coals of the Carboniferous. It seems as the Flood waters rose, sands on the shorelines of the world were picked up and redeposited to cap the Paleozoic rocks in what is known as the sands of the Permo-Triassic. In these sediments and above, the rising Flood waters picked up the land animals of the world (including its dinosaurs) and buried them in the famous Mesozoic rocks of the Triassic, Jurassic, and Cretaceous.

A New World

The transitions between the Creation Week and the Edenian epoch and then between the Edenian and antediluvian epochs primarily involved a substantial change in how God related to His creation. The different rules instituted after the Flood suggest that God relates differently to the current world than He related to the antediluvian world.

However, the primary difference between the antediluvian world and the world after the Flood, was a physical one. When the ark's inhabitants stepped off the ark, the world was radically different than it was when they entered the ark. The Flood completely remodeled

the antediluvian world. Land masses were moved around, thousands of feet of sediment were apparently shaved off the top of continents and redeposited, mountain ranges existed where none did before, the ocean appears to have been 20 degrees Centigrade (68 degrees Fahrenheit) warmer, and the earth's climate was warmer and wetter. All humans and land animals—except those on the ark—were destroyed. Most of the land plants and sea creatures were probably destroyed as well. When the ark organisms disembarked, at most only a few thousand land animals representing something like a thousand species existed on a single mountain. Land animals were nowhere else. The world at the end of the Flood was truly a new world.

A Command to Recover

Only an extremely small percentage of the organisms alive at the beginning of the Flood survived the Flood. Although the ark preserved representatives of every baramin of land animal, the number of species descended from them in the present suggests something like one tenth of one percent of the number of species before the Flood survived the Flood to repopulate it. Since God chose the survivors of the Flood (*e.g.* Gen. 6:20), it is likely that He selected them not just to preserve baramins, but also to illustrate His character.

However, such a low diversity could not possibly illustrate His character as effectively as a world full of organisms. It was therefore imperative that the earth be quickly re-filled and re-populated. The full diversity of life was intended to illustrate the invisible nature of God so that *every* person would be without excuse (Rom. 1:18-20). Thus, it was also necessary that the biological recovery from the Flood occur well within a single human lifetime. The first human listed after the Flood was Arphaxad, Shem's son. He was born two years after the Flood and lived for 438 years (Gen. 11:10-13).

Biological recovery from the Flood must have occurred, then, in less than four centuries. This explains the commandment given to Noah while he was still on the ark that all organisms on the ark were to leave the ark 'that they may breed abundantly on the earth, and be fruitful, and multiply upon the earth' (Gen. 8:16-17). That this recovery was actually accomplished in that amount of time is confirmed by the existence of late-arriving species of lions and camels only a few centuries after the Flood (Job 1:3, 17; 4:10-11, 38:39-40; Gen. 12:16; 24:10-22). Since this recovery occurred during the lifetime of Arphaxad, the period of Flood recovery is referred to as the **Arphaxadian epoch**.

Geologic Recovery

It is perhaps not intuitively obvious that the rocks of the earth should have to recover from the Flood, but the Flood did not just displace organisms, it moved lots of sediment and rock. In some places, more rocks were piled up than had been there before the Flood, and gravity would force these areas to sink into the earth. In other places rock was taken away and gravity would force these areas to rise. Some

Arphaxadian epoch—first few centuries after the Flood, when the earth began recovering from the Flood

rocks were buried so deeply that the earth's heat would heat them up and melt them. But the Flood often moved things around more quickly than gravity could raise or lower them or more quickly than the earth's interior could heat them.

In fact, although a majority of the recovery was probably completed in a few centuries, the earth has not yet fully recovered and it may take many more thousands of years to completely recover. This recovery explains why we experience volcanoes and earthquakes today—especially earthquakes in places that should have no earthquakes if the mountains were actually formed millions of years ago (*e.g.* the three to four earthquakes a day that occur in the area of Cleveland, Tennessee in mountains that are radiometrically dated at hundreds of millions of years old).

Immediately after the Flood, when geologic recovery was most rapid, the earthquakes and volcanoes would have been quite spectacular. This explains why older and older volcanoes experienced larger and larger eruptions. It also explains how the volcanic dust came to preserve so many fossils during the Arphaxadian epoch (in what are known as Tertiary sediments).

Climatic Recovery

The Flood also changed the earth's climate. Fossils indicate that the world's oceans heated up about 20°C (68°F) to an average temperature of 25°C (77°F) at the end of the Flood. Creationists hypothesize this heating occurred due to molten rocks from the earth's interior that erupted during the Flood. Warmer oceans would not only have made the earth warmer after the Flood, but it would have made the earth wetter as well. Since the continents cool rapidly at night, warm wet air from over the oceans would move across the continents and cool. Creationists reason that during the Arphaxadian epoch clouds dropped many times more rain over the continents than they drop over them today.

Although the warm, wet conditions would encourage plant communities to recover quickly from the Flood, the rain would also cause a lot of erosion, explaining the thousands of feet of Tertiary sediments found in many locations around the world. The rain would also fill lakes such as Lake Bonneville (the Great Salt Lake being what is left) and the lakes upstream from the Grand Canyon (which may have

spilled out catastrophically to carve out the Grand Canyon, perhaps in only a few weeks' time, a couple centuries after the Flood).

During the Arphaxadian epoch the evaporation which generated the rain would gradually cool the oceans back to antediluvian temperatures. During the Arphaxadian epoch in North America, the warm, wet forest ecosystem evidenced by fossils soon after the Flood, gradually transformed into a cool, dry temperate grassland in the interior, and a cold, dry tundra in the north. Some centuries after the Flood, temperatures should have fallen sufficiently in certain locations for rain to fall as heavy snow. In Antarctica, Greenland, Hudson's Bay, and high in many of the world's mountains this snow would accumulate into great thicknesses of ice. Some of this ice would get so thick as to flow down mountain valleys and across continents (such as in eastern North America). Climate models suggest that this **ice advance** occurred within centuries of the Flood and melted back to its current extent in only a few decades.

ice advance—development and spread of continental glaciers during the Arphaxadian epoch

Diversity Recovery

Since God had chosen the remnant that would survive the Flood (*e.g.* Gen. 6:20), it is reasonable to assume that He chose the organisms that would be best suited to restore the proper diversity of life. Then, during the centuries following the Flood, the remnant obeyed God's command, rapidly increasing in numbers and diversity.

Whatever the mechanism God placed in organisms that expresses the hidden diversity, it was activated in Arphaxadian baramins of the land, sea, and air to increase diversity about a thousand-fold in only a couple centuries. As is often the case with natural processes, rates of diversification probably declined exponentially during this period. At the peak of this intra-baraminic diversification, new species may have arisen hourly. Such diversification would explain evidences that modern species have arisen from other species. That such a diversification occurred immediately after the Flood only several thousand years ago would explain why the genetic evidence of the diversification has not yet been lost in most of the species of a baramin. This intra-baraminic diversification might also be the only way to explain the many examples of biological change evidenced by fossils.

Biogeographic Recovery

Since God had chosen the remnant that would survive the Flood (*e.g.* Gen. 6:20), He probably also chose organisms that would be best suited to spread out from the landing site of the ark and fill the earth. High sea level, warm oceans, and cool continents immediately following the Flood probably created rather effective migration routes along the shorelines of the continents. Whatever the temperature preference of an organism, it could move the appropriate distance from the ocean and follow that temperature along the shorelines of the world. This would permit organisms to spread across the globe—even more effectively when one considers that the ark landed near the geographic center of the earth's land-masses.

As for crossing oceans, billions of logs ripped up from antediluvian forests would be floating on the world's oceans for decades and centuries following the Flood. The kinds of surface winds that propel ocean currents of the present would blow these vast vegetation mats across oceans. Although these log mats would gradually disappear as the plant material became waterlogged and sank, for decades after the Flood they could rather effectively transport organisms across the oceans of the world.

Transportation by such log mats explains many **disjunct species** (species found in widely separated areas, such as Geochelone tortoises in South America, the Galapagos Islands of the Pacific, and the Aldabra Islands of the Indian Ocean), why many of the locations of highest

disjunct species—species with a geographic range in more than one widely separated area

diversity are places where ocean currents approach a continent, why organisms got to many locations from the direction that they did (*e.g.* Australia from South America), and why organisms of certain sizes are found on islands (such as the only elephants on islands being dwarf elephants).

Human History

In some ways the human story of the Arphaxadian epoch is different from that of the remainder of the biological world. The appearance of land animals in the oldest Tertiary rocks across the planet suggests that animals quickly and effectively obeyed God's command to spread out, multiply, and fill the earth. Humans, though, disobeyed God's command. They settled on the plain of Shinar (Gen. 11:2) and stayed together, apparently for at least five generations (Gen. 10:25), and began work to build a city and a tower (Gen. 11:2-4). It was only after God confused their language that humans finally scattered across the planet (Gen. 11:7-8).

For centuries, the designated rulers of the biological world did not even attempt to claim the breadth of their kingdom. By then, organisms had reached even the most inaccessible places, and most of their diversification was complete. The fossil record was already full of diversification evidence. Human fossils and artifacts first appear in post-Flood sediments at the beginning of the Quaternary (above the Tertiary). This is why some have looked at the fossil record and deduced that humans evolved from the primates and apes that left evidence of their diversification in Tertiary sediments *below* the first evidence of humans.

Among all organisms, it is very possible that most of the intra-baraminic diversification occurred in small populations as they spread across the planet. Several different mechanisms for rapid biological change occur only in small populations. The same is probably true for humans, so whereas humans increased in *number* at Babel, they probably did not develop different body forms (for example, races) until after they left Babel. Rapid independent diversification in each small family group as it left Babel probably generated the small differences we see between races in the present and the somewhat larger differences we see between human fossils in different locations in the fossil record.

It may have been because humans waited so long to spread out from Babel that they did not diversify as much as other organisms.

It should be noted that there is genetic evidence that modern humans had a common origin only thousands of years ago as Scripture indicates. The fact that there is no difference in some regions of the Y chromosome among men across the planet suggests that all human males descended from one man within the last 10,000 years—directly consistent with biblical history. Mitochondrial DNA (mtDNA) suggests all modern humans had a common ancestor in one female, but there is more variation than is found on the Y chromosome. If one assumes that after its origin the human population never experienced a **population bottleneck**, where it was reduced to a small population, then there is about 600,000 years of large population change.

population bottleneck—time when a population was reduced to a very small size (usually <10)

However, according to biblical history, the human population experienced *two* population bottlenecks—one through Noah's family on the ark and another through the families departing from Babel. Since change occurs so rapidly in small populations, each population bottleneck should reduce the estimated time of change by about 90% (10% of 10% of 600,000 years ≈ 6000 years). With the Y chromosome evidence, this suggests that all modern humans were descendant from one man and one woman who lived less than 10,000 years ago. This is not only consistent with biblical history, but provides independent evidence that human fossils are as young as biblical history would suggest.

Although the humans spreading out from Babel were intelligent and cultured, they struggled to survive once they were separated from a cultural center. Not knowing where the resources were located, they had to survive with what they had available, first gathering nuts and berries, hunting game, modifying sticks and stones, and seeking shelter in caves. In this first stage they left evidence of lithic (stone) culture. In the course of individual lifetimes, many people improved their tools ('paleolithic' to 'mesolithic' to 'neolithic' styles), in many cases skipping cultural stages when discoveries allowed it.

Some people located and mined metals and developed bronze and iron cultures. Some may have not only skipped cultural stages, but also progressed through the stages at different rates. Some people groups never developed past the original lithic culture, and still use that culture today. This explains why cultural sequences at different locations have to be associated with separate chronologies, and why stages are absent at many locations.

The Frontiers of Research

The Arphaxadian epoch is the one most similar to the modern world. In fact, it is part of the modern era, for after the Flood we have no indication of a change in how God interacts with the animals or plants of this planet. There is also no single catastrophe following the Arphaxadian epoch which should have destroyed much of the evidence from this time period—only the gradual erosion over the four thousand years or so since Arphaxad died. So, of the four epochs of earth history described above we know the most about the Arphaxadian. Yet so much happened in the Arphaxadian, that it could be said there is more that we do NOT know than any other epoch in history.

Research questions are many, and a few include the following: From where did land animals spread? What organisms actually survived the Flood and how did they thrive on a planet still experiencing large catastrophes? What paths did organisms travel? What stimulated intra-baraminic diversification and why did it stop? How did organisms change into the forms most appropriate for the environment they found themselves in? What caused the change in human longevity? How many humans and what size groups spread out from Babel? How much change did humans experience? Why is there less variation in modern humans than is evidenced in the fossil record?

Summary of Chapter

A. Regardless of their worldview bias, most historians agree on the timing of biblical events back to about the time of Abraham (~2000 B.C.). The history that Genesis offers from that time back (placing creation at about 4000 B.C.) is to be preferred over all other accounts of history because

   a. it is an eyewitness narrative account from a perfect eyewitness (Who saw it all perfectly, understood it all perfectly, can communicate perfectly, and cannot lie), so it is to preferred over any circumstantial inferences of history (*e.g.* theories of science); and

   b. at least three events in the history of life (creation, curse, and flood) involved processes that are not going on in the present, making inferences from present processes invalid.

B. The Creation Week was six days in time, during which God used creation processes He never used again (because He ceased from creating at the end of the Creation Week). He fashioned that creation in such a way that we could understand it and filled it with physical illustrations of his invisible attributes and spectra of perfections so that we could understand His nature and His infinity.

   When He created organisms (Days 3, 5, and 6), He created them with their associated biomatrix organisms, and He created them with great diversity, disparity, netted hierarchy of similarity, latent information (for survival through the curse, recovery from the Flood, and for us to reveal to God's glory), and He created them without any biological evil (no suffering, nephesh death, carnivory, or disease).

   a. On Day 1 God created the basic structure of the universe so life could thrive.

   b. On Day 2 God created the atmosphere and the water cycle for the support of life. About this day God created the biomatrix in the oceans so they could support life.

   c. On Day 3 God created the land with a biomatrix and biogeochemical cycles for the support of life (probably multiple continents; probably by raising them out of the oceans and forming most of the Precambrian sediments). Later on Day 3, God created the land plants (probably different plants on different continents plus a continent-sized floating forest, and apparently leaving Eden barren).

   d. On Day 5 God created the animals of the sea and air.

   e. On Day 6 God created the animals of the land (probably different animals on different continents, and apparently none in Eden), then (in sequence) God created man, planted the Garden of Eden, created animals for the Garden, had Adam name the Eden animals, and created Eve.

C. During the Edenian epoch (between the end of the Creation Week and the Fall; somewhere between a few days and few score years in length) there was no biological evil.

D. The Edenian epoch ended when God cursed the creation in response to Adam's sin. With the curse came biological evil, both degenerative biological evils (*e.g.* degenerative aging, disease) and evil-minimizing biological evils (*e.g.* nephesh death, natural selection, carnivory).

E. The antediluvian epoch (between the Fall and Flood; approximately 16 centuries in length) differed *at the very least* from the modern world in the following ways:

   a. climate (probably wider polar and temperate climatic zones; no cooler than cold temperate at the poles)

   b. diversity (probably more continents, more biological diversity and disparity)

   c. longevity (humans lived for nine centuries)

   d. diet (humans did not eat meat)

e. 2nd curse (murder punished with cursing of soil) → degeneration of antediluvian world

F. The Flood was a one-time, global catastrophe that destroyed all land animals and humans except the 8 humans and animals (2 of each baramin of unclean animals; 7 or 14 of each baramin of clean animals) on Noah's ark. The Flood substantially refashioned the surface of the earth, deposited Paleozoic and Mesozoic sediments, and warmed the world's oceans. It purified the earth of the effects of the 'second' curse and permitted a new start for man and animals.

G. The Arphaxadian epoch is a few centuries immediately following the Flood when the earth recovered from most of the effects of the Flood:

   a. (on-going) Geologic recovery is restoring rocks to their proper positions. This has resulted in gradually weaker and less frequent earthquakes and volcanoes.

   b. Evaporation cooled oceans to modern temperatures, resulting in

      i. extremely high rainfall following the Flood, gradually decreasing to the present

      ii. accumulation (after a couple centuries) of thick sheets of ice in mountains and at the poles and catastrophic (decades-long) melting. Slow recovery continues even today with the melting of the remnants of that ice.

   c. Intra-baraminic diversification restored the diversity of organisms.

   d. Migration (probably aided by floating log mats) allowed organisms to fill the earth.

H. Humans disobeyed God and remained at Babel until God dispersed them by confusing their language. Genetic drift in the small populations leaving Babel probably generated the modern races. Humans struggling to survive in a Flood-refashioned world lived initially in caves ('cave men'), and rapidly (decades or at most centuries) developed 'culture' (stone-based, then metal-based) according to the resources they discovered.

Test & Essay Questions

1. List two/three/four/five examples of the history of organisms being tied to history of humans.

2. Describe how time is dealt with differently in the Old and New Testaments.

3. List two/three/four examples of the importance of time in the Old Testament.

4. According to biblical chronology when did the creation begin?

5. Explain two reasons why the biblical chronology is followed rather than a chronology deduced from the physical world alone.

6. Why does the history/chronology deduced from the physical world alone differ so radically from biblical history?

7. How/When did the organisms of the biomatrix come to be and how do we know?

8. What is the nature of the Precambrian/Paleozoic/Mesozoic/Tertiary fossil record and how did it come to be?

9. Describe the 'day three regression' and how it might explain a portion of the fossil record.

10. What is the nature of the 'hidden information' in organisms, when did it arise, and what is its purpose?

11. What did God do on the seventh day, and what is the significance of that?

12. What change(s) occurred between the Creation Week and the Edenian epoch / between the Edenian and Antediluvian epochs / between the Antediluvian and Arphaxadian epochs?

13. What events mark the beginning and end of the Edenian / Antediluvian / Arphaxadian epoch and how long did the epoch last?

14. Why was a particular period in earth history labeled 'Edenian' / 'Antediluvian' / 'Arphaxadian'?

15. How does the shalom/'very good' nature of the Edenian epoch differ from the present?

16. Describe the history of biological evil in earth history.

17. How and when did biological evil arise?

18. List two/three examples of antediluvian communities that are unknown in the present world.

19. How was human longevity different at different periods of earth history and what caused the changes?

20. How was human diet different at different periods of earth history?

21. Why was 'all flesh' destroyed with a global flood and why did God choose a flood to do it?

22. Why are Paleozoic and Mesozoic rocks thought to be deposited in Noah's Flood?

23. List two/three/four things about the earth that the Flood changed.

24. What is the cause of Carboniferous coals / of Permo-Triassic sands?

25. How do we know that biological recovery after the Flood took only a couple centuries to occur?

26. In what way(s) did the climate of the world / of North America change during the Arphaxadian epoch?

27. What caused the ice advance and how long did it last?

28. How did diversity increase after the Flood? / How much did diversity increase after the Flood? / How fast did diversity increase following the Flood?

29. How did land animals cross oceans to get from the ark to distant continents or islands?

30. List two/three evidences for organisms rafting across oceans soon after the Flood.

31. How are disjunct species explained?

32. Why was Australia populated from distant South America rather than the much closer Southeast Asia?

33. How is it that the species most similar to the Galapagos tortoise on the Galapagos Islands of the eastern Pacific is the Aldabra tortoise on the Aldabra Island of the Indian Ocean?

34. Why does rafting become less and less effective in the years following the Flood?

35. Why are human fossils and artifacts found only above rocks containing the diversification of the animals most similar to humans?

36. What does the variation in modern mtDNA / Y-chromosomes indicated about human history?

37. Why are the oldest evidences of human culture so primitive? / What is the origin of the lithic culture?

APPENDIX

EVOLUTION

Perspectives on Evolution

Introduction

Because of the popularity of evolutionary theory in our world today, it is important that all students—even (and perhaps most importantly) students who do not believe that it is true—understand the concept sufficiently to actively enter the marketplace of ideas and therein shine with the light of Christ. Evolution is also a common stumbling block preventing people from accepting or even hearing the gospel, so the believer ought to know how to calm the fears and alleviate the concerns another person might have about evolution so that that person might come to know the Lord.

Evolution is also used to challenge the faith of believers, undermine the authority of the Bible, and assault the very character of God. Believers need to understand enough about evolution to understand and know that evolution is no threat to the truth of the Christian faith, not because evolution is somehow true, but because the symphony of the biological world proclaims quite soundly that it is not true.

Actually, many of the claims of evolution as it relates to biology have been addressed throughout the chapters of this textbook. However, the claims of evolution have been dealt with individually. Consequently, it would be very difficult for a student to understand what the argument for evolution looks like when all the pieces are considered together (rather than separately). Without the perspective of the entire argument, it would be difficult for the student to grasp how powerful the argument for evolution really is—and the argument for evolution really is a powerful one. It is because of the reasonableness and power of the argument for evolution that it has become so widely accepted in our culture and why it has such a stranglehold on the thinking of so many individuals.

Thus, it would be wrong for a biology textbook not to include an introduction to evolutionary theory. Yet, a discussion of evolutionary theory of this nature has no logical place in a biology textbook written from the perspective of young age creationism. Consequently, this topic is included as an appendix rather than included in the main text of the book.

As for the content of the text to follow, only two modern theories of naturalistic evolution are discussed. Out of logical necessity, the worldview of naturalism has generated naturalistic evolutionary

abiogenesis—naturalistic theory for the origin of the first cell from non-living material

theories for the origin of everything physical. Because this is a biology textbook, only two of these theories will be discussed here—abiogenesis and biological evolution. **Abiogenesis** is the naturalistic evolutionary theory for the origin of life itself, or more accurately the origin of the first organism or organisms from non-living matter. Biological evolution is the naturalistic evolutionary theory for the origin of life's diversity (*i.e.* how the first organism—or organisms—came to generate all the other organisms that existed in the past and exist in the present).

As for the presentation of the material, since a theory is best understood when it is presented in its entirety without distraction, each evolutionary theory is first presented without critique. The young-age creationist critique of each theory follows the presentation of the theory (as 'perspectives'), rather than being included in the discussion of the theory.

As frustrating as this might to be to some who wish to know *immediately* what might be wrong with an evolutionary claim, such interruptions to the logic of evolution prevent a person from fully comprehending the power of the theory. Also, to help clarify the distinction between fact and theory, a synopsis of the evolutionary theory is presented before—and separately from—the evidence marshaled in favor of that theory.

Abiogenesis

Since he does not believe in a God Who created, the advocate of the naturalistic worldview must develop another explanation for the origin of things. Over the years, evolutionary theory has been developed to fill that role. In the case of anything and everything that exists, some sort of evolutionary theory has been developed to explain its origin. Because life and non-life are so radically different, the origin of the first organism from non-living things is treated separately from how life diversified once the first organism was in existence.

This evolutionary theory for life arising from non-life is variously called abiogenesis [the beginning of life from non-life: *a* meaning 'non' or 'no' + *bios* meaning 'life' or 'living' + *genesis* meaning 'beginning'], abiotic evolution [the evolution of *a* meaning 'non' or 'no' + *bios* meaning 'life' or 'living'], or chemical evolution [since in the naturalistic worldview the entire process involves chemistry and only chemistry].

In the evolutionary theory for the origin of the earth's oceans and atmosphere, water and gases of the atmosphere were released from molten rocks by volcanic eruptions. Although molten rocks of the present release nitrogen, carbon dioxide, argon, and other gases now found in the atmosphere, they do not release molecular oxygen (O_2). Rather, all oxygen atoms are combined with other atoms in minerals of the rock, water (H_2O) for the oceans, and gases in the atmosphere like carbon dioxide (CO_2).

According to abiogenesis theory, natural energy sources (such as UV radiation because it was unblocked by ozone, or lightning, or geothermal energy) provided the energy necessary to unite the gases of the atmosphere into biomonomers. These biomonomers are thought to have accumulated in water where further energy joined the biomonomers together to produce biological molecules. Finally, the capture of some of these biological molecules into a naturally-forming plasma membrane generated the first cell.

Evidence for Abiogenesis

In 1952, Stanley Miller, a graduate student at the University of Chicago working in the lab of Harold C. Urey created an experiment to test the theory of abiogenesis. He released a spark (to simulate lightning) in a mixture of gases thought to have been in the earth's early atmosphere and collected the molecules that resulted.

The Miller-Urey experiment quickly generated amino acids, the monomers for proteins in living systems. The experiment has been run with different mixtures of gases and using different sources of energy and the results are similar. Such experiments have generated the 20 amino acids of living systems, with the four most abundant amino acids in the same order of abundance as the four most abundant amino acids in living systems. The same experiments have generated all five of the nitrogen bases and both of the sugars that make up the nucleotides of DNA and RNA.

Some of these organic molecules have even been found in outer space, confirming that these reactions do occur naturally. For example, thirty different organic molecules have been detected in interstellar clouds, including ribose sugar, formic acid and monoaminomethane which together can make one of the nitrogen bases of RNA. Seven of the biologically important amino acids have been found in the Murchison meteorite, the four most abundant of them in the same order of abundance as is found in organisms and experiments like those of Miller and Urey.

Other experiments show that under various conditions amino acids join together in chains of amino acids and, in the presence of certain clays, only left-handed forms join together (as in organisms). Other experiments demonstrate that nitrogen bases and sugars can unite to produce nucleosides, and nucleosides and phosphates can unite to produce nucleotides.

Other experiments show that the most stable string of three nucleotides is the same string that codes for the most abundant amino acid in living systems, and the second most stable string of three nucleotides is the same string that codes for the second most abundant amino acid in living systems. Still other experiments show that RNA catalyzes its own replication.

Finally, chains of amino acids in cool water tend to form microspheres that have the potential of capturing any organic molecules in that water. These microspheres are double-layered, about the size of cells, are stained by the same stains that stain living cells, allow osmosis, allow diffusion of some but not all molecules, and can divide or grow by accretion.

Perspectives on Abiogenesis

The origin of life experiments, like the Miller-Urey experiment, have revealed several important details about life. First, the proteins in living systems have been designed to be as energy efficient as possible. They most commonly utilize the amino acid that is easiest to make, then the amino acid that is next easiest to make, and so on through the fourth most common amino acid in organisms. That is really a very clever design.

Second, as encouraging as origin of life experiments have been to abiogenesis theory, a living cell has never been generated from non-living material. In fact, even when scientists set aside any constraint to be like the 'real world', put all their ingenuity into the process, and try over and over again for decades, a living being has not been constructed from non-living materials. Whatever ingenuity is necessary to generate life, it seems to exceed the ingenuity of the scientists who have worked on the problem. This is quite consistent with the special creation of life by God.

Third, phospholipids that make up cell membranes have never been simulated in origin of life experiments. The microspheres that have been formed that are supposed to function like cell membranes are made of chains of amino acids. No cell membrane is made of amino acid chains, so abiogenesis experiments have never generated cell membranes.

Fourth, origin of life experiments have never created any real proteins. Proteins are not simply chains of amino acids, they are amino acids linked by a particular kind of bond—a peptide bond. Origin of life experiments have never simulated a natural situation which would join amino acids with peptide bonds to produce actual proteins.

Fifth, the problem of homochirality has never been solved. Most organic molecules can take on more than one chiral or mirror-image form, rather like a right-handed glove is a mirror image of a left-handed glove. Therefore, one form of a chiral organic molecule is designated right-handed and the mirror image is designated left-handed. When organic molecules are created in origin of life experiments, both left- and right-handed forms are created in about the same amount. Proteins and sugars in organisms, however, are homochiral. Proteins are made of only left-handed amino acids and sugars are only right-handed. Origin of life experiments have never solved the problem choosing only left-handed proteins or right-handed sugars once amino acids or sugars have been formed. Although naturally occurring crystals, like some clays, can attract one chiral form, by the very nature of crystals they only attract one kind of amino acid. A string of identical amino acids has no biological meaning or use.

Sixth, even though a long string of different chemical reactions must occur in order to progress from natural inorganic molecules to the first cell, no origin of life experiment has ever generated any two of these steps in a row. An experiment that generates amino acids cannot also join those amino acids into proteins. An experiment that generates nitrogen bases cannot also join those nitrogen bases to sugars to make nucleosides. An experiment that joins nitrogen bases and sugars to generate nucleosides cannot also join the nucleosides to phosphates to produce nucleotides.

In every case, once a particular molecule is generated, other molecules react with it and prevent any further reactions. A few centuries ago it was thought to be easy to generate simple organisms from inorganic materials. Origin of life experiments have demonstrated that there is in fact a very great discontinuity between non-living materials and the biological world, a discontinuity much too wide to be crossed by natural process. Origin of life via special creation by the God of life is a much more reasonable explanation for the origin of life.

Biological Evolution

In the naturalistic worldview, the similarities found among all organisms strongly suggest that the full disparity and diversity of organisms descended from one ancestral cell. But the process which diversified that first cell into all the other organisms was not the same process that brought that first cell into being. Abiogenesis is thought to be the process whereby the first cell came to be; biological evolution is thought to be the process whereby that first cell generated all other organisms.

Different theories of biological evolution have been proposed at various times to explain the origin of particular organisms or particular categories of organisms in the biological world. For example, the kind of biological change that can be observed in a biologist's lifetime usually only creates only small modifications of a species. Theories developed at this level explain changes *within* species and are often referred to as theories of **microevolution**.

Charles Darwin (1809-1882), for example, is most famous for his theory of natural selection, which is a theory of microevolution. The favorite subject of evolutionary theorists, though, is how new species come to be. These are theories of speciation. Theories that focus on the origin of new genera, families and above can be described as theories of **macroevolution**.

Currently, the most popular theory of macroevolution is that the higher groups arose by the processes of microevolution and speciation operating over very long time periods of time. The most popular theory of microevolution at present is Neo-Darwinism—a theory that combined Darwin's theory of natural selection with the field of genetics as it was developed in the early part of the twentieth century.

The basic idea is that **mutations** create new character traits and natural selection eliminates the traits that harm the organism, leaving the organism only with the traits that allow an organism to survive

---

microevolution—change within species

macroevolution—the origin of taxa above species or the process that generates new taxa above species

mutation—genetic mistakes that occur as DNA is copied. Types: point mutation replaces one nucleotide with a different one; frame-shift mutation inserts or deletes a nucleotide

better in a particular environment. In this way, a species gradually changes over time.

The most popular speciation theory in the present suggests that one species can generate another species. This occurs when a few individuals from the parent population get isolated from the rest of the population, then change by mutation and natural selection until they can no longer interbreed with the rest of the population.

Once a new species arises, mutation and natural selection continue to alter it, and the original species, until they have become different enough to become new genera, families, orders, and so on. By this process, the first species on this planet changed continually through time like the trunk of a tree growing upwards.

From time to time, that species spun off a new species like branches coming off a trunk. Each new species also changed through time, spinning off other species. In the life of a tree, one tiny sprout can grow into a mighty tree, reaching scores of feet upwards and outwards in all directions, and supporting millions of leaves on thousands of branches. In the same way, in macroevolution, one tiny species can, along thousands of branching lineages, generate a huge disparity of millions of species.

Microevolution Evidence

Beneficial Mutations

Mutations that bring an advantage to organisms provide the raw material for change for all forms of modern theories of biological evolution. Most mutations, however, are harmful. If all mutations were harmful, evolution would not be possible, so evidences of beneficial mutations are important for evolution.

Examples of different kinds of beneficial mutations include the following: 1) Since sickle-shaped red blood cells tend to harm the malaria-causing parasite, people who have a mutation that leads to sickle-shaped red blood cells tend to be resistant to malaria; 2) many varieties of viruses, bacteria, mosquitoes, and rust fungi have arisen over the years by mutations that give them the ability to survive chemical control agents; and 3) In 1975, Japanese scientists discovered a variety of *Flavobacterium* capable of consuming nylon, a manmade substance invented in 1935. It is thought that a mutation gave *Flavobacterium* the ability to consume nylon.

Natural Selection

From the variety created by mutation, most forms of modern theories of biological evolution claim that natural selection selects those character traits that help an organism survive better than other organisms. Natural selection is therefore another critical element for modern evolutionary theory. Examples of natural selection include the following:

1. To explain why a higher frequency of insect species on islands are wingless, it was suggested that wind kills flying insects by blowing them off the island and into the ocean. The idea was tested by raising flies in a terrarium that had a fan set up at one end of a terrarium and flypaper at the other end of the terrarium. When the experiment was run, the number of wingless flies increased each generation, verifying that selection against winged insects tends to increase the number of wingless insects;

2. *Drosophila melanogaster* fruit flies in breweries tend to be able to tolerate alcohol whereas populations of the same species that live outside breweries tend to be killed by alcohol. To demonstrate that the brewery flies may have descended from normal fruit flies, an experiment was run, placing normal fruit flies in the presence of alcohol. After 28 generations, the surviving fruit flies were as tolerant of alcohol as the fruit flies in breweries, showing that selection can increase the tolerance of alcohol in fruit fly populations;

3. many examples of resistant varieties of viruses, bacteria, protozoa, fungi, and insects increase in commonness as chemical controls are used on them (for example, resistant forms of influenza increase in commonness with the use of flu vaccines, resistant forms of gonorrhea increase in commonness with the use of penicillin, resistant forms of wheat rust fungi increase in commonness with the use of fungicides, and resistant forms of mosquitoes increase in frequency with the use of DDT; and

4. to explain why over 200 insect species have exhibited **industrial melanism** (dark forms becoming common during a time of industrial pollution), it was suggested that predators find it easier to see light-colored insects against pollution-darkened surfaces. The most popular example concerns the normally white peppered

---

industrial melanism—dark forms of species that became common during the industrial revolution

British moth *Beston betularia*. A dark form was first observed in Manchester, England in 1848, and as coal dust pollution increased, the dark form became more common, making up 98-99% of Manchester peppered moths by 1895. Experiments with peppered moths verified that birds tend to eat more white moths than dark forms when the moths are resting on dark surfaces.

Speciation Evidences

Peripheral Isolate Theory

Ernst Mayr was one of the leading biologists of the twentieth century. In Mayr's opinion, of all the theories that had been developed to explain how species might arise, he believed the 'peripheral isolate theory of allopatric speciation' was the one most capable of allowing a species to spin off a new species. According to this theory, the process of speciation begins when a small group of organisms, living toward the edge of the geographic range of a species, gets isolated from the rest of the species.

A population near the edge of the geographic range is likely to be in a slightly unfavorable environment. As a result, it should experience considerable pressure (natural selection) to change. With this pressure (and the fact that lots of change can occur rapidly in small populations), this population can change until it is no longer able to interbreed with the original species. Now more successful in its environment, the new species would increase its range and population size until it became large enough to stop changing, being in equilibrium with its environment.

That speciation has occurred (and is occurring) by Mayr's peripheral isolate theory of allopatric speciation is evidenced by geographically separated populations of organisms in the natural world which display different degrees of interbreedability:

1. those that freely interbreed when placed together artificially (*e.g.* grizzly bears and polar bears; Abert and Kaibab squirrels from opposite sides of the Grand Canyon; Australian whipbirds species *Psophades olivaceous* and *P. nigrogularis*);

2. those that tend not to interbreed when placed together artificially, even though they are *capable* of interbreeding (*e.g.* the fruit fly species *Drosophila melanogaster* and *D. pseudoobscura*, even when raised in the same small terrarium);

3. those that, when artificially mated, produce offspring that are not fully fertile (*e.g.*, the fruit fly species *Drosophila pseudoobscura* and *D. persimilis* which not only mate only 4-6% of the time, but when they do, they produce sterile males and fertile females);

4. those that, when artificially crossed, produce completely infertile offspring (*e.g.* females of a particular Columbian population of *Drosophila pseudoobscura* which produce only sterile male offspring when mated with males from any other population of *D. pseudoobscura*; female horses and male donkeys that produce sterile mules; swamp wallabies (*Wallabia bicolor*) and agile wallabies (*Macropus agilis*) that produce only sterile offspring); and

5. those that have not been successfully crossed even artificially (*e.g.* marine animals on the Pacific and Gulf sides of Central America, where 10% of the species exist on both sides, but most of the rest of species have similar counterpart species on the other side with which they cannot breed).

Finally, ring species (where all but one pair of adjacent populations of a ring of populations can interbreed) would be another type of evidence suggesting that speciation can and does occur (whether or not that is done by Mayr's peripheral isolate theory).

Fossil Evidence

Since speciation events occur so rapidly (Mayr and others speculated maybe in 1000 generations), most of the history of life is a story of species persisting for millions of years in equilibrium with its environment, but once in a while spinning off a new species from the edge of its range. The long-standing equilibrium of a species is occasionally punctuated by brief speciation events.

In 1972, Stephen Jay Gould and Niles Eldredge speculated what such a story of **punctuated equilibria** would look like in the fossil record. Considering that the fossil record preserves hundreds of millions of years of life, fossilization is an extremely rare event. Fossilization is so rare that a peripheral isolate population would be too small and last for too short a time to expect that we might have any fossils from it in the fossil record.

Fossils are likely to be preserved only from large populations—in other words only from the established species. Since established species

punctuated equilibria—naturalistic evolution theory explaining stasis and abrupt appearance of species are in equilibrium with their environment and do not change much, the fossil record would only preserve unchanging species. Species will appear abruptly, and then not change for as long as they are found in the fossil record. If the peripheral isolate theory of allopatric speciation is really the way most species come to be, then Gould and Eldredge argued that fossil species should exhibit stasis and abrupt appearance.

Since the theory was proposed, stasis and abrupt appearance has been confirmed for fossil species of all types. As expected, very few fossil examples exist of a gradual transformation of one species to another. Some of the examples include the primate *Cantius*, some mollusks in sediments of Lake Turkana in Africa, the single-celled radiolarian *Theocorythium*, and the single-celled foraminiferan *Globorotalia*.

Macroevolution Evidences

Homology

Children in the same family tend to be more similar to each other than they are to anyone else. This is because they inherited many of the same traits from their parents. In general, the more closely related two people are, the more similar they are, and they have the similarities because they inherited those similarities from the closest ancestor they all have in common. The similarities between two people can be used to demonstrate that they are related, and the similarities in a group of people can be used to reconstruct a family tree.

If all species are related to each other in a tree of life as macroevolution claims, then similarities between species demonstrate that they are related and those same similarities can then be used to reconstruct the evolutionary tree—the family tree for all organisms. A **homology** is a similarity between two organisms that is there because the two organisms inherited the similarity from a common ancestor. Since characters that do not give an advantage to an organism are not likely to have evolved twice, homologies that provide no advantage to the organism make the most convincing evidence for common ancestry. Similarities between and among organisms abound, and they can be grouped into three categories:

homology—similar structure in a similar position in different organisms, but having a different function (due to common ancestry in naturalistic evolution; due to a common Creator in creationism)

1. *adult similarities* (similarities that exist between and among adults of different species, such as milk and hair possessed by all mammals, the opposable thumb possessed by all primates, and so on). The

most celebrated adult similarity is the "1-bone, 2-bone, multiple-bones, finger bones" structure of the vertebrate limb found in the wings of birds and bats, the fins of whales and porpoises, and the legs of amphibians, reptiles, and mammals. Since this particular sequence of bones provides no particular advantage to organisms that have it, there is good reason to believe the organisms that have it do so merely because they inherited it from a common ancestor;

2. *embryological similarities* (such as the blastula stage of development in all mammals and the virtually identical development of humans and chimps);

3. *molecular similarities* in many different molecules (such as DNA and RNA in all cells, all the molecules of chlorophyll a-b photosynthesis in green algae and plants, all the molecules of Krebs cycle aerobic respiration in micrococci bacteria and mitochondria, 98% of the DNA sequence in human and bonobo chimps being similar, and so on).

Not only do these similarities exist, but in each case the similarities allow the reconstruction of an evolutionary tree, just as macroevolution expects.

Concordance of Evolutionary Trees

An evolutionary tree can be developed by using only adult similarities (in other words using no embryological or molecular similarities). Another evolutionary tree can be developed using only embryological similarities (in other words using no adult or molecular similarities). And, even more evolutionary trees can be developed from any number of a host of different biological molecules (in other words using no adult or embryological similarities or similarities from any other molecule). The evolutionary trees produced in this manner are very similar to each other, suggesting that organisms really are related to each other in the fashion indicated by any one tree, and really did evolve according to the branching pattern of that tree.

Nested Hierarchy

On a tree, the smallest twigs branch off a slightly larger branch, and several of these larger branches branch off an even larger branch, and so on until the largest branches branch off the trunk of the tree. If organisms evolved along the path of an evolutionary tree, then it ought to be possible to classify the species that evolved most recently from a common ancestor into one genus, and the genera that evolved most recently from a common ancestor into one family, and so on, until all organisms are classified into the very largest groups.

If organisms arose by macroevolution, organisms will be classifiable into such a nested hierarchal pattern. The success of taxonomists at classifying organisms into the nested hierarchy of species within genera, genera within families, and so on, suggests that organisms did in fact evolve by the process of macroevolution.

Embryological Recapitulation

In the process of development, organisms pass through a series of developmental stages. Sometimes one or more of these developmental stages preserves something that was found in an evolutionary ancestor of that organism. When that happens, an organism's development (or 'embryology') is said to recount (or 'recapitulate') that organism's evolutionary history (or 'phylogeny'). Cases when 'embryology recapitulates phylogeny' are examples of **embryological recapitulation**. An example of embryological recapitulation is a tracheid (water-carrying) cell in a flowering plant which, as it develops, passes through the kinds of tracheid cells found in non-flowering plants.

embryological recapitulation—developmental stages that look like the stages of that species' evolution

Suboptimal Improvisations

The pain and suffering we see in the world makes it clear that the world in which we live is not perfect. However, when we focus on how things *normally* work in the biological world (when not messed up by DNA copying errors or by pathological organisms), the world works very well. When they are functioning normally, there are few biological systems that humans can improve upon.

Consequently, although the world is not perfect, it appears to be near-optimal. Natural selection explains why this is. Natural selection chooses that which provides the most benefit to the organism. Since natural selection is also the primary process guiding evolutionary change, macroevolution should generate biological structures that are at or very close to optimal design.

Yet, because macroevolution is a physical process unguided by a mind, it would not be expected to always provide an organism with the *best* possible design, so that explains why the world is not perfect. Natural selection will choose the best design from the variation made available to it by mutation, but mutation might not generate the best design. Or, it may not be possible to get the organism from whatever state it is in to the best possible design.

Under these situations, natural selection does not choose the best of all possible designs, but rather the best structure dealt to the organism by its evolutionary history. When the modification on the structure inherited from ancestors is not the best possible design it is called a **suboptimal improvisation**.

> **suboptimal improvisation**—structure of poor design, thought to be evidence of naturalistic evolution

The most well-known example is the panda's thumb. Both the lesser panda and the greater panda have five claws and five fingers or toes on each paw, just like the remainder of the bears of the world. However, when a panda eats bamboo, it lays the bamboo between its palm and a thumb, curls the five fingers around the bamboo, and pulls the branch through the closed hand. The panda then eats the leaves stripped off in this process. But what is this 'thumb'? It is obviously not like our thumb—one finger opposite four others—because the panda's thumb is opposite all five fingers. It turns out that the panda's thumb is not a thumb at all, but a lengthened radial sesamoid bone—one of the bones of the wrist. The panda's thumb is suboptimal, for there are much better thumbs in the world. Whereas a wise designer might have given the panda a better thumb, the panda inherited five non-opposable fingers from its bear ancestor. Without an opposable finger to work on, natural selection could not choose an optimal thumb. Instead, it chose the best available improvements upon what macroevolution had provided for it. Therefore, the suboptimal improvisation known as panda's thumb is not only evidence that pandas are descendant from bears, but that whatever brought pandas into being was not capable of generating the best of all possible structures for the panda.

Vestigial Structures

At any given time in the process of macroevolution, an organism probably contains multiple structures that were inherited from ancestors, but which are not as useful to the present organism as they were to the ancestors. Some of these structures might have lost some or all of their function due to mutations. Other structures might be able to function but they function only in an environment in which the present organism no longer lives. Still other structures might be in the process of being replaced by a structure that is more beneficial to the organism.

A structure is said to be a vestigial structure if it either has no function at all, or if it has a reduced function in that organism compared to the function that that structure had in an ancestor of that organism. Examples of vestigial structures include: **pseudogenes** (segments of DNA that have start and stop codons like genes, but which do not code for a working protein); partially developed teeth in the embryos of baleen whales; wings of ostriches, emus, and rheas that can no longer be used to fly; remnants of hind leg and hip bones under the skin and scales of pythons; remnant toe bones in horses; hind leg and hip bones in whale embryos; and non-functional, partially-developed eyes in Mexican cave fish.

> **pseudogene**—DNA segment with gene-like start and stop codons, but not coding for a working protein

In the case of the horses, the vestigial nature of the splint bones is confirmed by the oldest genera of fossil horses having three toes on each foot rather than the single toe of modern horses. In the case of the whales, the vestigial nature of the embryonic hind limbs is confirmed by the hind legs on the *adults* of the oldest known families of fossil whales. In the case of the Mexican cave fish, the vestigial nature of the eyes is confirmed by fully functional eyes in populations of the same species found outside caves.

Genetic Throwbacks

Occasionally, an organism is born that displays a character trait of an ancestor of that species. Such a structure is known as a **genetic throwback**. Occasionally, for example, a chicken embryo begins to develop teeth. Although this is apparently caused by a mutation that kills the chicken while it is still in its shell, it indicates that chickens possess at least some of the information for building teeth in their DNA. That early birds had teeth is indicated by the fact that the oldest order of fossil birds possessed a full set of teeth. This suggests chickens have descended from an order of birds with teeth.

> **genetic throwback**—occasional organism manifesting a complex trait otherwise only known in fossils

Another genetic throwback concerns the toes of horses. Occasionally, a horse is born with three fully developed toes on its feet. Modern horses walk on a single toe and have two more vestigial toes in the form of splint bones. That horses once actually had three toes is indicated by the fact that the oldest fossil horse genera had three to four toes on each foot. The horse toe genetic throwback suggests that modern horses are descendant from a genus of horses that had three toes.

Biogeography

biogeography—discipline of biology that studies where organisms live and how they got there

It is common for the most similar species in the world to be located next to each other, just as might be expected if they evolved from one another or a common ancestor in that area. Islands, for example, are often inhabited by species found nowhere else in the world, but most similar to species found on the nearest mainland. This, for example, is true of the Cape Verde Island off the coast of South Africa and the Galapagos Islands off the coast of South America. Very often the converse is also true.

The more distant a land mass is from everything else, the greater the number of unique species that are found in that region, as might be expected if organisms changed as they travelled from one place to another. This can be seen in the many unique species found in Hawaii, New Zealand, and Madagascar. And, when distant land masses are the size of continents, entire groups of organisms are restricted to those areas—as if the origin and entire evolution of the group occurred on that continent. All Australian marsupials, for example, both living and fossil, are found only on the island continent of Australia. Likewise, all living and fossil species of new world monkeys, marmosets, rat possums, caviomorph rodents and a few other groups are known only from the continent of South America (which did not become connected to Central America until only a couple million radiometric years ago).

Tertiary Fossils

According to macroevolution, new species are coming to be all the time. This would mean that all the species alive today are of different ages. Some species are young, having arisen very recently. Other species are older, having arisen a bit farther back in time. Others are even older, and so on. As a consequence, when older and older rocks are considered, fewer and fewer of the species in the rock will be living species, and more and more of them will be extinct species. Most of the species found in young rocks will be found alive in the modern world, and few of them will be extinct.

In older rocks, fewer of the species will be still alive today and more of them will be extinct species. If rocks are old enough, none of the fossils will be from living species. All of them will represent extinct species. Not only is this the case, but the first divisions of the Cenozoic were made based on what percentage of fossils in the rocks were species still living today.

First Appearance of Fossils

According to the evolutionary tree created from homologies, the first organisms on this planet were bacteria or archaea. From bacteria, protozoa and algae evolved. From protozoa, fungi and animals were evolved. From algae, plants were evolved. Among the plants, non-vascular plants evolved first, then plant divisions evolved in the following order: *Horneophytopsida*, *Rhyiopsida*, *Zosterophyllopsida* & *Lycopsida*, *Filopsida* & *Equisetopsida*, *Progymnospermopsida*, *pteridosperms*, *Pinopsida*, *Cycadopsida*, and *Gnetopsida* & *Magnoliophyta*. Among the animals, invertebrate animals evolved first, then some invertebrates evolved into jawless fish, then some jawless fish evolved into jawed fish, then some fish evolved into amphibians, then some amphibians evolved into reptiles, and then some reptiles evolved into mammals and birds. Among the mammals, insectivores evolved first, and then some insectivores evolved into primates. Among the primates, monkeys evolved first, then some monkeys evolved into apes, and then some apes evolved into humans.

The order that major groups of organisms appear in the fossil record is consistent with that expected by macroevolution. For example, the oldest rocks on the earth that contain any fossils at all, date at about 3.5 billion years of age and contain only fossils of bacteria. Tens of thousands of feet of sediments contain no evidence of any other kind of organism at all—no plants, no animals, no algae, no protozoa, no fungi, and no spores or pollen or molecules made by any of those organisms—only bacteria.

The oldest known eukaryotic fossils are protozoa and algae, dating at about 1.4 billion years. The oldest fossil evidence for plants dates at about 450 million years ago and vascular plant divisions appear in exactly the order predicted by macroevolution. The oldest fossil animals are invertebrates and they date at about 600 million years ago. The oldest vertebrates are jawless fish dated at 490 million years, the oldest jawed fish are dated at 424-408 million years, the oldest land animals are amphibians dated at 377-362 million years, the oldest non-amphibian land animals are reptiles dated at 322-303 million years, the oldest mammals are dated at 208-178 million years, the oldest primates are dated at 88-65 million years, the oldest apes are dated at 22 million years, and the oldest human-like fossils are dated at about 4 million years.

Stratomorphic Intermediates

If macroevolution is true, organisms continually evolve from one type into another through a series of intermediate forms. Since fossils are forming all the time, some of the intermediates should have been preserved in the fossil record. Older rocks should preserve fossils from the ancestral group, younger rocks should preserve the intermediates, and the youngest rocks should preserve fossils from the descendant group. Intermediate forms (or morphological intermediates) should be found in intermediate-aged rocks (or stratigraphic intermediate rocks). A morphological intermediate fossil found in a stratigraphic intermediate rock is called a stratomorphic intermediate. Stratomorphic intermediates can be classified into the following categories:

- *A stratomorphic intermediate taxon* is a biological group of organisms located in a stratomorphic intermediate position (*e.g.* mammal-like reptiles between reptiles and mammals; anthracosaurs between amphibians and reptiles; phenacodontids between Mesozoic mammals and horses; and *Baragwanathia* between rhyniophyte and lycopod plants).

  The same thing is evidenced by the oldest chordate *Pikaia*, the oldest bird *Archaeopteryx*, the oldest amphibian *Ichthyostega*, the oldest primate *Purgatorius*, the oldest whale *Pakicetus*, the oldest hominids *Aegyptopithecus* and *Proconsul*. In the case of *Pikaia*, evolutionists even predicted what it would look like before it was found.

- *A stratomorphic intermediate series* is a sequence of more than one stratomorphic intermediate between an ancestor and descendant in the correct order. Examples include a mammal-like reptile series formed by genera in each of several families of mammal-like reptiles, a suture series formed by ammonite genera, a whale series formed by early whale genera, a bird series formed by early bird genera, a Cantius series formed by species in the genus *Cantius*, a Plesiadapus series formed by species in the genus *Plesiadapus*, a horse series formed by genera in the horse family, an elephant series formed by genera in the elephant family, a camel series formed by genera in the camel family, a titanothere series formed by genera in the titanothere family, and a hominid series formed by species between apes and man.

- A *stratomorphic intermediate structure* is an intermediate structure located in a stratomorphic position. The fossil feather Praeornis is intermediate between the scales of reptiles and the feathers of birds. The very small jaw bones near the jaw joint in mammal-like reptiles are intermediate between the three jaw bones and one inner ear bone of reptiles and the single jaw bone and three inner ear bones of mammals.

Summary of Macroevolution Evidence

Looking over the entire argument for macroevolution, the evidence for it is quite strong. This is especially the case when one considers the power of arguments using consilience of inductions. Prosecutors in criminal court cases have a difficult task. In our court system the defendant is supposed to be presumed innocent until *proven* guilty, and for conviction, the jury is supposed to be convinced 'beyond the shadow of a doubt'. Yet, rarely is there an eyewitness or a taping of the crime. The evidence is often merely circumstantial. Nonetheless, convictions are not uncommon. Most of the time the reason for the conviction is that the prosecutor makes an argument by consilience of inductions. From each piece of circumstantial evidence an inference is made—a conclusion is arrived at by induction. When induction from more than one piece of circumstantial evidence arrives at the same conclusion we have consilience of inductions. When there is consilience from several, very different, unrelated evidences, the power of the consilience of induction is strong enough to convince a jury beyond a reasonable doubt.

The circumstantial evidence of adult homologies leads by induction to a conclusion of macroevolution, but so does the circumstantial evidence of embryological homologies, and each of numerous molecular homologies. The consilience of inductions between them *and* the similar evolutionary trees deduced from them argues strongly for macroevolution. But induction leads to the same conclusion from nested hierarchal classification, embryological recapitulation, suboptimal improvisations, vestigial structures, genetic throwbacks, biogeography, the percent living fossils in Tertiary rocks, the order of first appearance of fossils, and stratomorphic intermediates, not to mention the evidences from microevolution and speciation. The abundant evidence from comparative anatomy, embryology, molecular biology, taxonomy, genetics, biogeography, and paleontology, all leading to the same conclusion makes a powerful argument for macroevolution.

Perspectives on Biological Evolution

Perspectives on Mutation

Mutational Load

Most mutations are harmful. And there is no doubt that harmful mutations play an important role in the biological world—a source of considerable biological evil. Some mutations cause problems in the organism in which they occur. Other mutations alter the DNA in the germ line, so it is passed on to future generations. One mutation of this nature can cause problems in thousands or millions of descendants. The more that is learned about DNA, the more diseases are found to be caused by mutations. Even many of the mutations that are not linked to disease destroy genetic information, such as seems to be the case in pseudogenes.

Once a mutation occurs in a germ line, it is very difficult to eradicate from all its descendants. Consequently, organisms tend to accumulate mutations, increasing their overall **mutational load** (a measure of the number of harmful mutations carried by an organism or a species). In fact, unless there is a process that rids organisms of mutations that is completely unknown to biologists, organisms cannot have been here for millions of years or their mutational load would have killed them by now.

Although we cannot yet directly measure an organism's mutational load, the general health of organisms suggests they have accumulated only thousands of years of mutations. The low mutational load of organisms is inconsistent with the millions of years required for evolution to occur. It is also inconsistent with a long lineage that takes an organism back millions of generations to the first bacterium.

---

mutational load—the number of harmful mutations carried by an organism or species (aka genetic load)

Biological Information

For biological evolution to be true, mutations are the source of whatever DNA information in the present world was not carried on the DNA of the first bacterium. There are a couple problems with this claim. The first is that no random process (like mutation) is known to generate any real information, and the amount of information in the world's DNA is truly extraordinary. Secondly, randomly changing the alphabet letters of a message always destroys the meaning of that message. It never creates a new message that has any sort of meaning.

Since the information on DNA appears to be conveyed in the form of a language, not only should mutations be incapable of adding information, they should also be incapable of altering information in any meaningful way. Mutations are incapable of providing the biological information that biological evolution requires them to provide.

Lack of Beneficial Mutations

As critically important as beneficial mutations are to evolutionary theory, it has been extremely difficult to find good examples of truly beneficial mutations. The only examples that have been reported to date are beneficial to an organism only when the organism is under considerable duress. For example, the only benefit that the sickle-cell anemia mutation brings to someone is some resistance to malaria. When a person is not in the presence of malaria, the mutation is harmful.

The remaining examples of beneficial mutations are reported in organisms that were about to die—bacteria that would have starved unless they learned to derive energy from something they have never been able to consume before, or viruses that would have been destroyed by an immune system, or bacteria that would been exterminated by antibiotics, or mosquitoes by insecticides, or rust fungi by fungicides. It was under these conditions of duress that mutations supposedly occur that allow the organism to survive[a]. In each case, the mutant form is beneficial to the organism only in the conditions under which it arose. When the conditions of duress are lifted, the non-mutant forms outcompete the mutant varieties. Closer examination of a couple different organisms has revealed that mutations become more common in the organism while the organism is under duress.

---

[a] Mutations are difficult to prove, so few of them have actually been proven to be due to mutations.

As soon as a mutation arises that allows the organism to thrive under those conditions, mutations in that organism become uncommon again. This research suggests that these organisms actually control their mutation rate, probably by controlling the processes that check the DNA for mistakes and then correct the mistakes. This in turn suggests that this is part of a higher design. It is possible that this is a survival mechanism placed in organisms at the creation in order to give those organisms a chance to survive when faced with situations that threaten to destroy them. All known beneficial mutations might actually be examples of a survival mechanism built into organisms to permit their survival through life-threatening situations. If so, such a clever mechanism is almost certainly the design of a wise Creator desiring to preserve His organisms. Furthermore, we are left with no known examples of the kind of beneficial mutations that biological evolution requires.

Perspectives on Natural Selection

Evil Minimization

Natural selection is usually described as a process that chooses the organisms that are better fit for a particular environment. Technically, natural selection is the opposite—the process that eliminates the organisms that are *less* fit for a particular environment. When 'selection' or 'selection pressure' is put into mathematical form, it is the fraction of the population that *fails* to pass on its genetic material to the next generation. This fact points to a function of natural selection in the biological world that is rarely acknowledged by evolutionary biologists—the minimization of biological evil.

In experiments where natural selection is taken away from a population, the overall health of the populations drops. Disease spreads through more of the population, and disease becomes more common. Natural selection takes out the weaker organisms, including the unhealthy ones. Natural selection, then, tends to take out disease. Natural selection keeps the natural evil in the world to a minimum.

Reversible

Natural selection determines what fraction of a population possesses particular character traits. In an environment where Character Trait A is favorable and Character Trait B is unfavorable, natural selection tends to take out B, causing A to be more common and B less common.

In another environment where Character Trait A is unfavorable and Character Trait B is favorable, natural selection tends to take out A, causing B to be more common and A less common.

What natural selection changes in one direction, it is equally capable of changing back in the other direction. Studied by Peter and Rosemary Grant over the course of three decades, Darwin's finches on the Galapagos Islands are probably the best studied example of natural selection known. The commonness of particular character traits in Darwin's finches changes with El Niño, a cyclical climatic effect. Since the climate is cyclical, finch character traits change cyclically. All known examples of natural selection seem to work the same way. The percentage of winged flies increases again when the fan is turned off and the fly paper taken out of the terrarium. The percentage of flies killed by alcohol increases again when the flies are raised outside the brewery. The percentage of pests killed by chemical controls increases again when the chemical controls are taken away. The percentage of dark peppered moths returned to less than 10% again 50 years after they became most common in the 1890's.

According to how natural selection is supposed to work in theory, and in the case of every known biological example, natural selection causes organisms to change as the environment changes. According to biological evolution, natural selection is what causes organisms to change in a particular direction. If so, then examples of speciation in the fossil record should show a strong correspondence between the environment and biological change. In fact, in not a single one of the claimed evidences of microevolutionary change in the fossil record (for example, in *Cantius*, or Lake Turkana mollusks, or in *Theocorythium*, or in *Globorotalia*) is there any relationship between the biological change and any change in the environment (as might be evidenced by different sediments). The changes in biology seem to be occurring without any corresponding changes in the environment.

Furthermore, it is most common for evolutionary theorists to explain why a species does *not* change by suggesting that the species is already fit to the environment and the environment does not change. For as long as the environment does not change, natural selection would keep the species from changing.

Unfortunately for this hypothesis, species remain unchanged in the fossil record through huge changes in sediment (usually interpreted as changes in environment). Species seem to remain in stasis even though the environment changes radically. Consequently, although examples of natural selection do operate in the *modern* world, when applied to the fossil record natural selection seems to explain neither biological change (because it is usually not associated with evidence of environmental change) nor stasis (because it is often associated with evidence of environmental change).

Too Slow

Early in the twentieth century, one of the developers of the theory of **neo-darwinism**, J. B. S. Haldane, expressed concern over whether natural selection could be responsible for the amount of biological change required in evolutionary theory. Because natural selection depends on the death of organisms, the speed of natural selection determines how many organisms must die. Fewer organisms die if natural selection operates slowly, but for it to occur faster more organisms must die.

For a species to survive, it must be able to produce young faster than natural selection takes them out. The number of young that a species can produce puts a limit on how fast natural selection can operate. Haldane calculated that given the average number of young produced by organisms, at least 300 generations would be required to 'fix' a particular trait (To fix a trait is to raise its commonness to 100%). This is assuming only one trait is worked on at a time, so two traits would require at least 600 generations whether worked on one right after another or whether worked on at the same time. Three traits would require 900 generations, four would require 1200 generations, and so on. This is extremely slow change. Humans and bonobo chimps show about 98% similarity in the nucleotide sequences of their DNA. 2% difference in 3 billion nucleotides amounts to 60 million nucleotide differences.

If these differences are all fixed and considered separately, and if natural selection is responsible for fixing them, by Haldane's estimate it would require 9 billion generations in each lineage, or some 180 billion years. Even with the great ages deduced from radiometric dating, there is only about 4 million years of time available for this transition. In 4 million years there is time for natural selection to fix only about 1300 nucleotides—1/45,000 of the number of differences between

---

neodarwinism— naturalistic evolution theory of the 1920's to 1940's, updating evolution with genetics

humans and chimps. Haldane's concern (which is sometimes referred to as 'Haldane's dilemma') has never been solved by evolutionary theorists. Natural selection seems to be too slow to explain the amount of biological change necessary to change one species into another.

Timing

According to evolutionary theory, natural selection changes an organism *after* the organism experiences a change in environment. Either the environment changes and the organism changes in response, or the organism moves into a new environment and changes in response. Charles Darwin, for example, claimed that organisms migrated from South America to the Galapagos Islands and then migrated from island to island, adapting by natural selection to each island.

Modern research, however, is able to measure the (unseen) genetic differences among Galapagos species. The large amount of genetic difference in many Galapagos species suggests that these species were most probably established *before* they migrated to the Galapagos Islands. Other organisms in other locations are showing the same thing. Again, although natural selection does operate in the modern world, it is becoming more and more doubtful that it is responsible for any substantial biological change, especially that which leads to new species.

What we know for sure about natural selection is that it minimizes biological evil and allows species to change in minor ways to track the environment. Evidence seems to argue against natural selection being capable of more substantial biological change. What is most likely is that all substantial biological change is due to information built into organisms at the creation.

Perspectives on Speciation

Isolation Without Infertility

All theories of speciation must explain the common feature of the biological world that species tend to be distinct and stable. Most species are easy to identify and distinguish from other species, and species tend to remain unchanged for long periods of time. The ancient Greeks observed this about species and explained it by saying that as an individual organism develops, it strives to achieve the perfect form for that species—a form that exists in the world of perfect, invisible, unchanging forms and ideas known as the logos.

Crosses between species would violate the order of the logos, so the natural order is hard on the offspring (imposing such things upon it as banishment, ugliness, sterility, or even death). The sterility of the mule (a cross between a female horse and a male donkey) was brought up as one of the most popular proofs of this characteristic of the natural order.

In the Middle Ages, as Greek ideas were incorporated into Christianity, the distinctiveness of species was explained to be a consequence of God (the mind of perfect forms and ideas) creating species ('kinds') that were to remain distinct by divine command. Again, reflective of Greek thought, crossing between species was generally considered against God's order and punishable by God by banishment, ugliness, sterility, or even death. And, once again, the mule was a favorite proof.

More recently, advocates of naturalism, not able to appeal to either God or the logos, explain the distinctiveness of species by populations of organisms losing the ability to interbreed with other species. And, once again, the mule is used as evidence. As biological evolution has developed, all speciation theories are dependent on species being unable to interbreed with other species. After all (so the reasoning goes), if species *can* interbreed, then it seems that they *would* interbreed, and if they *did* interbreed, then species would not be distinct. Mayr's peripheral isolate theory of allopatric speciation, for example, points to examples of various degrees of infertility as evidence.

Unfortunately for these theories, instances of partial or full infertility are uncommon in the biological world. It appears that a majority of species in any given family can interbreed and produce fertile offspring. Even considering the horse family, there are nine species of horses—three zebra species, three ass species, and three horse species—and either directly or indirectly every species combination has produced fertile offspring. Even in the case of the horse and donkey, the hinny (a cross between a female donkey and a male horse) is fertile. Even one in a thousand or so mules (crosses between a male donkey and a female horse) is fertile.

It would appear that a majority of species are capable of inter-specific hybridization in biological families throughout the biological world and infertility in the offspring of these crosses is uncommon[a]. This suggests that infertility is not the reason for species distinctiveness (contrary to the expectation of speciation theories). Also, contrary to the belief of Ernst Mayr and most evolutionary biologists, geographic isolation is not necessary for the generation of new species.

Stasis and Abrupt Appearance

Eldredge and Gould reasoned that if Mayr's peripheral isolate theory of allopatric speciation were true, then most species would appear suddenly in the fossil record, without intermediate forms from previous species, and species would remain unchanged for the duration of their fossil record. Mayr's theory could provide an explanation for stasis and abrupt appearance of most *species*, but stasis and abrupt appearance is not just the pattern of species. Most genera appear abruptly in the fossil record and maintain their form throughout the fossil record. The same is true for families, orders, classes, phyla, kingdoms, and domains. In fact, it is even true for communities of organisms in the fossil record.

Yet, since Mayr's theory only deals with speciation (the origin of new species) it cannot explain stasis and abrupt appearance of anything but species. This suggests that stasis and abrupt appearance of species is *not* the result of Mayr's theory at all, but rather the consequence of a much bigger pattern of life, something that includes all groups of organisms at all taxonomic levels.

Species Designs

Plants commonly produce new species using a process that is uncommon in animals. During meiosis it is fairly common in plants for all the homologous chromosomes to remain attached in anaphase I. This results in offspring receiving extra sets of chromosomes, thus doubling the chromosome number, or tripling it, and so on. Rather than the offspring being diploid (each gene carried on two homologous chromosomes), the offspring are triploid (each gene carried on three homologous chromosomes), or tetraploid, and so on. Not only does such a **polyploid** usually look different from its parents, but it often cannot interbreed with them either.

polyploidy—cell state when it contains more than two homologous chromosomes

---

[a] The fact that this is not widely known appears to be due to the lingering impact of Greek thought in our culture.

In this manner, a new species is produced immediately—in the course of a single generation. Counting the numbers of chromosomes in plants suggests that at least 70% of plant speciation may have occurred by such a process. The result in the fossil record is the same stasis and abrupt appearance for plants as is seen for other organisms. But this also suggests that another reason for the lack of intermediates in the fossil record of all organisms is that they never existed.

God created baramins in a netted hierarchal pattern to illustrate His unified hierarchal nature. If God also hid information in those original organisms that He wished humans to express to His glory, then it stands to reason that He would create that information in the same netted hierarchal pattern. Furthermore, if that same hidden information was also there to restore intra-baraminic diversity after the Flood, then intra-baraminic diversification would yield the same netted hierarchal pattern among the species that arose after the Flood. In fact, this *should* have happened so that a similar netted hierarchal pattern was seen by people after the Flood as was seen by people before the Flood (thus illustrating the same nature of the same God). This suggests that each species is a result of a **species design** (a package of information capable of generating and maintaining a species) and the species designs of the world are themselves arranged in the same netted hierarchal pattern as are the baramins.

species design— package of information capable of generating and maintaining a species

Furthermore, God hid many of these species designs in the organisms of the original creation for man to reveal to the glory of God and to restore biological diversity after the Flood. Such species designs would allow the sudden appearance of species without intermediates (even when such intermediates are possible) and they would explain the stability of species (even when inter-specific hybridization is possible). The hierarchal arrangement of the species designs would explain why the hierarchal arrangement of created groups matches the hierarchal arrangement of groups produced by intra-baraminic diversification. It would also explain the stasis and abrupt appearance in the fossil record of both communities and taxa at every taxonomic level.

Recent Diversification

Inter-specific hybrids argue rather persuasively that groups of modern species and even genera are closely related to each other and could have been descendant from a common ancestor. Vestigial hind legs in fetal whales combined with fossil whales with legs provides fairly strong evidence that modern whales are descendant from whales with legs, or at least contain the information for developing legs. Vestigial toes in horses, combined with three-toed genetic throwbacks, a stratomorphic trajectory of horses in the Cenozoic, and three-toed fossil horses makes an extremely strong case that modern horses are descendant from a three-toed horse.

However, the same evidences suggest that if these things happened, they did so very recently. It is easy to lose the ability to interbreed. The many couples not able to have children can testify to that. If scores, hundreds, or even thousands of species in a particular biological family can still interbreed and produce fertile offspring, then the diversification that produced the species in that family did not occur that long ago, or else many of these species would have lost their ability to interbreed. The intra-baraminic diversification evidenced by inter-specific hybridization fits much better with diversification after the biblical Flood than it fits the multi-million-year scenario of biological evolution.

Similarly, the ongoing process of mutation destroys genetic information. Although natural selection does tend to eliminate *expressed* harmful mutations (as less fit expressions of a gene), natural selection cannot eliminate *unexpressed* mutations. Since the average nucleotide will be mutated every million generations or so, a million generations should alter every single nucleotide of a gene, rendering it completely unusable—even unrecognizable as a gene. In a few million years, there should be absolutely no genetic information left in unused genes. Vestigial organs and genetic throwbacks suggest that the genetic information has not been unused for very long, and if modern whales are descendant from the legged whales in the fossil record and modern horses are descendant from the three-toed horses in the fossil record, then this has happened recently—probably since the Flood only several thousand years ago.

This, in turn, would suggest that the Cenozoic fossil record is not recording millions of years of time, but only thousands of years. This, in turn, would suggest that organisms diversified much too rapidly to be explained by any theory of speciation currently proposed by biological evolution. Rapid revelation of hidden species designs is the only reasonable explanation for such rapid speciation.

Perspectives on Macroevolution

Common Hierarchal Design

When God created baramins, He created them in a netted hierarchal pattern as an illustration of the unified hierarchy of His own nature (Chapter 11). When He peppered organisms with biological similarity so as to demonstrate that there is only one God (Chapter 9), He created that similarity in the same netted hierarchal pattern. A portion of this similarity is interpreted in biological evolution as homology and is used to create concordant evolutionary trees and classify organisms in a nested hierarchy. The remainder of the similarity is the 'netted' aspect of the 'netted' hierarchical pattern. This similarity explains why a number of organisms are not easily classified into a nested hierarchy (such as the pronghorn and the lesser panda) and why every classification generates so many homoplasies. It would also explain why very similar organisms can be found in very different groups (such as both marsupials in Australia and placentals in North America having examples of skunks, flying squirrels, moles, saber-toothed tigers, wolves and many others).

Whereas a unified hierarchal creator can explain the netted hierarchal similarities of life, macroevolution has difficulty explaining the same thing. The netted hierarchal pattern that God created, not only explains the homology, concordance of evolutionary trees, and nested hierarchy marshaled as evidence for macroevolution, but also explains the 'netted' aspect of biological similarity not readily explained by macroevolution.

Optimal Design

Although the world was perfect when it was created, our present world is not perfect. Since the Fall, biological degeneration has been accumulating in the creation. This is seen, for example, in increasing mutational load, diversity of disease, and biological suffering. Mutations after the Fall have also destroyed biological information. It might be that pseudogenes have something to do with the information God hid in organisms, but it could also be that some vestigial structures, including some or many pseudogenes, are evidence of that degeneration.

Setting aside the biological evil that has entered the creation since the Fall, the creation is astonishingly close to optimal. Most of the time humans have been unable to design anything capable of even equaling the design of the biological world, let alone something with a better design. Claims of suboptimal design have been made, but most of them involve making the claim that there is a better structure somewhere else in the biological world (such as humans having a better thumb than pandas). But optimal design does not necessarily involve giving an organism a perfect form of a given structure. When God created the biological world, he created biological *systems* (Chapter 8). A multiplicity of parts (reflective of His plurality—Chapter 10) were put together in complex cooperative relationships (reflective of His personhood—Chapter 6) to accomplish unified purposes (reflective of His unity—Chapter 9).

Every biological structure was created for the biological system into which it was placed. Therefore, it was the *whole* that was optimally designed. For the sake of the whole system, no part of the system—no single organism, no single organ system, no single organ, no single cell, no single organelle, no single strand of DNA, and so on—was created infinitely capable. Consequently, the optimality of a component should be measured *within* the system in which it was placed. Time after time claims of suboptimal design have not been found to be truly suboptimal upon consideration of imperfection that has entered the creation since the Fall and (most importantly) upon close examination of the biological system in which the structure is found. For example, although the panda's thumb would not be the optimal thumb for a human, for thousands of years it has worked quite well for pandas. Furthermore, after Stephen Jay Gould made it famous as an example of suboptimal design, the panda's thumb was discovered to be designed rather like vice-grips to be better than any other known thumb for precision pinching. So much for suboptimal design.

It is the optimal design of embryology that provides a better explanation for embryological recapitulation than is provided by macroevolution (Chapter 14). An organism's phylogeny is the best organism-to-organism process that humans can imagine in order to generate that organism from a single celled bacterium. It is an *optimal* sequence.

The very reason that an organism's (optimum) phylogeny is compellingly similar to the same organism's embryology is because the embryology is the *optimal* sequence for developing that organism from a single cell. In fact, the similarity is so compelling that embryological recapitulation continues to be used as evidence for macroevolution even though just about every embryological recapitulation claim that has ever been made has proven to be only a visual similarity. It is extremely difficult to find any embryological stage that could in any way be related to the adult form of any ancestor.

God's optimal design goes beyond explaining the near-optimality, the so-called suboptimal improvisations, and embryological recapitulation evidences marshaled for macroevolution. It even goes beyond providing a *better* explanation for suboptimal improvisations and embryological recapitulation. God's design can explain what macroevolution cannot even begin to explain, namely the ubiquity and abundance of exquisite designs in the biological world.

This is true at every level of biological design: from complex individual molecules such as the motor-like design of ATP synthase[a], to systems of molecules such as that which clots blood, to organelles such as the mitochondrion, to cells such as extremophiles that live in boiling water, to organs such as the jackhammer design of the woodpecker head, to organisms such as cheetahs which can accelerate from 0 to 62 miles per hour in less than three seconds, to organism societies such as that of the honey bee, to mutualisms such as lichens, and to communities such as the vent communities at the bottom of the ocean. Examples of exquisite biological design are too many to enumerate.

Intra-baraminic Diversification

In the original creation, God hid 'species designs' (explained above) in organisms with a process that could reveal that information so that the organisms which left the ark could multiply and refill the earth in the course of a couple centuries (Chapter 14). The existence of hidden species designs would explain the large number of different breeds, varieties, and cultivars of plants and animals that have been developed by breeders.

Other evidence for hidden information is found in vestigial organs and genetic throwbacks[b]. Since the Paleozoic and Mesozoic sediments are thought to be sediments of the Flood (because of their

---

[a] Although ATP synthase is considered a single protein, it is constructed from 8 protein subunits constructed from information on at least 13 different genes.

[b] Until we know for sure that modern birds actually descended from birds with teeth, ratite birds actually descended from birds that could fly, and snakes actually descended from snakes with hind limbs, the information for building teeth that chickens seem to have, and the reduced wings of ratites and hind limbs of pythons should be considered nothing more than of this hidden information.

transcontinental scale), Cenozoic sediments preserve a record of the intra-baraminic diversification that occurred immediately following the Flood. This intra-baraminic diversification would explain:

- stasis and abrupt appearance in the Cenozoic because new species are generated quickly and tend to persist in the 'species design' of species as well as genera and higher taxa, for which macroevolution has no explanation;
- the increase in percentage of living species in younger and younger sediments of the Cenozoic because species were generated continuously throughout the Arphaxadian;
- stratomorphic intermediate series within families in the Cenozoic because intra-baraminic diversification generated a string of organisms throughout the Arphaxadian (examples including the *Cantius*, *Plesiadapus*, horse, elephant, camel, and titanothere stratomorphic series);
- parallel evolution in North American mammal families in the Cenozoic because similar designs were placed in different baramins of increased body size (as in the case of horses, titanotheres, elephants, and camels), and of hypsodonty (as in the case of rabbits, horses, elephants, and camels), which is extremely difficult to explain in macroevolution;
- vestigial toes and three-toed genetic throwbacks in horses because modern horses are descendant from a three-toed horse species early in the Arphaxadian, also evidencing *recent* ancestry, which is contrary to macroevolution;
- sudden appearance of C4 photosynthesis in 16 different plant families in the Cenozoic because God placed the design for C4 photosynthesis in many plant baramins, something which is extremely difficult to explain in macroevolution.

The fossil and living whales are classified in several biological families in the order Cetacea. Some creationists believe that there might be one baramin of whales. It is possible that only one walrus-like whale species existed before the Flood and the whales diversified following the Flood to replace the water-dwelling reptile groups that were exterminated in the Flood (such as plesiosaurs, ichthyosaurs, mososaurs). If so, then the post-Flood diversification of whales would explain the stratomorphic intermediate *Pakicetus* species, the stratomorphic whale series, the vestigial hind limbs in whale embryos, and the vestigial teeth in baleen whale embryos.

Simultaneous with the intra-baraminic diversification, organisms after the Flood were also filling the earth. Land animals, for example, had to migrate from the ark to all the continents of the earth. Vegetation mats blown around the oceans by wind may have aided in this migration in the centuries immediately following the ark (Chapter 14). Intrabaraminic diversification occurring simultaneously with migration and rafting would explain the following:

- for most species, the most similar species to it has an adjacent geographic range …explained by macroevolution
- for many island species, the most similar species is found on the nearest mainland …explained by macroevolution
- for many species with limited geographic ranges, the most similar fossil species has a similar restricted geographic range …explained by macroevolution
- the more distant an island is from everything else, the greater the number of unique species …explained by macroevolution
- areas with the largest number of unique species correspond with places where currents approach continental coasts …not explained by macroevolution
- many disjunct species …unexplained by macroevolution
- all fossil and living species of Australian marsupials are only found on Australia (if Australian marsupials are one baramin) …explained by macroevolution
- Australia was colonized from South America …not explained by macroevolution
- many similar species between Australian marsupials and North American placentals …extremely difficult to explain by evolution

Waters On and Off the Continents

The pre-Abraham earth history outlined in Genesis, provides a means of explaining many of the characteristics of the fossil record (Chapter 15). If the continents were created early on the third day by being raised up, letting the water run off them, then this could be the source of most of the Precambrian sediments. This would explain why they contain only bacteria and archaea fossils because that may have been all that had been created at that point. Since a full disparity, high diversity biomatrix was deposited in the course of a single day, there was no time for the development of intermediate forms or for change in species. This not only explains the stasis and abrupt appearance of all taxa, but it also explains the 'Archaean explosion'—the full disparity of Archaea and bacteria appearing in the oldest sediments of the earth. Because most of the Paleozoic and Mesozoic sediments are deposited on trans-continental or global scale, they are thought to be the sediments deposited during the Flood. Similar to the day three regression, the Flood was a rapid burial of organisms from a full disparity and diversity of organisms.

Since the Flood only lasted a bit over a year, there was not enough time for many organisms to change. And, when the Flood began there as a *complete* complement of baramins on the planet—not just the biomatrix that was there at the time of the day three regression. This again would explain the stasis and abrupt appearance of taxa from all taxonomic levels as well as the 'Cambrian explosion'—a full disparity of animals living on the bottom of the ocean. The movement of the Flood onto the land would explain why invertebrates on the ocean bottom would get buried before swimming organisms, swimming organisms before floating organisms (for example the floating forest ecosystem), and floating organisms before land organisms.

The remarkable order of Paleozoic plants is because they were living in that same order from the outside to the inside of the floating forest—from plants that live in the water to plants that need standing water, to plants that need less water and so on. And that was the order that the flood destroyed the floating forest. The destruction of the floating forest also explains the order of the animals living in the environments of the floating forests (from fish along the shore to fish-amphibian intermediates on the thin vegetation mat at the edge of the forest to amphibians in a heavily ponded region, to amphibian-reptile intermediates in a thicker part of the mat, and true land animals where the forest mat was thickest.

On the land, dinosaur and gymnosperm plant communities appear to have been separate from mammal and angiosperm plant communities, and nearer to the ocean. In between them, God apparently placed an intermediate plant community with intermediate organisms such as mammal-like reptiles and reptile-like birds. Burying the communities in this order explains the sequence of fish before fish-amphibian intermediates, before *Eusthenopteron* and other amphibians, before anthracosaurs, before reptiles, before mammal-like reptiles and *Praeornis* and *Archaeopteryx* and the early bird series, before mammals and birds. It would also explain the similar organisms found in different families of the mammal-like reptiles and in different early birds as one steps up the Mesozoic sediments (interpreted as very specific examples of parallel evolution, that are extremely difficult to explain in macroevolution).

Human Disobedience

Following the Flood, humans disobey God and refuse to disperse across the earth. This explains why they first appear very late in the Cenozoic, after the intra-baraminic diversification of most organisms is complete.

Major Features of Life

There are a number of characteristics of life that are not explained at all by evolution but are explained by the God of the Bible. They include:

- biological beauty (explained by creation by the God of glory: Chapter 3)
- biological systems (explained by creation by the God of unity: Chapter 9)
- animal behavior (explained by creation by a personal God: Chapter 6)
- animal souls (explained by creation by a personal God: Chapter 6)
- diversity (explained by creation by a God of plurality: Chapter 10)
- discontinuity (explained by creation by a God Who is like none other: Chapter 4)
- mutualism (explained by creation by a good God: Chapter 9)
- the language of life and DNA (explained by creation by the Word: Chapter 13)

- each new organism's non-physical life (explained by creation by the living God: Chapter 14)

Summary of Appendix

A. If macroevolution has a strong argument by consilience of inductions, then by consilience of inductions creation has an argument several times stronger. Almost all the evidences of macroevolution can be explained by creation. In fact, most of the macroevolution evidences are *better* explained by creation. But most of the macroevolution evidences are unusual, subtle, or rare features of life. They do not represent the major features or themes of life.

B. Advocates of naturalism are arguing for an explanation that is foreign to this world. The creation has to be scoured high and low for any evidence that might be explained by naturalistic hypothesis. When it comes to the major characteristics of life, very few of them are in any way addressed by macroevolution. The world truly is the Lord's.

Test & Essay Questions

1. Explain the steps of abiogenesis.
2. Why was the Miller-Urey experiment performed and what sort of things have been created in this and similar experiments?
3. What is the significance of the four most common amino acids generated in Miller-Urey experiments?
4. Comment on the success of origin of life experiments at creating life.
5. What have origin of life experiment *not* been able to create?
6. What is homochirality and what relevance does it have to origin of life experiments?
7. Comment on origin of life experiments and the fact that there are several chemical steps involved in abiogenesis.
8. What does it mean to say that there is very deep discontinuity surrounding life?
9. Compare and contrast microevolution, speciation, and macroevolution.
10. What is neo-darwinism?
11. What role does natural selection / do mutations play in biological evolution?
12. List two/three examples of beneficial mutations / of natural selection.
13. Explain Mayr's peripheral isolate theory of allopatric speciation.
14. List three/four/five evidences for Mayr's peripheral isolate theory of allopatric speciation.

15. What did Stephen Jay Gould and Niles Eldredge predict to be true if Mayr's peripheral isolate theory of allopatric speciation were true?

16. Explain the theory of punctuated equilibria. / What kind of evidence do you expect for punctuated equilibria?

17. What does stasis and abrupt appearance mean? / How does biological evolution explain stasis and abrupt appearance of species / of genera and higher taxa?

18. What is a homology? / Give two/three/four examples of homologies.

19. Compare and contrast adult, embryological, and molecular homologies. / Give one example of each of the following: adult homology, embryological homology, molecular homology

20. How are evolutionary trees constructed from homologies and what does the ability to construct one suggest?

21. What is the significance of concordance of evolutionary trees?

22. How is the nested hierarchy of life indicated? / What is the significance of being able to classify organisms in a nested hierarchy?

23. Provide an example of embryological recapitulation and what it means.

24. Why does macroevolution expect life to be near-optimal? / Why does macroevolution expect suboptimal improvisations? / Give an example of a suboptimal improvisation and why it is a suboptimal improvisation.

25. Explain what a vestigial structure / what a genetic throwback is and what it means. / List two/three/ four examples of vestigial structures. / List two examples of genetic throwbacks.

26. Compare and contrast vestigial structures and genetic throwbacks.

27. List two/three different kinds of biogeographic evidences for macroevolution.

28. Explain the relevance of the biogeography of Australian marsupials to macroevolution.

29. What does macroevolution expect about the commonness of living species in various rocks in the fossil record.

30. Explain how the order of first appearance of fossils relates to macroevolution.

31. What is a stratomorphic intermediate? / List two/three different types of stratomorphic intermediates. / Give one/two example(s) of each of the following: stratomorphic intermediate taxon, stratomorphic intermediate series, stratomorphic intermediate structure.

32. To what does consilience of inductions refer? / What does consilience of inductions got to do with macroevolution? / Present a consilience of inductions argument for macroevolution.

33. What is mutational load and what does the mutational load of organisms indicate?

34. Explain why biological information is a problem for mutations.

35. What does it mean to say that examples of beneficial mutations are only found under conditions of duress? / What might beneficial mutations suggest about God's design in biology?

36. What role does natural selection play in the minimization of biological evil? / When did natural selection begin?

37. What does it mean to say that natural selection is reversible?

38. What is natural selection not able to explain about species stasis / about gradual species changes in the fossil record?

39. Why did J. B. S. Haldane believe that natural selection might be too slow to explain macroevolution?

40. What is the significance of a Galapagos Island species being established *before* it arrived in the Galapagos Islands?

41. What are two ways that inter-specific hybrids pose a challenge to theories of speciation in biological evolution?

42. What is a species design?

43. How does polyploidy result in speciation?

44. What evidence indicates that intra-baraminic diversification has occurred recently? How is recent diversification a challenge to macroevolution?

45. What macroevolution evidences are explained by common hierarchal design / by optimal design / by intra-baraminic diversification, and how are they explained?

46. How is homology / concordance of evolutionary trees / nested hierarchy / near-optimality / suboptimal improvisations/ embryological recapitulation / explained by creationism?

GLOSSARY

abiogenesis—naturalistic theory for the origin of the first cell from non-living material

abrupt appearance—status of a fossil organism when it is not preceded in the fossil record by intermediates to any other fossil organism

acid rain—rain made more acidic than normal due to acid-causing pollutants in the atmosphere

aging, degenerative—gradual wearying of an organism—a negative effect of the curse

alga (pl. *algae*)—eukaryotic organism that is generally a one-celled, aquatic producer

allele—one of multiple character states for a particular gene

amino acid—monomer for proteins

anabolism—chemical reactions in cells that build molecules

anaphase—mitosis substage when sister chromatids separate to opposite ends of the cell

anaphase I—meiosis substage when homologous chromosomes separate to opposite ends of the cell

anaphase II—meiosis substage after cytokinesis I, when sister chromatids separate in each cell

animal—(multi-cellular) eukaryotic organism whose cells lack cell walls (also possesses biblical life)

antediluvian epoch—period of earth history between the curse and the Flood (at least 15 centuries long)

anthropic principle—the universe has characteristics that suggest it is was fashioned with man in mind

Arphaxadian epoch—first few centuries after the Flood, when the earth began recovering from the Flood

atom—an electrically neutral arrangement of proton(s) and an equal number of orbiting electrons(s)

atomic number—number of protons in an atom's nucleus (determining the properties of that atom)

autotroph—organism that gets energy from the non-biological world. Types: photoautotroph gets energy from sunlight; chemoautotroph gets energy from inorganic molecules

bacterium (pl. *bacteria*)—single-celled organisms small enough not to need organelles (aka prokaryote)

baramin—recognizable group of similar organisms surrounded by deep discontinuity that persists by members interbreeding and producing similar offspring (aka biblical kind)

baraminology—the science of the discovery and study of baramins; a creationist biosystematics method

beauty—the attractive holistic fit of an entity's attributes; deep beauty is beauty at multiple scales

bioaccumulation—accumulation and storage of pollutants by an organism

biodegradable—decomposable into components that organisms can use (vs. non-biodegradable)

biodiversity—the number of taxa (usually species)

biogenesis, law of—a general observation that a given type of organism only arises from another organism of the same type, sometimes briefly summarized by 'life only comes from life'

biogeochemical cycle—organism-driven process continuously supplying organisms with a required element by fixing the element from an inorganic reservoir, passing it through all organisms and then returning the element to the reservoir

biogeography—discipline of biology that studies where organisms live and how they got there

biological evil—anything in the biological world that causes suffering of humans or animals

biology—a science that studies organisms

biomatrix—system of bacteria, algae, protozoa, and fungi required for plants and animals to live on earth (aka organo-substrate)

biome—flora and fauna found in a particular climate and across a large area of the earth's surface

bioremediation—transformation of a harmful substance into a less toxic form

biosystematics—the science of classifying (grouping) and naming organisms

breeding—human activity of controlling what sorts of organisms are produced in the next generation by selecting which parents are mated (or 'crossed')

breeds—distinguishable group of animals that remains distinct only with breeding

Calvin cycle—chemical reactions in photosynthesis' light independent reactions that fix carbon

capillary action—natural tendency of a liquid's surface to rise where the surface intersects a vertical wall

carbohydrate—organic molecule with carbon, hydrogen, and oxygen atoms in a ratio of 1:2:1

carnivore—an animal that eats another animal (aka predator)

carnivory—the eating of animals by other animals (aka predation), an evil-minimizing effect of the curse

carrying capacity—the number of organisms a particular environment can support over the long term

catabolism—chemical reactions in cells that break down molecules

catastrophic die-off—the death, due to harsh conditions, of a large percentage of a population—usually because that population exceeds the carrying capacity of the environment

cell—tiny, membrane-bound structures that make up all organisms; smallest unit of biological life

cell cycle—stages and substages in the life of a cell. Stages: interphase when the nucleus is not dividing; mitosis or meiosis when the nucleus is dividing

cell theory—1: every organism is made of one or more cells; 2: the cell is the smallest unit of biological life; 3: every cell comes from the division of a pre-existing cell; 4: all metabolism occurs within cells; 5: heredity information is located in cells; 6: every cell has the same basic chemical ingredients

cell wall—structure outside the cell membrane strengthening cells of bacteria, algae, plants, and fungi

cellular reproduction—cell process that divides a (parent) cell into two or more daughter cells.

cellulose—complex carbohydrate making plant structure (fiber); earth's most abundant organic molecule

centromere—section of a chromosome containing proteins for attachment

chemoautotroph—organism that gets its energy from inorganic molecules

chemosynthesis—cellular process that extracts energy from chemicals in the physical environment

chitin—complex carbohydrate making up cell walls of fungi and exoskeletons of arthropods

chloroplast—triple-membraned organelle where photosynthesis occurs

Christian theism—a worldview that begins with the existence of the Christian God, Creator of the physical world and its organisms as well as the non-physical world and its non-physical beings

chromosome—(in a eukaryotic cell) a linear segment of DNA and its associated proteins

citric acid (or Krebs) cycle—aerobic respiration reactions that extract energy from carbon-hydrogen bonds

class—taxonomic level above order and below phylum

climate—average temperature (and rainfall) at a given location (*e.g.* tropical; temperate; polar; arid)

climax community—the stable community for a particular area; the last in community succession

clone—an organism with identical DNA as another organism

cloning, reproductive—producing from one organism, another organism with identical DNA. Types: horticultural cloning by rooting a plant cutting; embryonic splitting by separating the first cells of a developing animal; adult cloning by getting an adult cell to develop into a mature organism

codominant—allele type that is fully expressed when both homologous alleles are codominant

codon—sequence of three nucleotides on DNA or RNA that represents a 'letter' of the genetic code

commensalism—symbiosis where only one organism benefits; the other(s) neither harmed nor helped

community—(biological) system of plant, animal, and biomatrix species found in a particular location

community succession—a sequence of communities in a particular area leading to the climax community in that area, each community altering the environment for the community to follow

competition—the struggle of organisms to survive when resources are limited; caused by overproduction, leads to natural selection, and designed as an evil-minimizing effect of the curse

compound—a combination of elements that exhibits properties different from the component elements

condensation—the dense packing of a chromosome that occurs at the beginning of mitosis or meiosis

conditioned response—behavior learned through repeated rewards or punishments

coniferous forest—forest land biome found between the temperate and polar zones of the earth

conservation biology—discipline of biology that seeks to preserve organisms and biological communities

consumer—an organism that gets its energy by consuming organic molecules (vs. producer)

coral reef—high diversity, low-latitude marine biome living in limestone structures built by organisms

covalent bond—(strong) attraction between atoms caused by sharing electrons

Creation Week—first six days (~6000 years ago) of time, during which God created the physical world

creation, young-age—the belief that God created the entire universe about 6000 years ago (vs. naturalistic evolution)

crossing over—process in prophase I when homologous chromosomes exchange genetic material

cultivars—distinguishable group of plants that remain distinct only with breeding (aka varieties)

curse—natural evil allowed into the creation by God in response to Adam's disobedience. Types of changes: negative effects: truly negative changes, probably imperfections in cycles and repair processes; evil-minimizing effects: changes introduced to minimize natural evil in a cursed creation

cytokinesis—cell process that divides a cell in two

cytokinesis I—meiosis substage, after homologous chromosomes separate, when the cell first divides

cytokinesis II—meiosis substage after the separation of sister chromatids, when the two cells divide

cytoplasm—water filling a cell, which is thick with a host of dissolved molecules

cytoskeleton—network of structural proteins that give a cell its shape and allow the cell to move

Day Three regression—theory that water running off rising continents on Day Three of the Creation Week formed most of the Precambrian rocks and fossils

death, biblical—cessation of biological life in animals & humans as an evil-minimizing effect of the curse

deciduous—when a tree drops its leaves at a certain point in the year and then regrows them at another

decomposer—organism that gets its energy from breaking down dead organisms and recycling the molecules so they can be used again in the community

deep discontinuities— large gaps that seem unbridgable that divide life into thousands of baramins

defenses—protections that God provided to organisms as evil-minimizing effects of the curse so as to protect them from being eaten

desert—land biome designed for the earth's arid zones (less than 10 inches of rainfall per year)

diffusion—spontaneous motion of a molecule towards areas of lower concentration of that molecule

diploid—cell state when it contains both chromosomes of each pair of homologous chromosomes

direct development— animals developing directly into a miniature version of the adult and grow in size from there

discontinuity—gap in form between or among organisms. Different types of discontinuity: very shallow discontinuity is bridgeable, persists only with breeding, and separates breeds and varieties and cultivars; shallow discontinuity is bridgeable, persists naturally, and separates species; deep discontinuity is unbridgeable and separates baramins

disease—degeneration of an organism due to genetic error(s) and/or harmful organism(s); a negative effect of the curse (aka pathology)

disjunct species—species with a geographic range in more than one widely separated area

disparity—measure of how different organisms are when compared with one another

diversification—increase in number of species

DNA—large nucleic acid that stores genetic information

DNA myth—common (false) belief that DNA fully determines the nature of an organism

domain—taxonomic level above kingdom

dominant—allele type that is fully expressed no matter the homologous allele

dominion—divinely-assigned role of humans to rule over the creation, enhancing its glorification of God

Edenian epoch—period of earth history between the Creation Week and the curse (<100 years)

egg cell—(haploid) gamete of a sexual (usually female) organism that grows into an adult after fertilization

electron—small negatively-charged subatomic particle often found orbiting the nucleus of an atom

electron shell—a region of space around an atomic nucleus where electrons might orbit the nucleus

electron transport system—cell membrane molecules that extract energy from an excited electron

elegance—beautiful simplicity of a design that accomplishes a complex result

element—atoms and ions having similar properties because they have the same number of protons

embryological recapitulation—developmental stages that look like the stages of that species' evolution

emergent properties—properties of an entity unaccounted for by the entity's component parts

endangered species—a species whose population is declining at a rate that puts the species at risk

endomembrane system—membrane organelles found between the nucleus and the cell membrane

endoplasmic reticulum (ER)—nucleus-connected, reticulated membrane of the endomembrane system where the cell builds macromolecules. Types: rough ER builds proteins; smooth ER builds lipids

enzyme—protein molecule that speeds ups (catalyzes) biological chemical reactions

estuary—ocean outlet of river where water saltiness transitions from fresh to marine

eukaryotic—type of cell that contains organelles

evolution, naturalistic—the belief that all things originate by spontaneous or natural change from previously existing physical things (vs. young-age creationism)

exotic species—an organism introduced into one biological community from another

extinction—death of every member of a taxon

faculative mutualism—a symbiosis when each organism is benefited by being with the other but can survive apart from each other

family, taxonomic—taxonomic level above genus and below order

fatty acid—long carbon chains with hydrogen atoms; monomer for non-steroid biological lipids

fauna—all the animal species at a given location

fermentation—chemical reactions a cell uses to get energy from organic molecules without using oxygen

fertilization—union of a sperm and an egg when the egg accepts the DNA of the sperm

fixation—extraction of an element from its physical world reservoir and 'fixing' it in an organic molecule

Flood—unique, global, year-long catastrophe that destroyed all land animals except those on Noah's ark

flora—all the plant species (and biomatrix species, when broadly defined) at a given location

fossil fuels—fuels like coal, oil, and natural gas which are made of fossils of dead organisms

frame shift mutation—a nucleotide is added or omitted in the process of copying DNA

fungi—(generally multi-cellular) eukaryotic organism whose cells have cell walls of chitin

gamete—haploid cell generated by meiosis in sexual organisms (aka sex cell). Types: a sperm, at fertilization, contributes only DNA; an egg cell, after fertilization, grows into an adult organism

gene—segment of DNA that codes for a protein (and some simple character traits)

genetic load—the number of harmful mutations carried by an organism or species (aka mutational load)

genetic throwback—occasional organism manifesting a complex trait otherwise only known in fossils

genetics, Mendelian—simple rules of inheritance that apply to a few dominant/recessive characters

genotype—set of alleles possessed by an organism (2 per gene, even if the same or unexpressed)

genus (pl. *genera*)—taxonomic level above species and below taxonomic family

geographic range—the area of the earth where an organism lives

global warming—rise of Earth's surface temperature due to greenhouse gas increase (especially CO_2)

glycogen—carbohydrate macromolecule for storying energy in animals

glycolysis—respiration reaction in the cytoplasm that breaks a glucose molecule in half to extract energy

Golgi body or apparatus—reticulated sac of the endomembrane system that labels macromolecules

greenhouse gas—atmospheric gas that absorbs sunlight, thus heating the earth's surface

groundwater—water that flows underground on its way to the ocean

gut flora—biomatrix microorganisms aiding digestion by living in mutualism in animal intestines

habitat—space in a biological community in which species can thrive

habituation—an organism's lack of response to experiences it has learned neither hurts nor helps the organism

haploid—cell state when it contains only one of each pair of homologous chromosomes

heat of fusion—heat required to melt 1 gram of a solid that is already at its melting point

heat of vaporization—heat required to evaporate 1 gram of a liquid that is already at its boiling point

hetcrotroph—organism that gets its energy from organic molecules

heterozygous—genotype with different alleles on the two homologous chromosomes

hierarchal coding— a few symbols code for a larger number of other symbols, which in turn code for an even larger number of other symbols

hierarchy—distinct levels. Types of hierarchy: unified hierarchy has both distinct levels and equality; nested hierarchy has lower level groups fully within higher level groups; netted hierarchy is a nested hierarchy with substantial cross-hierarchy similarity

holism—a perspective that seeks to understand something by looking at its larger context, and discovers more to something than is accounted for by a sum of its parts (vs. reductionism)

holobaramins—groups of known organisms surrounded by deep discontinuity hypothesized to correspond to baramins

homologous chromosomes—two chromosomes containing the same genes in the same sequence

homology—similar structure in a similar position in different organisms, but having a different function (due to common ancestry in naturalistic evolution; due to a common Creator in creationism)

homoplasy—character trait that does not fit the pattern of classification suggested by other characters

homozygous—genotype with the same allele on the two homologous chromosomes

host—the organism in parasitism that is harmed (by the parasite organism)

hot spot, biodiversity—a region that, if altered, would result the greatest extinction of organisms

hybrid zone—geographic region located between adjacent ranges of different organisms, which region contains intermediates between the different organisms

hydrogen bond—(weak) attraction between slight electrical charges on molecules

hydrophilic—water-'loving' (because molecule is polar)

hydrophobic—water-repelling (because molecule is non-polar)

ice advance—development and spread of continental glaciers during the Arphaxadian epoch

identical twins—two people with identical DNA; arise from embryo cells separated soon after fertilization

image of God—the divinely-declared, special status of each and every human—from conception to death—as a representative of God

imitative learning—behavior an organism copies from the behavior of another organism

immune—resistant to an infection or toxin

immunize—inject someone with a chemical identification tag of a pathogen so that that person's immune system develops the ability to fight that pathogen (aka vaccinate)

imprinting—organismal behavior based upon early life experiences of that organism

incompletely dominant—allele type partially expressed when both homologous alleles are of that type

indirect development—animals developing through more than one radically different stage

industrial melanism—dark forms of species that became common during the industrial revolution

instinctive behavior—behavior performed perfectly the very first time it is tried (vs. learned behavior)

interphase—stage in the life of a cell when the cell is not dividing

inter-generic hybrids—hybrids between different genera

inter-specific hybrids—offspring generated by parents from two different species

intra-baraminic diversification—proliferation of species within baramins, especially soon after the Flood

in-vitro fertilization—medical procedure fertilizing an egg cell and inserting it into a womb to develop

ion—an electrically charged arrangement of proton(s) with a different number of orbiting electron(s)

ionic bond—attraction between oppositely-charged ions

ionic compound—a compound composed of ions bound together by ionic bonds

irreducible complexity—characteristic of a system where it cannot function with <3 component parts and thus cannot function while assembled one component at time (*e.g.* by naturalistic evolution)

IVF—see *in-vitro fertilization*

karyokinesis—nuclear division; eukaryotic cell process that divides a cell nucleus into two nuclei

keystone species—species that may increase the diversity in a community

kind, biblical—recognizable group of similar organisms surrounded by deep discontinuity that persists by members interbreeding and producing similar offspring (aka baramin)

kingdom—taxonomic level above phylum and below domain

kingship—divinely-assigned role of humans to rule over the creation, serving both God and creation

Krebs (or citric acid) cycle—aerobic respiration reactions that extract energy from carbon-hydrogen bonds

learned behavior—behavior which is performed imperfectly the first time, but more perfectly as the organism repeats it (vs. instinctive behavior)

life—non-physical source of vitality. Different types of life: divine life is part of the natural essence of God and creature life is created by God in spirit beings and organisms; biblical life is possessed by God, spirit beings, and living humans and animals; nephesh life is possessed by living humans and animals; biological life is possessed by all organisms.

light-dependent reactions—photosynthesis reactions that require sunlight

light-independent reactions—photosynthesis reactions that do not require sunlight

lipid—organic molecules that, because they are non-polar, do not dissolve in water

lysosome—vesicle in some cells with enzymes for breaking down damaged or ingested macromolecules

macromolecules—organic molecules many times larger than typical molecules found outside organisms

macroevolution—the origin of taxa above species or the process that generates new taxa above species

macroevolutionary theory—naturalistic evolution theory explaining the origin of taxa above species

magnification, biological—more rapid accumulation and storage of pollutants by higher consumers

meiosis—(in generating sex cells) cell cycle stage yielding four nuclei with unique, haploid DNA sequences

mental map—memory of the environment that an organism develops so as to move through it

metabolism—cell processes storing & releasing energy and building & breaking down macromolecules

metaphase—mitosis substage when chromosomes align and attach to centrioles via spindle fibers

metaphase I—meiosis substage when homologous chromosome pairs align and attach to spindle fibers

metaphase II—meiosis substage after cytokinesis I: chromosomes align & attach to spindles in each cell

microevolution—change within species

microorganisms—organisms too small to be seen without magnification

mitochondrion—double-membraned cell organelle where aerobic respiration occurs

mitosis—cell cycle stage yielding two nuclei containing identical DNA (to one other and to the parent cell)

molecule—atoms bound together by covalent bonds

monomer—molecular building block of (biological) macromolecule

monosaccharide—5- or 6-carbon sugar molecule that is the monomer for biological carbohydrates

multi-cellular—organism made up of more than one cell

mutation—genetic mistakes that occur as DNA is copied. Types: point mutation replaces one nucleotide with a different one; frame-shift mutation inserts or deletes a nucleotide

mutational load—the number of harmful mutations carried by an organism or species (aka genetic load)

mutualism—symbiosis where two (or more) organisms benefit. Types of mutualism: facultative mutualism organisms that can survive apart; obligate mutualism organisms cannot survive apart

mycorrhizal fungi—fungi in mutualism with plant roots that provide water and nutrients to the plants

native species—an organism that is normally part of a particular biological community

natural evil—anything in the physical world that causes suffering of humans or animals

natural law—a regularity of the universe

naturalism—a worldview that accepts the existence of only physical things (vs. Christian theism)

neodarwinism—naturalistic evolution theory of the 1920's to 1940's, updating evolution with genetics

netted hierarchy— a widespread network of similarity that unites all the organisms

neutron—large, chargeless subatomic particle usually found in the nucleus of an atom

niche—role of an organism in a biological community

non-biodegradable—not degradable into components that organisms can use (vs. biodegradable)

non-polar substance—a substance with electrical charges (vs. polar substance)

non-renewable resource—a substance needed by organisms that is continually made available or continually generated (vs. renewable resource)

nuclear envelope—double membrane surrounding a cell's nucleus

nuclear pores—holes in the nuclear envelope that allow macromolecules to pass in and out of the nucleus

nucleic acid—(biological) macromolecule constructed of nucleotides, designed for carrying information

nucleolus—dense region in the nucleus of a cell where ribosome components are assembled

nucleus—double-membrane-surrounded organelle that houses the cell's DNA

obligate mutualism— a mutualism when each organism not only benefits from the relationship, but also cannot exist without the other

order—taxonomic level above family and below class

organ—biological system of tissues in multi-cellular organisms; make up organ systems in large organisms

organ system—biological system of organs that interact with other systems in a large organism's body

organelle—membrane-bound biological system of molecules, found in eukaryotic cells; the substructure of a cell

organism—an animal, plant, or single-celled life form

organo-substrate—system of bacteria, algae, protozoa, and fungi required for plants and animals to live on earth (aka biomatrix)

osmosis—the spontaneous motion of water across a semi-permeable membrane towards areas of lower concentration of water

overproduction—organisms producing more offspring than will survive to reproduce; leads to competition and natural selection, and introduced by God after the curse to replace dead organisms

parasite—the organism in parasitism that benefits, thus harming the host organism

parasitism—symbiosis where one organism benefits and one or more organism is harmed; a negative effect of the curse

pasteurization—process of heating or boiling that kills microorganisms which might spoil beverages

pathogen—a disease-causing organism

pathogenesis—the cause of a disease

pathology—degeneration of an organism due to genetic error(s) and/or harmful organism(s); a negative effect of the curse (aka disease)

periodic table—tabular arrangement of elements by atomic number, which places elements with similar characteristics in the same column

phenotype—character(s) exhibited by an organism

phospholipid—lipid with hydrophobic and hydrophilic ends; make up biological membranes

photoautotroph—organism that gets its energy from sunlight

photosynthesis—cellular process that fixes carbon and stores sunlight energy in organic molecules

phylum (pl. *phyla*)—taxonomic level above class and below kingdom

pigment—that which absorbs some color and reflects the rest; photosynthesis molecule that collects sunlight energy

pioneer community/species—the first organisms (in community succession) to settle in an area

plant—(multi-cellular) eukaryotic organism whose cells have cell walls of cellulose

point mutation—a simple change in one nucleotide, analogous to changing one alphabet letter in a sentence

polar substance—a substance with electrical charges (vs. non-polar substance)

pollutant—harmful chemical introduced into the environment

pollution—harmful chemicals accumulated more rapidly than remediated by the biomatrix

polyploidy—cell state when it contains more than two homologous chromosomes

population bottleneck—time when a population was reduced to a very small size (usually <10)

Precambrian—oldest sediments on the earth; fossils in all but the uppermost layers are only bacteria

predation—the eating of animals by other animals (aka carnivory), an evil-minimizing effect of the curse

predator—an animal that eats another animal (aka carnivore)

prey—an animal eaten by another animal

priest—divinely-assigned role of humans to use the creation to better know God, worship God, and bring others into the worship of God

producer—an organism that gets its energy from the physical environment (vs. consumer)

prokaryote—evolutionary term for a cell lacking organelles (aka bacterium)

prometaphase—mitosis substage some place after the prophase, when the nuclear envelope dissolves

prometaphase I—meiosis substage some place after prophase I, when the nuclear envelope dissolves

prophase—first mitosis substage, when chromosomes, centrioles, and spindle fibers appear and nuclear envelope and nucleoli disappear; see *prometaphase*

prophase I—first meiosis substage: chromosomes, centrioles, & spindle fibers appear, nuclear envelope & nucleoli disappear, homologous chromosomes pair, and crossing over occurs; see *prometaphase I*

prophase II—meiosis substage, after cytokinesis I, when centrioles and spindle fibers appear in each cell

protein—(biological) macromolecule constructed of amino acids

proton—large positively-charged subatomic particle usually found in the nucleus of an atom

protozoan (pl. *protozoa*)—eukaryotic organism that is generally a one-celled consumer

pseudogene—DNA segment with gene-like start and stop codons, but not coding for a working protein

punctuated equilibria—naturalistic evolution theory explaining stasis and abrupt appearance of species

Punnett square—matrix method used to calculate genotypes of offspring in Mendelian genetics

radiometric date—age calculated from amount and rate of decay of radioactive elements

rainforest—forest land biome designed for the wetter part of the earth's tropical zone

reason—the memory, processing, and rearranging of information by an organism

recessive—allele type that is not expressed at all when the homologous allele is dominant

recycle—clean and reform trash (especially non-biodegradable trash) so as to reuse it

reductionism—a perspective that seeks to understand something by looking at its component parts (vs. holism)

renewable resource—a substance needed by organisms that is continually made available or continually generated (vs. non-renewable resource)

replication—cell process that makes a copy of the cell's DNA

reservoir (elemental)—location where an element essential to organisms is stored in sufficient quantity to provide all organisms with the element

resource partitioning—division of a biological community's resources among different organisms

respiration, aerobic—chemical reactions a cell uses to get energy from organic molecules using oxygen

respiration, cellular—chemical reactions a cell uses to get energy out of organic molecules. Types: fermentation is done without using oxygen; aerobic respiration is done using oxygen

ribosomes—protein-making machines found in the membrane walls of the endoplasmic reticulum

ring species—circle of populations where all but adjacent population pair can interbreed successfully

savannah—grass land biome designed for the dryer part of the earth's tropical zone

science—something humans do to understand the physical world by proposing tentative truths as theories of explanation and valuing fit with the physical world

second law of thermodynamics—the tendency of everything in the universe to spontaneously move towards areas where that something is in lower concentration

selection, artificial—human choice of some available organisms (*e.g.* for use or breeding)

selection, natural—natural process resulting in death of (the less capable) part of a population; caused by overproduction and designed and introduced by God as an evil-minimizing effect of the curse

sister chromatids—(following replication) two identical DNA molecules attached at their centromeres

sister species— species recently descended from the same ancestor

solvent—a substance into which other substances dissolve. Types of solvents: polar solvent dissolves polar substances; non-polar solvent dissolves non-polar substances

special creation—direct action of God to bring something into being

speciation—origin of a new species or the process that generates a new species

species—smallest recognizable group of similar organisms that persists naturally by the production of similar offspring; taxonomic level below genus

species design—package of information capable of generating and maintaining a species

specific heat—heat required to raise the temperature of 1 gram of a substance $1°$ C

spectrum of perfection—a range of degrees of perfection of a God-illustrating trait, created in a suite of different physical things so that humans extrapolate to God's infinite manifestation of that trait

sperm—(haploid) gamete of a sexual (usually male) organism that contributes only its DNA at fertilization

spontaneous generation—the natural (not created) origin of life from non-living things; once believed of all lower life forms, now only believed in naturalistic evolution for the first organism(s)

starch—carbohydrate macromolecule for storing energy in plants

stasis—status of a fossil taxon when it shows no change through successive layers of the fossil record

steroid—lipid built from four connected rings of carbons; act as chemical messengers in organisms

stewardship—divinely-assigned role of humans to be stewards of God's creation, preserving and enhancing its glorification of God, being ready always to give an account to God

stratomorphic intermediate—a fossil having attributes appearing intermediate between two other organisms, and which is also located in strata (rock layers) located between the same two organisms; a stratomorphic series is a sequence of stratomorphic intermediates

stratamorphic series— a sequence of fossil genera found in successive layers that step through a sequence of forms

suboptimal improvisation—structure of poor design, thought to be evidence of naturalistic evolution

sustainable development—preserving biological communities as we transform the environment

symbiosis (pl. *symbioses***)**—long-term interaction between two or more organisms

system, biological—multiple parts that interact in such a way that the whole has emergent properties

systems biology—a discipline of biology that studies biological systems

taxon (pl. *taxa***)**—group of organisms united by similarity, distinguished from other taxa by discontinuity

taxonomic level—the hierarchal level of a particular taxon (*e.g.* species, genus, family, order…)

taxonomy—the science of naming organisms

telophase—last mitosis substage, when, in both cells, chromosomes and centrioles disappear, and nuclear envelope and nucleoli appear

telophase I—meiosis substage, after homologous chromosomes separate, when centrioles disappear

telophase II—last meiosis substage, when, in all four cells, chromosomes and centrioles disappear, and nuclear envelopes and nucleoli reappear

temperate deciduous forest—forest land biome found in the wetter part of the earth's temperate zone

temperate grasslands—grass land biome designed for the dryer part of the earth's temperate zone

threatened species—a species that is likely to become endangered in the near future

tier—vertical zone of a forest characterized by a unique flora and fauna

tissue—biological system of cells in multi-cellular organisms that make up organs in larger organisms

transcription—cell system that makes an RNA copy of genetic information on DNA

translation—cell system that makes an amino acid sequence (for protein) from a nucleotide sequence

trophic structure—the flow of energy through a community (from producers through consumers)

tundra—biome found in the earth's polar zone

unicellular—organism that is made up of only one cell

unified hierarchy—simultaneous unity and diversity, equality and distinction of a hierarchy

vaccinated—when a person is injected with a vaccine that contains an identification tag for a pathogen, making the person immune to the pathogen

vaccine—the injection, with a chemical identification tag of a pathogen, given so that the recipient's immune system develops the ability to fight that pathogen (aka immunize)

vacuole—fluid-filled organelle in some cells which stores nutrients, water, or waste

varieties—distinguishable group of plants that remain distinct only with breeding (aka cultivars)

vesicles—endomembrane system spheres that transport macromolecules among a cell's organelles

vestigial organ—organ in a modern species having a reduced function from that in an ancestral species

water cycle—continuous natural process supplying organisms with water, by evaporation from the oceans, precipitation to the earth's surface, and flow back to the oceans

worldview—a person's perspective (set of beliefs) that colors or influences the way that person interprets everything that person perceives (*e.g.* naturalism; Christian theism)

INDEX

abiogenesis. *See* evolution: abiogenesis
Adam
   names animals. 109, 115, 443
   sin of. 7, 261, 446, 451
aging, degenerative. *See* curse, the Adamic: evil-minimizing effects: degenerative aging
aging, developmental. *See* development, biological
algae. 56, 247, 251, 352, 353
   biomatrix. 246
   in lichens. 248
anaphase. 404
   I & II. 408
animals. 54, 77, 193, 198, 352, 359
Antediluvian Epoch. 442, 452
anthropic principle. 207, 238
archaea. *See* bacteria
Arphaxadian Epoch. 462, 499
atomic number. *See* elements
bacteria. 56, 116, 326, 345, 419, 449, 501
   biomatrix. 246, 267, 282, 305
   chemosynthetic. 248, 305, 352
   gut flora. 249, 254, 282
   mineral-fixing. 252
   nitrogen-fixing. 93, 252, 257
   photosynthetic. 247, 251, 352
   remediators. 252, 268
baramin. 115, 122, 129, 413, 441, 494, 496, 501
   biblical kind is defined. 110
   holobaramin approximates. 115, 121
   not the genus. 113, 494
   not the species. 112, 115, 413
baraminology. *See* biosystematics: baraminology
beauty
   definition. 85
   human appreciation of. 97
   of biology. 95, 502
biodiversity. 298, 308
   hot spot. 315
biogenesis, law of. 66, 69
biogeochemical cycle. *See* cycle: biogeochemical
biogeography. 417, 459, 482

biological systems. *See* systems, biological
biology
   conservation biology. 312
   definition of. 28
   limitations of. 56
   reasons for its study. 45
   systems biology. 285
biomatrix. 245
biomes. 299, 305
bioremediation. 253
biosystematics. 337
   baraminology. 115, 122
bond, chemical
   hydrogen bond. 224, 358
breeds. 102, 117, 126, 130, 131, 316, 414, 420
carnivory. *See* curse, the Adamic: evil-minimizing effects: predation
carrying capacity. 78, 426, 427
cell. 172, 281, 352, 359
   organelles. 281, 342
   theory. 328
chemosynthesis. 248
chloroplast. 353, 362
choice. *See* will
chromosome. 402
   homologous. 406
class (taxonomic). 339
communication, animal. 371
community, biological. 92, 282, 285, 290, 305
   succession. 92
condensation. 402
conditioning. *See* intelligence: learned behavior
consumers. 247, 249, 256, 264, 305, 352
counting. *See* intelligence: learned behavior

creation
    illustrates God's nature. 14, 51, 74, 85, 88, 98, 107, 128, 170, 198, 207, 233, 245, 262, 273, 289, 297, 312, 324, 345, 351, 363
    knowability traits of. 20
    week. 444
    young-age creationism. 7, 97, 128, 193, 261, 287, 311, 344, 362
cultivars. 102, 117, 126, 130, 131, 316, 414, 420
curse, the Adamic. 7, 36, 186, 261
    evil-minimizing effects. 442
        death. 6, 67
        defenses. 152, 178, 255
        natural selection. 96, 126, 420, 448, 475, 479, 491, 495
        predation. 448
    negative effects. 448
        degenerative aging. 6, 445
        parasitism. 451
        pathology. 78, 282, 445, 448, 451, 463
        suffering. 6, 445, 496
    origin of natural evil. 7, 448
cycle
    biogeochemical. 255, 259
    water. 210, 304
cytokinesis. 401
cytoplasm. 360
Darwin, Charles. 96, 472, 491
Day Three regression. 441
death. *See* curse, the Adamic: evil-minimizing effects: death
decomposers. 250, 253, 277, 305
defenses. *See* curse, the Adamic: evil-minimizing effects: defenses
desert biome. 300
development, biological. 171, 175, 280, 339, 371, 372, 412, 422,
    direct. 174
    indirect. 175
discontinuity, biological. 114, 115, 124, 472, 502
    challenge to evolution. 127
    deep. 121, 124
disease. *See* curse, the Adamic: negative effects: pathology
disparity. 308, 312, 337
diversification
    intra-baraminic. 417, 461, 500
diversity. *See* biodiversity

DNA. 117, 171, 278, 275, 316
    myth. 371, 373
domain (taxonomic). 339
Edenian Epoch. 446
egg cell. 171, 372
electron. *See* subatomic particles
elements. 218, 250
embryological recapitulation. 479, 497
emergent properties. 5, 280, 285, 287, 342
emotions. 77, 190
ethics. 42
    abortion. 77
    acid rain. 265
    antibiotics & -icides. 76, 263, 266, 283, 289
    breeding. 102, 131, 200, 292, 318
    cloning. 318
    DNA. 394
    euthanasia. 77
    exotic introductions. 292
    global warming. 238
    gluttony, wasting, hoarding. 76
    humane treatment of animals. 199
    hunting & fishing. 74, 199
    ozone protection. 236
    pollution. 235, 267
    preserve beauty. 103
    preserve diversity. 130, 198, 319
    reproductive ethics. 427
evolution
    abiogenesis. 68, 472
    naturalistic evolution. 6, 96, 98, 128, 193, 261, 287, 311, 344
extinction. 130, 307
family (taxonomic). 114, 338
fixation. 251, 252
    carbon. 251
    nitrogen. 252
forest
    coniferous. 302
    floating. 449, 454, 501
    stromatolite. 449
    temperate deciduous. 301
fossil
    communities. 442, 449
    disparity. 309
    fuel. 236
    intermediates. 123, 416, 484, 485
    species. 308
    stasis. 119, 477, 489, 493

fossils. 320
  human. 460
  in Paleozoic & Mesozoic. 450, 454, 483
  in Precambrian. 441, 501
  in Tertiary. 457, 482, 486, 489, 495
fungi. 56, 174, 178, 274, 339, 352
  biomatrix. 250, 246, 305
  lichen. 248
  mycorrhizal. 254, 259, 261, 282, 357
gamete. 380
gametes. 406
gene. 373, 376, 392
genetic load. *See* mutational load
genetics, Mendelian. 381
genetic throwback. 415, 420, 481, 495
genus/genera. 112, 121, 338
  not the biblical kind. 113
God's
  beauty. 85, 288
  desire to be known. 20, 51, 74, 170, 207
  life. 51
  love. 73, 95, 207, 233, 245, 261
  personhood. 170, 273, 311, 344
  provision. 207, 233, 245, 262
grassland
  savannah. 300
  temperate. 301
greenhouse gas. 209, 235
growth, biological. 173, 281
gut flora. 249, 254, 259, 282
habitat. 305
habituation. *See* intelligence: learned behavior
hierarchy
  nested. 479
  netted. 341, 342
  of similarity. 340
  unified. 324, 496
holism (holistic perspective). 5, 29, 85, 440
holobaramin. *See* baramin: holobaramin approximates
homology. 477
homoplasy. 340
horse series. 416, 484, 495

humans
  as image. 41, 54, 75, 101, 130
  as kings. 34
  as priests. 31
  as stewards. 39
  king responsibilities. 40, 80, 103, 131, 200, 238, 293, 319
  priest responsibilities. 33, 74, 101, 130, 198, 233, 262, 289, 312, 345, 363
hybrid
  inter-generic. 112, 124
  inter-specific. 112, 114, 124, 415
  zone. 112, 128
imitation. *See* intelligence: learned behavior
imprinting. *See* intelligence: learned behavior
intelligence, organismal. 186
  instinct. 181, 192, 245
  learned behavior. 181, 194
  reason. 177, 186
interphase. 403
ions. *See* elements
irreducible complexity. 286, 343
Jacob. 102
karyokinesis. 401
kind, biblical. *See* baramin
kingdom (taxonomic). 339
life
  biblical. 54, 193
  biological. 123, 328
  divine. 51
  nephesh. 411
  non-physical nature of. 193
  origin of. 69
life cycles. 174
Linnaeus, Carl. 114
macroevolution. 344, 486, 503
Marsh, Frank. 115
medicine, modern. 283
meiosis. 403, 409
melanism, industrial. 474
membranes. 215, 276, 326, 342
memory. *See* intelligence: learned behavior
Mendeleyev, Dmitry. 219
Mendel, Gregor. 385
mental maps. *See* intelligence: learned behavior
metabolism. 172, 326, 342, 363
metaphase. 404
  I & II. 407

microevolution. 472, 475
mitochondrion/mitochondria. 360
   DNA. 461
mitosis. 403, 404
molecules. 216, 230, 277, 285, 325
   macromolecules. 123, 171, 273, 288
   monomers. 274, 277
monera. See bacteria
monomers. See molecules: monomers
moral sense. See will
mutation. 126, 377, 379, 386, 393, 415, 416, 420, 472, 480, 488, 495, 496
   beneficial. 473, 488
   frame-shift. 378
   point. 377
mutational load. 486, 496
mutualism. 498, 502
mycorrhyzal fungi. See fungi: mycorrhyzal
natural evil. See curse, the Adamic: origin of natural evil
natural law. 213, 217
natural selection. See curse, the Adamic: evil-minimizing effects: natural selection
neodarwinism. 472, 490
neutron. See subatomic particles
niche. 305
Noah. 108, 109
nucleus. 402
order (taxonomic). 339
organo-substrate. See biomatrix
orientation. 181
overproduction. 448
parasitism. See curse, the Adamic: negative effects: parasitism
pathogen. 448, 451
pathology. See curse, the Adamic: negative effects: pathology
periodic table. 219
phospholipids. 276, 326
photosynthesis. 212, 247, 359
phylum/phyla. 339
plants. 54, 247, 316, 352
   design of. 228, 359
population bottleneck. 461
Precambrian fossils. 441, 501

Precambrian fossils
   in Precambrian. 454
producers. 246, 247, 249, 251, 255, 257, 305, 352
prokaryote. See bacteria
prophase. 404
   I & II. 407
proton. See subatomic particles
protozoa. 56, 249, 352
   biomatrix. 246
pseudogenes. 481, 496
punctuated equilibria. 476
Punnett squares. 384
radiometric dating. 438
rainforest. 300
Redi, Francesco. 66
reductionism (reductionistic perspective). 4
remediation. See bioremediation
replication. 402, 470
reproduction. 179, 280
ribosomes. 388
savannah. See grassland: savannah
science
   definition of. 23–27
   origin of. 20
self-awareness. 190
speciation. 415, 472, 477, 489, 495
species. 111, 116, 121, 338
   abrupt appearance of. 119, 477, 493, 499
   designs. 494
   disjunct. 459, 500
   endangered. 315
   exotic. 284, 290, 292
   from hidden information. 130, 494, 498
   keystone. 293
   native. 290
   not the biblical kind. 112, 115, 413
   pioneer. 92
   ring. 416
   sister. 125, 128
   stasis. 119, 477, 493, 499
   threatened. 315
spectrum of perfection. 72, 98, 116, 123, 128, 171, 181, 188, 197, 279, 287
sperm. 171, 371, 406
stratomorphic
   intermediate. 112, 114, 128
   series. 113, 114, 122, 416, 484, 499
   structure. 485

subatomic particles. 218, 273, 325, 355, 360

suboptimal improvisations. 480, 498

suffering. *See* curse, the Adamic: negative effects: suffering

systems, biological. 196, 279, 285, 287, 342, 362

taxonomy. 114, 129, 338

taxon/taxa. 111, 338

telophase. 405
  I & II. 408

trophic structure. 305

tundra. 302

varieties. 102, 117, 126, 130, 131, 200, 316, 414, 420, 498,

vestigial structures. 415, 420, 481, 495

viruses. 123, 328

vitamins. 254

water. 210, 233

waterbond, chemical
  hydrogen bond. 230

will, organismal. 77, 188, 193, 195

Wise, Kurt. 115

worldview. 3
  Christian theism. 4
  naturalism. 3, 4, 6, 8, 9, 19, 28, 68, 97, 124, 193, 231, 260, 287, 309, 362

young-age creationism. *See* creation: young-age creationism